STORAGE CARBOHYDRATES IN VASCULAR PLANTS

DISTRIBUTION, PHYSIOLOGY AND METABOLISM

Edited by

D. H. LEWIS

Professor, Department of Botany, University of Sheffield

The right of the University of Cambridge to print and sell all manner of books was granted by Henry VIII in 1534. The University has printed and published continuously since 1584.

CAMBRIDGE UNIVERSITY PRESS

Cambridge

London New York New Rochelle

Melbourne Sydney

Published by the Press Syndicate of the University of Cambridge
The Pitt Building, Trumpington Street, Cambridge CB2 1RP
32 East 57th Street, New York, NY 10022, USA
296 Beaconsfield Parade, Middle Park, Melbourne 3206, Australia

First published 1984

Printed in Great Britain by the University Press, Cambridge

Library of Congress catalogue card number: 84-3167

British Library Cataloguing in Publication Data

Storage carbohydrates in vascular plants. –
(Society for Experimental Biology seminar
series; 19)

1. Carbohydrates – Metabolism 2. Plants – Metabolism
I. Lewis, D.H.
581.1′33 QK898.C3

ISBN 0 521 23698 3

UP

CONTENTS

CONTRIBUTORS

ap Rees, T.
Botany School, University of Cambridge, Downing Street, Cambridge, CB2 3EA, UK

Drew, E. A.
Australian Institute of Marine Science, Townsville, Queensland, Australia

Duffus, C. M.
Department of Agricultural Biochemistry, School of Agriculture, University of Edinburgh, West Mains Road, Edinburgh, EH9 3JG, UK

Giaquinta, R. T.
E. I. du Pont de Nemours and Company, Central Research and Development Department, Experimental Station, Wilmington, Delaware 19801, USA

Halmer, P.
Germain's (U.K.) Ltd, Hansa Road, Hardwick Industrial Estate, King's Lynn, Norfolk, PE30 4LC, UK

Herold, A.
Department of Botany, The University of Sheffield, Sheffield, S10 2TN, UK

Hopf, H.
Botanical Institute, University of Munich, Menzinger Strasse 67, D-8000 München 19, Federal Republic of Germany

Kandler, O.
Botanical Institute, University of Munich, Menzinger Strasse 67, D-8000 München 19, Federal Republic of Germany

Lewis, D. H.
Department of Botany, The University of Sheffield, Sheffield, S10 2TN, UK

Pollock, C. J.
Welsh Plant Breeding Station, Aberystwyth, Dyfed, UK

Stitt, Mark
Lehrstuhl für Biochemie der Pflanze der Universität Göttingen, Untere Karspüle 2, D-3400 Göttingen, Federal Republic of Germany

Thorne, J. H.
E. I. du Pont de Nemours and Company, Central Research and Development Department, Experimental Station, Wilmington, Delaware 19801, USA

PREFACE

Animals possess a limited range of compounds which are used as storage carbohydrates. Most are based on glucose [e.g. free glucose itself, trehalose (di-glucose), glycogen (poly-glucose), glucitol (reduced glucose)]. By contrast, vascular plants contain a vast range, based on several monosaccharides. Despite this variety in plants, most attention has been devoted to sucrose and starch. This concentration of research on one soluble and one insoluble storage carbohydrate, which reflects their dietary and commercial importance, is in turn reflected in the space (over five chapters) that is specifically devoted to them here. They both also receive mention in most of the other chapters, which principally concern less well-known compounds (sucrosyl and other oligosaccharides, cyclitols, alditols and 'hemicelluloses'). There is a thoroughly deserved, increasing interest in the latter, which it is hoped this book will foster.

The book aims to bridge the gap between the statutory chapter or two given to storage carbohydrates in standard texts of plant physiology and biochemistry, and encyclopaedic monographs. Hopefully, this has been done in a way which will be intelligible to students and will act as a detailed introduction to others interested in the field – the terms of reference of the SEB Seminar Series. The book is the outcome of the International Seminar held in Oxford in December, 1980. I must accept the blame for the long delay in publication. However, published information which, since 1980, has changed concepts rather than merely increased the detailed knowledge of specific aspects has been incorporated in particular chapters to keep the book up-to-date in broad principles. Where necessary, authors have highlighted areas about which there is insufficient information and pin-pointed specific problems.

The book begins with an overview of the inter- and intra-specific distribution of storage carbohydrates in plants. This chapter has been compiled from contributions by several of the authors of subsequent ones. Two chapters, one on the metabolism and one on the transport of sucrose, follow. This naturally leads to the two chapters on sucrosyl fructans and sucrosyl galactans. The latter chapter which also concerns the biosynthesis of oligosaccharides not based on a sucrosyl group, links to the next via their common interest in *myo*-inositol, a pivotal compound in aspects of plant metabolism outside its role in synthesis of other carbohydrates. The review of soluble reserves ends with a discussion of the physiology and metabolism of alditols. Starch merits three chapters – synthesis in leaves, degradation in leaves, and metabolism in storage tissues, and the book ends with a survey of the synthesis and degradation of the immense variety of extracellular storage polysaccharides, earlier tabulated in Chapter 1.

Abbreviations for compounds which are referred to frequently throughout the book are listed immediately after this Preface. Plants, especially agricultural, horticultural or otherwise domesticated species, are most often referred to by their vernacular names to avoid cluttering the text with Latin binomials. However, by way of a taxonomic glossary, the binomials are included after the vernacular names in the Index. For less familiar plants, binomials are used in the text.

It is a pleasure to acknowledge the help of the following, all staff of Cambridge University Press, in the preparation of this book: Robin Pellew, Marion Jowett, Audrey Smith and Martin Walters.

Department of Botany
University of Sheffield

D. H. Lewis
December 1983

ABBREVIATIONS

ADP Adenosine diphosphate
ADPG, ADPGlc Adenosine diphosphate glucose
Ara Arabinose
CDP Cytidine diphosphate
DHAP Dihydroxyacetone phosphate
Dol-P Dolichol phosphate
DP Degree of polymerization
DPGA Diphosphoglyceric acid
EDTA Ethylenediaminetetra-acetic acid
edwt Extracted dry weight
f Furanose
F, Fru Fructose
*F6P, Fru6P Fructose 6-phosphate
FBP, Fru1,6 P_2 Fructose 1,6-bisphosphate
G, Glc Glucose
*G1P, Glc1P Glucose 1-phosphate
*G6P, Glc6P Glucose 6-phosphate
*G3P Glyceraldehyde 3-phosphate

* With or without hyphens
Some abbreviations are used in combinations (especially for sugar nucleotides).

Gal Galactose
GalA Galacturonic acid
GlcA Glucuronic acid
GlcNAc *N*-acetyl glucosamine
GDP Guanosine diphosphate
HP Hexose monophosphate
Man Mannose
MI, *myo*-Inos *myo*-Inositol
myo-Inos-1-P *myo*-Inositol phosphate
NAD (or NAD^+), NADH, Nicotinamide adenine dinucleotide, nicotinamide
NAD (or $NADP^+$), NADPH adenine dinucleotide phosphate and their reduced forms
OPP Oxidative pentose phosphate
p Pyranose
P Phosphate (in organic esters)
Pi Inorganic orthophosphate
PPi Inorganic pyrophosphate
PGA Phosphoglyceric acid
Rha Rhamnose
RPPP Reductive pentose phosphate pathway
RuBP Ribulose bisphosphate
SBP Sedoheptulose bisphosphate
SPP Sucrose phosphate phosphatase
SPS Sucrose phosphate synthetase
TDP Thymidine diphosphate
TP Triose phosphate
UDP Uridine diphosphate
UDPG, UDPGlc Uridine diphosphate glucose
UMP Uridine monophosphate
UTP Uridine triphosphate
Xyl Xylose

D. H. LEWIS
(*Compiled from contributions from other authors of this book*)

1 Occurrence and distribution of storage carbohydrates in vascular plants

General introduction

Vascular plants contain a wide range of soluble and insoluble storage carbohydrates. Their chemical natures and the amounts present vary in both space and time – from species to species, organ to organ, tissue to tissue, and cell to cell, as well as diurnally and with season. This chapter aims briefly to document this distribution for the major kinds of storage carbohydrate, whereas subsequent chapters will deal with the physiology and metabolism of synthesis and degradation of individual groups of compounds.

The existing literature on carbohydrates in plants is vast. The following books are especially relevant either because of their historical importance to the subject or because they are recent compilations (Staněk, Černý, Kocourek & Pacák, 1963; Staněk, Černý & Pacák, 1965; Pigman & Horton, 1970–72; Miller, 1973; Preiss, 1980; Loewus & Tanner, 1982). As well as in broad-spectrum, less specialized botanical journals, research papers and reviews concerning the carbohydrates of plants regularly appear in *Plant Physiology*, *Phytochemistry*, *Carbohydrate Research*, *Annual Review of Plant Physiology* and *Advances in Carbohydrate Chemistry and Biochemistry*.

More is probably known about the biology of starch and sucrose than of all other storage carbohydrates of plants grouped together. For this reason, this pair receives mention in almost all subsequent chapters and five are devoted specifically to them. An aim of this book, however, is to draw attention to the second league – sucrosyl-fructans, sucrosyl-galactans and other oligosaccharides, cyclic and acyclic sugar alcohols (inositols and alditols), and storage polysaccharides other than starch. The last are frequently referred to as 'hemicelluloses' but, since they are not related chemically or biosynthetically to cellulose, a β-D-($1 \rightarrow 4$)-glucan, this term is a misnomer and should be avoided (Davies, Giovanelli & ap Rees, 1964: 152 and see 'Other storage polysaccharides' below). These five groups warrant a chapter apiece and more space than sucrose and starch in this chapter. In addition, several monosaccharides (reducing sugars) occur in the uncombined, free state in plants. Glucose and fructose occupy the centre of the stage here with several others in the wings. These do not warrant a separate chapter since their metabolism is often intimately bound up with that of their oligo- and polysaccharides or their reduction products (alditols). They therefore receive appropriate mention in the chapters dealing with these.

A basic knowledge of the structure and chemistry of monosaccharides is, however, fundamental to an understanding of the physiology and metabolism of all storage carbohydrates since some monosaccharides, rarely found free in plant tissues in substantial amounts, are important building blocks of storage oligo- and polysaccharides.

The monosaccharides

Structure

Figure 1.1 shows the nomenclature and inter-relationships between the D-series of 3–6 carbon aldose and ketose sugars (see Guthrie & Honeyman, 1974 for some useful mnemonics for remembering these!). Table 1.1 records the 7–9 sugars which occur in vascular plants. The formulae are represented as both their open chain and ring configurations and, in Fig. 1.1, the abbreviations often used for the sugars are listed. (In some chapters, G and F are used for glucose and fructose, respectively.) These sugars exist in either their α- or β-form (anomers) (Fig. 1.1), an important consideration in relation to the formation of the glycosidic bonds of oligo- and polysaccharides. Pairs of sugars which differ in orientation of hydrogen and hydroxyl groups on other specific carbon atoms are termed epimers (e.g. glucose and mannose are epimers with respect to carbon atom 2). These and other elementary aspects of the chemistry of monosaccharides are clearly explained in such biological texts as Davies *et al.* (1964) and Hall, Flowers & Roberts (1974) as well as in the specialist books on carbohydrates mentioned above (see also Stoddart, 1971).

Fig. 1.1 The D-series of 3–6 carbon monosaccharides shown in both open-chain and ring configurations, aldohexoses as pyranose and aldopentoses and ketohexoses as furanose forms. The rings of pentoses and aldohexoses are in the α-configuration and those of the ketohexoses are β. Carbon atoms are numbered on the left-hand side for each set. Hydrogen atoms are omitted from all formulae so that each intersection represents =CH. Horizontal bars on open chains and vertical bars on rings represent hydroxyl groups such that those drawn to the *right* on open chains become *downwards* on the rings. Only those labelled in full capitals with abbreviations occur free or combined among the storage carbohydrates of vascular plants. The corresponding alditols to those marked * also occur in vascular plants (see below and Chapter 7). † Arabinose differs from the other aldoses by occurring as the L-sugar and by combining in polymers as the furanose form (see below and Chapter 11).

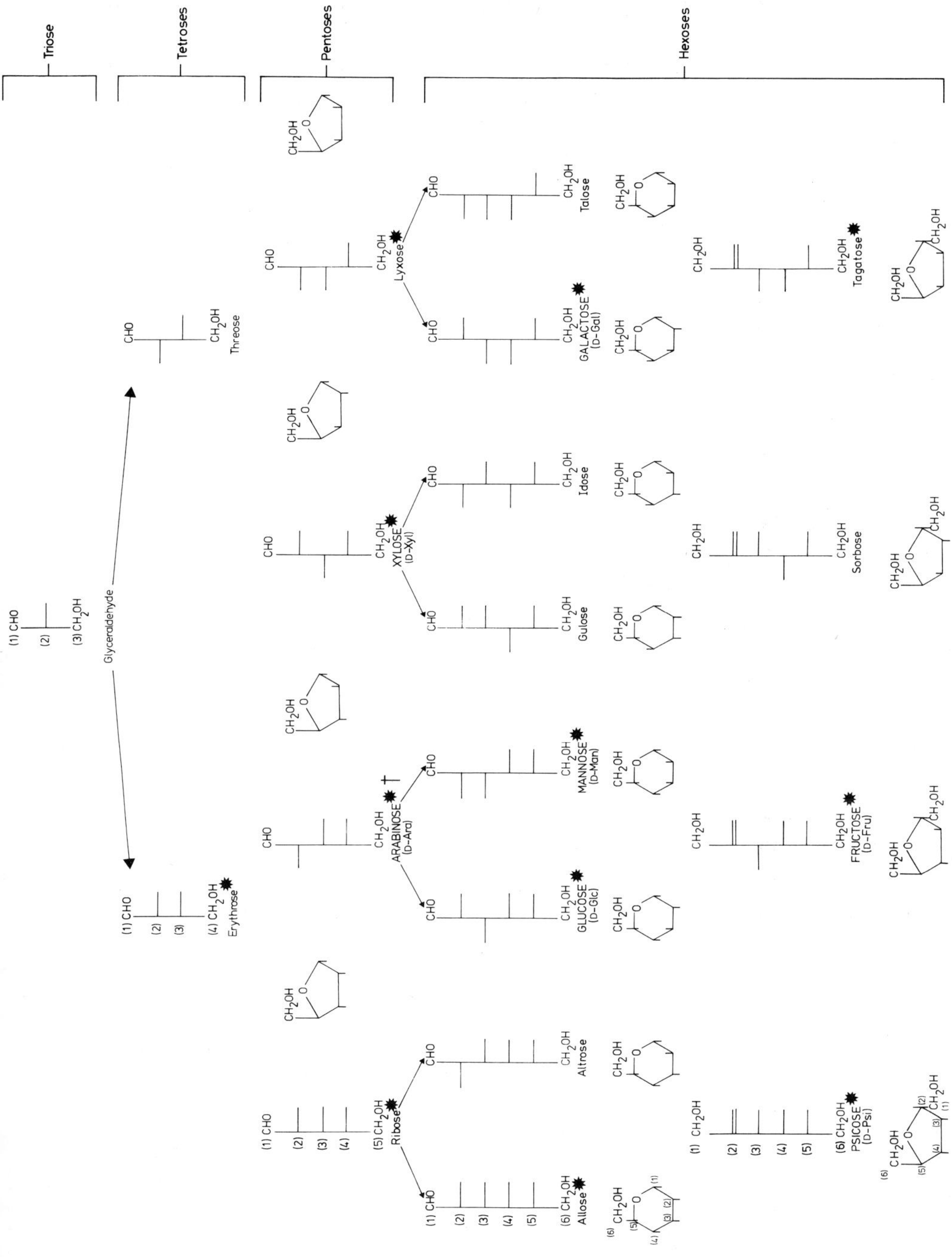

Triose
Tetroses
Pentoses
Hexoses
Glyceraldehyde
Threose
Erythrose
Ribose
ARABINOSE (D-Ara)
XYLOSE (D-Xyl)
Lyxose
Allose
Altrose
GLUCOSE (D-Glc)
MANNOSE (D-Man)
Gulose
Idose
GALACTOSE (D-Gal)
Talose
PSICOSE (D-Psi)
FRUCTOSE (D-Fru)
Sorbose
Tagatose

Table 1.1 *The occurrence of 7–9 carbon sugars in vascular plants (compiled from Schaffer, 1972)*

Sugar[a]		Trivial name	Formula[a]	Source[b]
Aldo-heptoses	D-*glycero*-D-*galacto*-heptose[c]		α 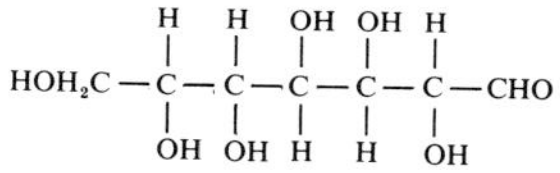 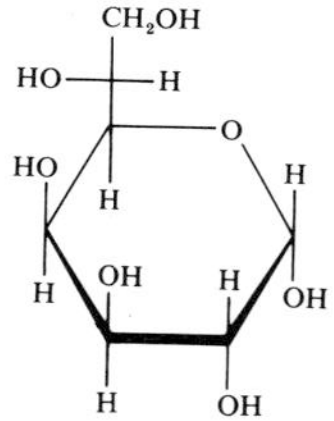	A
	D-*glycero*-D-*gluco*-heptose[c]		α 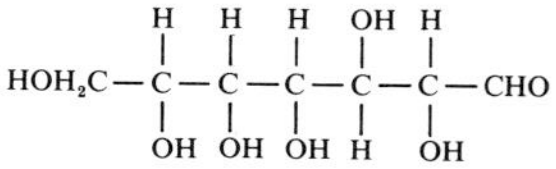 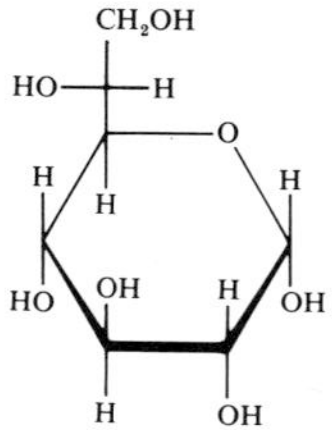	P
	D-*glycero*-D-*manno*-heptose[c]		α 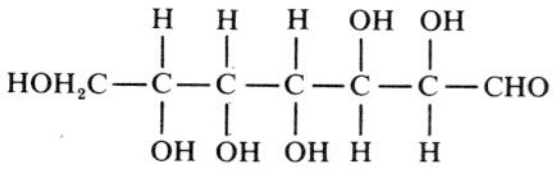 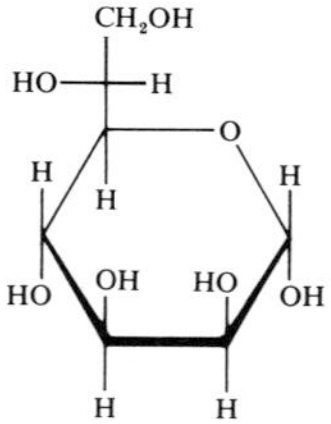	P

Keto-heptoses	D-*allo*-heptulose		α	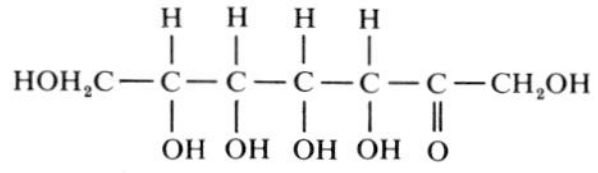	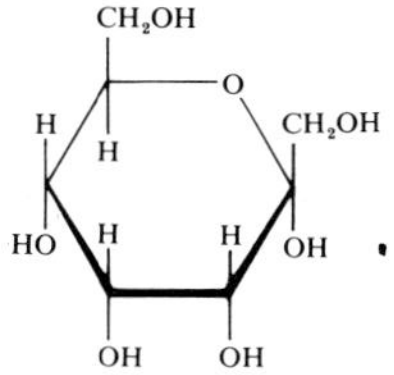	A, P
	D-*altro*-3-heptulose	Coriose	β	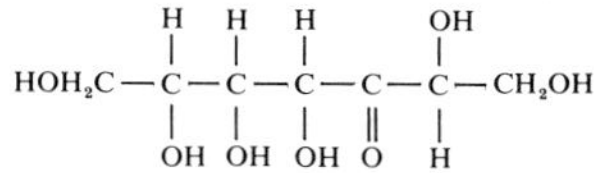	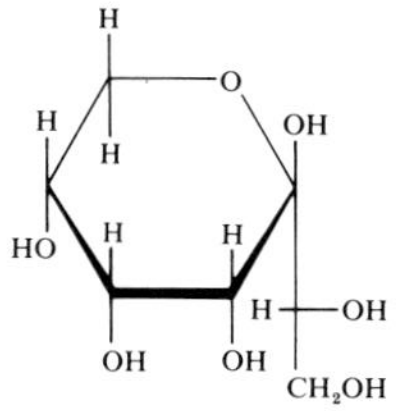	C, P
	D-*manno*-heptulose[c]	Perseulose	β	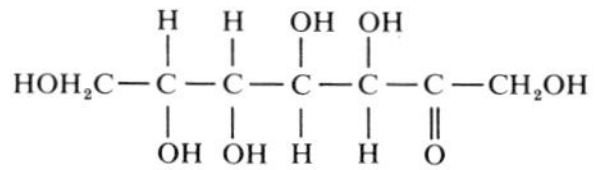	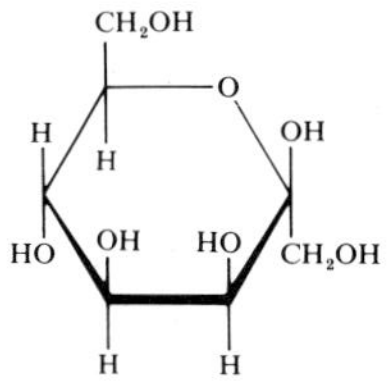	A
	D-*altro*-heptulose[c]	Sedoheptulose	α	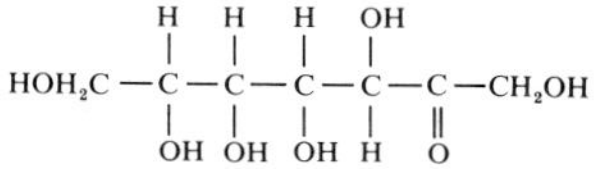	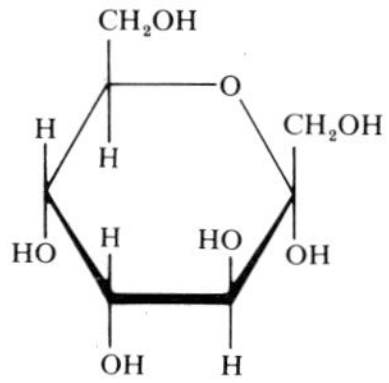	S

Table 1.1 (*cont.*)

Sugar[a]		Trivial name	Formula[a]	Source[b]
	D-*talo*-heptulose		α 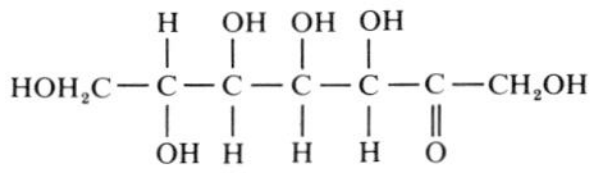 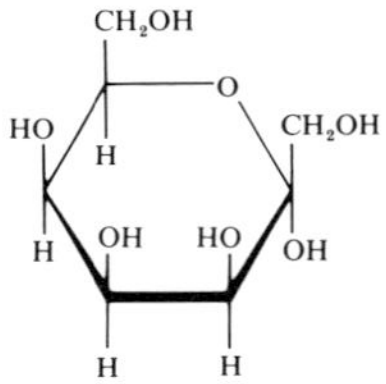	A
Keto-octoses	D-*glycero*-L-*galacto*-octulose		α 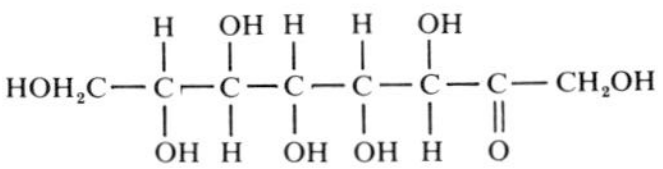 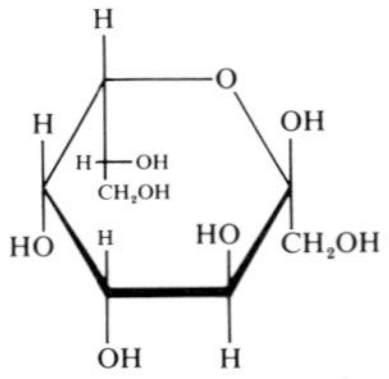	A
	D-*glycero*-D-*manno*-octulose[c]		α 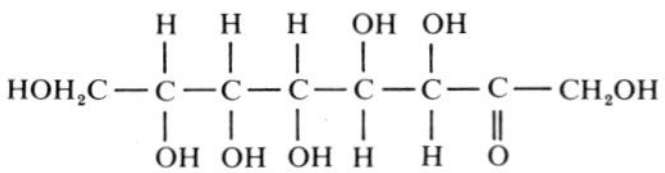 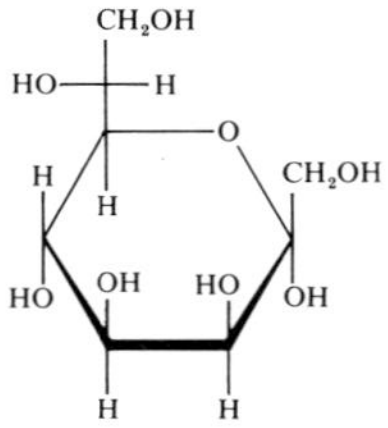	A, S

Keto-nonoses	D-*erythro*-L-*galacto*-nonulose	α	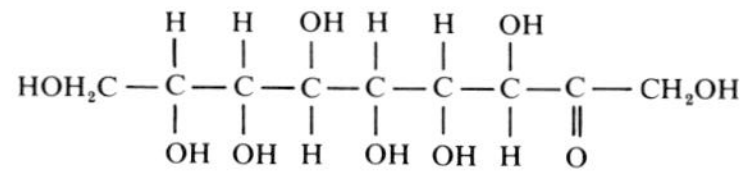	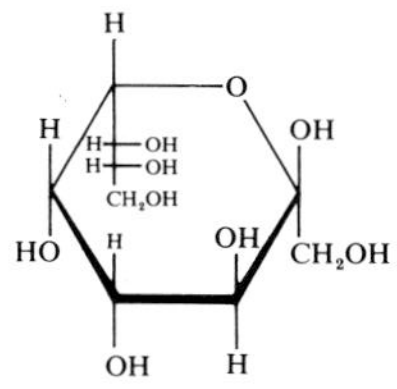	A, P, S
	D-*erythro*-L-*gluco*-nonulose	α	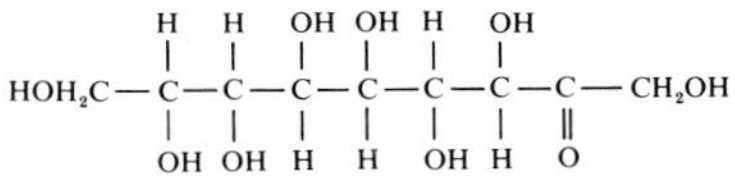	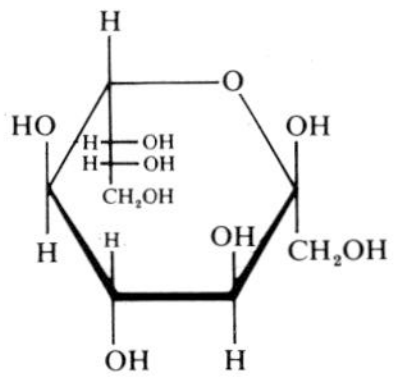	A, P, S

[a] Consult Fig. 1.1 to determine the logic of this nomenclature in relation to the formulae

[b] Sources: A = seed, fruit or leaves of avocado; C = all parts of *Coriaria japonica*; P = root or leaves of primroses; S = leaves of stonecrops and related plants

[c] Corresponding alditol also found in vascular plants (see below)

Occurrence

Glucose and fructose occur in all vascular plants as products of hydrolysis either of their phosphate esters or especially their disaccharide, sucrose (see Chapter 2). They can also be derived from their polymers, starch and fructan (see Chapters 4, 9 and 10). Concentrations of these reducing sugars vary widely but may constitute over 50% by weight of the edible portion in some fruits (e.g. date, Whiting, 1970). They are both substrates for the enzyme, hexokinase, which initiates their entry into metabolism, as sources of energy (via glycolysis, the oxidative pentose phosphate pathway and tricarboxylic acid cycle), as carbon skeletons for synthesis from intermediates of these pathways and as unchanged building blocks for the synthesis of oligo- and polysaccharides.

$$\begin{matrix}\text{Glc}\\\text{Fru}\end{matrix} + \text{ATP} \xrightarrow[\text{Mg}^{2+}]{\text{Hexokinase}} \begin{matrix}\text{Glc 6-P}\\\text{Fru 6-P}\end{matrix} + \text{ADP}$$

Other free monosaccharides have been found in substantial amounts much less frequently. Mannose and galactose occur transiently during degradation of their polymers e.g. in germinating legume seeds (see Chapter 11). D-Galactose has also been reported from ivy berries, L-arabinose from coniferous wood and a variety of aldoses from various fruits (Whiting, 1970). The only other recorded free hexose is D-psicose (= D-allulose) which occurs in the leaves of the saxifragaceous genus, *Itea*, along with its reduction product, allitol (see below and Chapter 7). Similarly, pairs of several 7–9 carbon ketoses and their corresponding alditols, as well as three aldoheptoses, concurrently exist in the vegetative and/or reproductive parts of a few, taxonomically unrelated species notably the stonecrops, primroses and avocado (Table 1.1). Although often present only in trace amounts, some e.g. D-*manno*-heptulose in some avocado varieties may reach 5% of the fresh weight. Note that, although its phosphate esters are universal compounds of both oxidative and reductive pentose phosphate pathways, the presence of substantial amounts of free sedoheptulose is restricted to a few vascular

Fig. 1.2 The open chain formulae of (*a*) apiose and (*b*) hamamelose.

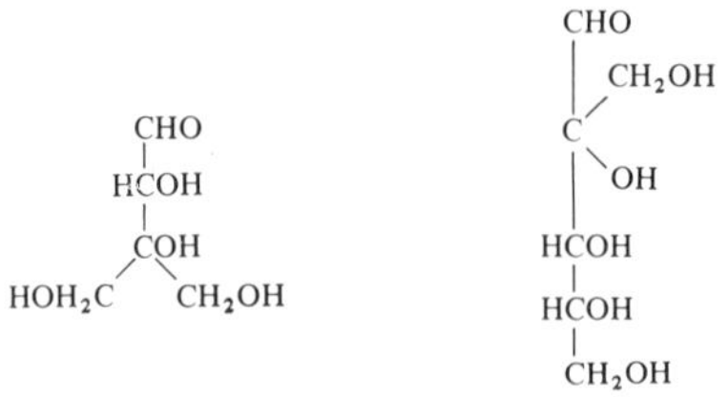

(*a*) (*b*)

plants, notably succulents of the Crassulaceae. The wider occurrence of this and other 7–9 carbon sugars in trace amounts will probably be revealed as sensitive gas chromatographic techniques (Holligan, 1971; Holligan & Drew, 1971) are applied to extracts from a wider range of plants than have hitherto been analysed.

Although branched-chain monosaccharides are common as components of many antibiotics (Hanesslan & Haskell, 1970), only two, apiose and hamamelose (Fig. 1.2), have been recorded in the free state in vascular plants. They are probably widespread in trace amounts (Watson & Orenstein, 1975; Sellmair, Beck, Kandler & Kress, 1977). Hamamelose, together with its alditol, occurs in many *Primula* spp. (Sellmair *et al.* (1977)). These sugars are better known as components of simple or complex glycosides, or polysaccharides, in a variety of plants e.g. parsley, duckweeds and witch-hazel.

The oligosaccharides

Introduction

With the exception of cyclic and acyclic sugar alcohols, all other carbohydrates to be considered in this book consist of various kinds, numbers and configurations of monosaccharide units linked together in various ways by glycosidic bonds. Carbohydrates containing from two to ten monosaccharide units, termed their *degree of polymerization* (DP), are termed oligosaccharides. Individual kinds of these are called di-, tri-, tetrasaccharides etc. Carbohydrates with larger degrees of polymerization comprise the polysaccharides.

In a glycosidic bond, the hydrogen atom of the hemi-acetal hydroxyl group (on carbon atom 1 of aldoses and 2 of ketoses) is replaced by an organic radical – another reducing monosaccharide or sugar alcohol in the examples to be discussed in this book. If the hemi-acetal hydroxyls of both units are linked, the resulting disaccharide is no longer a reducing sugar. When any other hydroxyl group is involved, one hemi-acetal configuration remains, so that the resulting glycoside retains reducing properties. For this reason, it is important to specify which hydroxyls are involved in the glycosidic bond. In order fully to characterize particular glycosides, three other features need to be known – whether the sugars involved are D- or L- (their chirality), whether the anomeric carbons are linked in the α- or β-configuration and whether the units are in pyranose or furanose forms. Four examples will illustrate these points.

Maltose, gentiobiose and trehalose are all disaccharides consisting of two glucose units; sucrose is a combination of glucose and fructose. The first three differ in that the glucose units are linked $\alpha 1 \rightarrow 4$ in maltose, $\beta 1 \rightarrow 6$ in gentiobiose and via both hemi acetal hydroxyls in trehalose ($\alpha 1 \rightarrow \alpha 1$). Thus, the first two are reducing sugars whereas trehalose is not. In all three, the glucose units are in the pyranose form. Sucrose also contains glucopyranose but its ketose unit, fructose, is in the furanose configuration. Since sucrose is also non-reducing, the linkage must be between C1 of glucose and C2 of fructose. It turns out that the hydroxyl of the former involved in the glycosidic bond is α and that of the latter is β (see Fig. 1.3 below). These structures may be summarized as follows.

Maltose: 4-*O*-α-D-glucopyranosyl-D-glucopyranose
(4-*O*-α-D-Glc*p*-D-Glc*p* **or** α-D-Glc*p* (1→4)-D-Glc*p*)
Gentiobiose: 6-*O*-β-D-glucopyranosyl-D-glucopyranose
(6-*O*-β-D-Glc*p*-D-Glc*p* **or** β-D-Glc*p* (1→6)-D-Glc*p*).
Trehalose: α-D-glucopyranosyl-α-D-glucopyranoside
(α-D-Glc*p*-α-D-Glc*p* **or** α-D-Glc*p* (1→1)-α-D-Glc*p*).
Sucrose: β-D-fructofuranosyl-α-D-glucopyranoside
(β-D-Fru*f*-α-D-Glc*p* **or** β-D-Fru*f* (2→1)-α-D-Glc*p*).

Note that, in the full names, those ending in *-ose* indicate that the disaccharides are reducing; those ending in *-oside* are non-reducing. Clearly, although informative, the full names rapidly become cumbersome as oligosaccharides increase in size. For simple description of groups of oligo- and polysaccharides, use of the suffix *-an* is a valuable shorthand. Glucans are chains of glucose; sucrosyl-fructans are chains of fructose connected to a sucrose end-group; sucrosyl-galactans are chains of galactose connected to a sucrose end-group; galactomannans are heteropolymers of galactose and mannose, etc. More useful descriptions can be formed by adding details of the anomeric and

Table 1.2 *Free energy of hydrolysis* ($\Delta G^{\ominus}$) *of glucosidic linkages* (*Dey, 1980*)

Compound	$\Delta G^{\ominus}$ (kcal mol^{-1})	$\Delta G^{\ominus}$ (kJ mol^{-1})
UDP-D-glucose (pH 7.4)	−7.6	−31.8
α-D-Glucosyl phosphate[a] (pH 8.5)	−4.8	−20.1
D-Glucose 6-phosphate (pH 7.0)	−3.3	−13.8
D-Fructose 6-phosphate (pH 7.0)	−3.8	−15.9
Sucrose	−6.6	−27.6
Maltose	−4.6	−19.3
α, α-Trehalose	−4.4	−18.4
Glycogen [(1→4)-α-D-glucosyl linkage]	−4.4	−18.0
Dextran [(1→6)-α-D-glucosyl linkage]	−2.0	−8.4
Levan [(2→6)-β-D-fructosyl linkage]	−4.0	−16.8

[a] Synonym of glucose 1-phosphate

positional nature of linkages, and of chirality, e.g. α-D-(1→4) glucan to describe the amylose portion of starch (see below).

When different disaccharides are hydrolysed to their constituent monosaccharides, the free energy of hydrolysis varies considerably. Table 1.2 compares values for the glycosidic bonds of some di- and oligosaccharides, and phosphate esters. The large values observed mean that synthesis of the bond cannot be conducted physiologically by a reversal of hydrolysis. Mechanisms for the synthesis of glycosidic bonds are discussed below, especially in Chapters 2, 4, 5, 10 and 11. The significance of the large value for sucrose will be mentioned in the next section and also in Chapters 4 and 5.

Structure, occurrence and location

Sucrose. The reasons why the non-reducing disaccharide sucrose (Fig. 1.3) must be regarded as the principal sugar of vascular plants are discussed in Chapter 2. Sucrose is found throughout plants, often in high concentrations. After comparing various biophysical and biochemical attributes of solutions of glucose and sucrose, Arnold (1968) concluded that, by

Fig. 1.3 The formulae of (*a*) sucrose, (*b*) trehalose, (*c*) maltose and (*d*) gentiobiose.

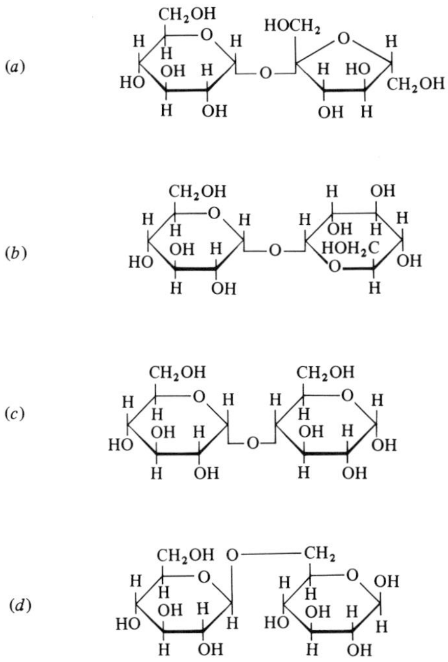

virtue of its non-reducing nature, sucrose could be regarded as a 'protected' hexose. In this role, sucrose can be moved around plants and stored without being readily metabolized until required. Following Pontis (1977), it should also be stressed that the free energy of hydrolysis of sucrose is almost as large as that of sugar nucleotides (Table 1.2). These are important intermediary metabolites in the interconversion of sugars and synthesis of polysaccharides (see Chapters 2, 5, 10 and 11). In reactions which conserve this energy, sucrose can act as a source both of glucosyl units for the synthesis of UDPGlc, probably in all vascular plants (see Chapter 2), and of fructosyl units for the synthesis of fructans in the most highly evolved angiosperms (see Chapter 4). However, in contrast to this conservation, hydrolysis by invertases, in which the free energy is lost, is also universally of metabolic importance (see Chapter 2). The metabolism of plants is not always organized in the theoretically most efficient thermodynamic manner!

Trehalose and selaginose. α,α-Trehalose (Fig. 1.3), a non-reducing disaccharide like sucrose, is a common sugar of heterotrophic eukaryotes especially many fungi and invertebrates and is also found occasionally in algae and bryophytes. It is uncommon in vascular plants and, until the mid 1960s, was only reliably known to occur in the pteridophyte lesser clubmosses, from which it had been isolated by Anselmino & Gilg (1913) and in which its metabolism had been briefly investigated (Quillet & Soulet, 1964). In the last twenty years, largely by studies in Kandler's laboratory, it has been found in leaves of several eusporangiate ferns as well as many further species of *Selaginella*, and in ripening fruits of umbellifers (Apiaceae) (see Hopf & Kandler, 1976; Kandler & Hopf, 1980). It co-exists with sucrose but often substantially exceeds this in concentration. Indeed when it occurs, it appears to replace sucrose as a translocatory and storage disaccharide. A trehalosyl-glucoside, selaginose (α-D-Glc*p* (1 → 1) α-D-Glc*p* (2 → 1) α-D-Glc*p*), has also been isolated from the clubmoss, *Selaginella kraussiana*, and related species (Fischer & Kandler, 1975). The metabolism of trehalose and selaginose is briefly discussed in Chapter 5.

Maltose, gentiobiose and other rare disaccharides. As noted above, the diglucosides, maltose and gentiobiose (Fig. 1.3) differ from sucrose and trehalose in being reducing sugars. This is also true of lactose [β-D-Gal*p* (1 → 4)-D-Glc*p*]. Of these three sugars, free gentiobiose and lactose are rarely found in plants. Gentiobiose, a component of the trisaccharide gentianose, discussed below, has been found in xylem sap of a birch and throughout the vegetative parts of plants containing gentianose (Franz & Meier, 1972). There are well authenticated reports of lactose, the characteristic disaccharide of mammalian milk, in chicle and anthers of *Forsythia*. Table XI of Staněk *et al.* (1965) lists these and further occurrences of gentiobiose, lactose and other rare disaccharides e.g. melibiose [α-D-Gal*p* (1 → 6)-D-Glc*p*] and planteobiose [α-D-Gal*p* (1 → 6)D-Fru*f*]. Like gentiobiose, these are also components of trisaccharides (raffinose and planteose, respectively). The trisaccharides are found free more frequently than their component, galactose-containing, disaccharides (see below).

Maltose is of much more frequent occurrence (again, see Table XI of Staněk *et al.* 1965). Indeed, it is now becoming clear that this sugar, either formed as a product of degradation of starch (see Chapters 9 and 10) or synthesized *de novo* from hexose phosphates (see Chapter 5), is probably of widespread occurrence in at least trace amounts. There are occasional reports of its presence in high concentrations e.g. 45, 38 and 14% respectively in petioles, stems and roots of soybean (Quillet & Bourdon, 1956; Miller, 1973). The possibility of its presence in enhanced amounts in phosphate-deficient photosynthetic tissues needs to be explored (Herold, McGee & Lewis, 1980).

Many other disaccharides, which will not be considered in this book, occur as components of glycosides etc. but rarely as the free sugar (see Staněk *et al.* 1965; Miller, 1973).

Sucrosyl-fructans. A number of species store reserve carbohydrate in the form of fructose-rich polysaccharides. The first to be described and characterized

Table 1.3 *A representative list of the occurrence of fructans in vascular plants*[a]

Group and family	Representative species	Type of fructan
Monocotyledons		
Gramineae (= Poaceae)	Timothy grass, Cocksfoot, Barley	Phlein in vegetative organs; branched in grains
Liliaceae	Onion	Inulin
Amaryllidaceae	Century plant	Branched
Iridaceae	Yellow flag	Branched
Dicotyledons		
Compositae	Jerusalem artichoke	Inulin
	Dahlia	Inulin
	Dandelion	Inulin
	Chicory	Inulin
Campanulaceae	Creeping campanula	Inulin
Boraginaceae	Comfrey	Inulin

[a] The material in this table has been derived from Archbold (1940), Hirst (1957), Aspinall & Das Gupta (1959), Akazawa (1965) and Scott (1968)

was obtained from tubers of Jerusalem artichoke and given the name 'inulin' (Archbold, 1940). Subsequently these polymers, referred to interchangeably as sucrosyl-fructans, fructans, fructosans or polyfructans, have been detected in a range of material including bacteria (Harrison, Tarr & Hibbert, 1930), algae (Meeuse, 1963), bryophytes (Maas & Craigie, 1964) and vascular plants.

Fructans have been isolated from both dicotyledons and monocotyledons (Table 1.3), Both fructans and starch (see below) can be isolated from the majority of these species, although they may occur in different organs. Accumulation of fructan therefore should be regarded as a complement to synthesis of starch, rather than as an alternative. The occurrence of these polymers in such distinct and highly evolved families as the Compositae and Gramineae indicates that accumulation confers a selective advantage. Their synthesis does not represent a minor pathway of declining evolutionary significance.

There is agreement on the major features of the structure of fructans but considerable uncertainty still exists concerning differences between material isolated from different sources. One of the problems involved in evaluating such work is the nomenclature employed by different workers. Often polymers have been assigned names based only upon the species from which they were extracted. Here, structural differences will be used to classify fructans, with the species cited where relevant.

Generally, fructans can be divided into two groups based upon the nature of the major linkage between adjacent fructose residues. The structure of inulin from Jerusalem artichoke, reviewed by Hirst (1957), is shown in Fig. 1.4*a*. The fructofuranose residues are linked $\beta(2 \rightarrow 1)$ and are terminated by a glucopyranose residue, giving a linear, non-reducing molecule. The mean DP of inulin from mature tubers of Jerusalem artichoke is around 35 (McDonald, 1946). Chromatographic separation shows that this inulin

Fig. 1.4 The formulae of (*a*) the 'inulin-type' and (*b*) the 'phlein type' of sucrosyl-fructan.

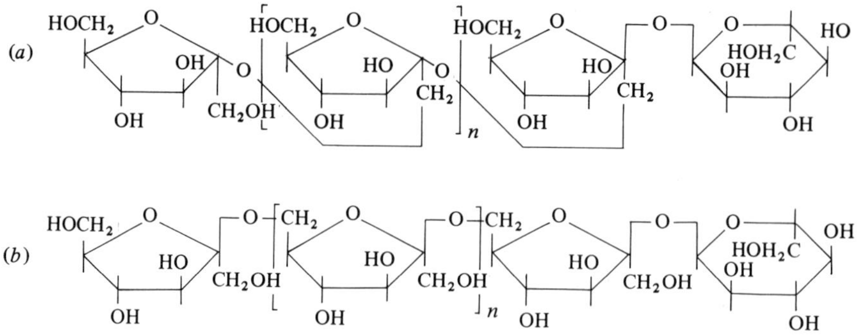

consists of a range of individual oligosaccharides and short-chain polysaccharides (Edelman & Jefford, 1968). Paper chromatography can resolve these components into an homologous series starting with sucrose (DP 2) and in which each successive member has one more fructose residue than the one eluted immediately before it (French & Wild, 1953). Such series are found in other members of the Compositae (Bacon & Edelman, 1951) and in onion (Darbyshire & Henry, 1978). This structure appears to be the principal one found in vascular plants except the Gramineae and collectively these polymers are referred to as the 'inulin type'.

The principal linkage between adjacent fructose residues of fructans from several temperate grasses is β (2→6) (Laidlaw & Reid, 1951; Bell & Palmer, 1952). Again the chains are terminated by a glucopyranose residue and are non-reducing (Fig. 1.4*b*). In at least some grasses there are no detectable oligosaccharides between sucrose and high-molecular-weight fructan (Pollock, 1979). Polymers of this general structure may be referred to as 'grass levans' by analogy with bacterial fructans which are also called levans, or as the 'phlein type' after the polymer isolated from *Phleum pratense* (timothy grass).

Significant variations exist on these basic structures. Glucose-free fructans have been reported but are probably artefacts caused by hydrolysis during extraction (Quillet, 1957). Recently, two oligo-fructans from asparagus roots with glucose residues contained within the chain have been characterized (Shiomi, 1981). There is also some evidence for branched-chain fructans, particularly among the low-molecular-weight oligosaccharides isolated from cereal grains. Two series have been described, those with backbones of the inulin type and 2→6 linkages at the branch-points, and those with phlein type backbones and 2→1 linkages at the branch-points (Axelrod, 1965). Indeed, there are reports of three distinct homologous series in developing wheat grains (Meredith & Jenkins, 1973).

There are also reports of branched fructans in the vegetative organs of grasses (Arni & Percival, 1951; Bell & Palmer, 1952) but these have not been supported by later studies using infra-red spectroscopy (Kühbauch, 1974), ^{13}C n.m.r. spectroscopy (Tomašić, Jennings & Glaudemans, 1978), and improved methods for methylation analysis (Pollock, Hall & Roberts, 1979). Use of these improved techniques for structural analysis on a wider range of material will be necessary to resolve this discrepancy.

The mean size of fructans isolated from different species or from different parts of the same species may vary markedly. Inulin from Jerusalem artichoke has been reported with a mean DP of 35 (mol. wt, 6 kD) (McDonald, 1946) or up to 50 (8.5 kD) (Bell & Palmer, 1952). In contrast, inulin-type polymers from onion scales had a maximum DP of only 11 (1.8 kD) (Darbyshire & Henry, 1978). In grasses, fructans show an even greater variation in size.

Estimates vary from DP 25 (4.3 kD) in meadow fescue to DP 50 (8.5 kD) in timothy (Pollock & Jones, 1979).

Although fructans may be detectable throughout the plant, accumulation is generally extensive in specific organs. These may be morphologically distinct storage organs such as the tubers of Jerusalem artichoke (Edelman & Jefford, 1968), modified non-photosynthetic organs such as the roots of dandelion (Rutherford, Griffiths & Wain, 1966), leaf bases of onion (Darbyshire & Henry, 1978) and swollen stem bases of *Hordeum bulbosum* and *Arrhenatherum elatius* (C. J. Pollock and D. H. Lewis, unpublished data) or the morphologically unmodified leaves and stems of many temperate grasses (Pollock & Jones, 1979). In addition accumulation of fructan can be initiated by biotrophic fungal infection in tissues where it is normally absent (Holligan, Chen & Lewis, 1973), by detaching the leaf (Tetley & Krivak, 1976), or by incubating leaf disks on solutions of sucrose (Chandorkar & Collins, 1972).

Reliable evidence on the intracellular compartmentation of fructans is lacking, mainly owing to the high solubility in water of the polymers and the absence of sensitive and specific histochemical methods. Based on the precipitation of inulin by ethanol within the vacuole after sectioning, it is generally believed that a substantial proportion of the accumulation must be within the vacuole (Edelman & Jefford, 1968). This is supported by the occurrence of fructan contents of up to 70% of tissue dry weight (Edelman & Jefford, 1968; Darbyshire & Henry, 1978). It is not known, however, whether fructans are accumulated selectively within the vacuole or are in equilibrium with the cytoplasmic component.

Sucrosyl-galactans and sucrosyl-glucosides. Sucrosyl-fructans, described in the last section, represent only one way in which the sucrose molecule can be used as a template for the elaboration of oligosaccharides. Vascular plants contain a wide range of sucrosyl-galactans in which the galactose units are joined by various linkages either to the fructose or glucose units of sucrose, or to both. In addition one sucrosyl-glucoside, gentianose, occurs in the Gentianaceae. Figure 1.5 shows the relationship of these carbohydrates to sucrose and to the fructans previously discussed. Tables 1.4 and 1.5 summarize the taxonomic distribution, and that within plants, of these oligosaccharides.

The raffinose series (Fig. 1.6), omitted from Table 1.4 because of its very wide distribution, consists of seven oligosaccharides in which successive units of galactose are joined $\alpha(1\rightarrow6)$ to the glucose moiety of sucrose. Raffinose occurs widely in gymnosperms and angiosperms in both vegetative and reproductive parts of the plants, and is often accompanied by its homologues of larger DP. As their trivial names suggest, these are abundant in genera of the Labiatae and Scrophulariaceae but are by no means confined to these.

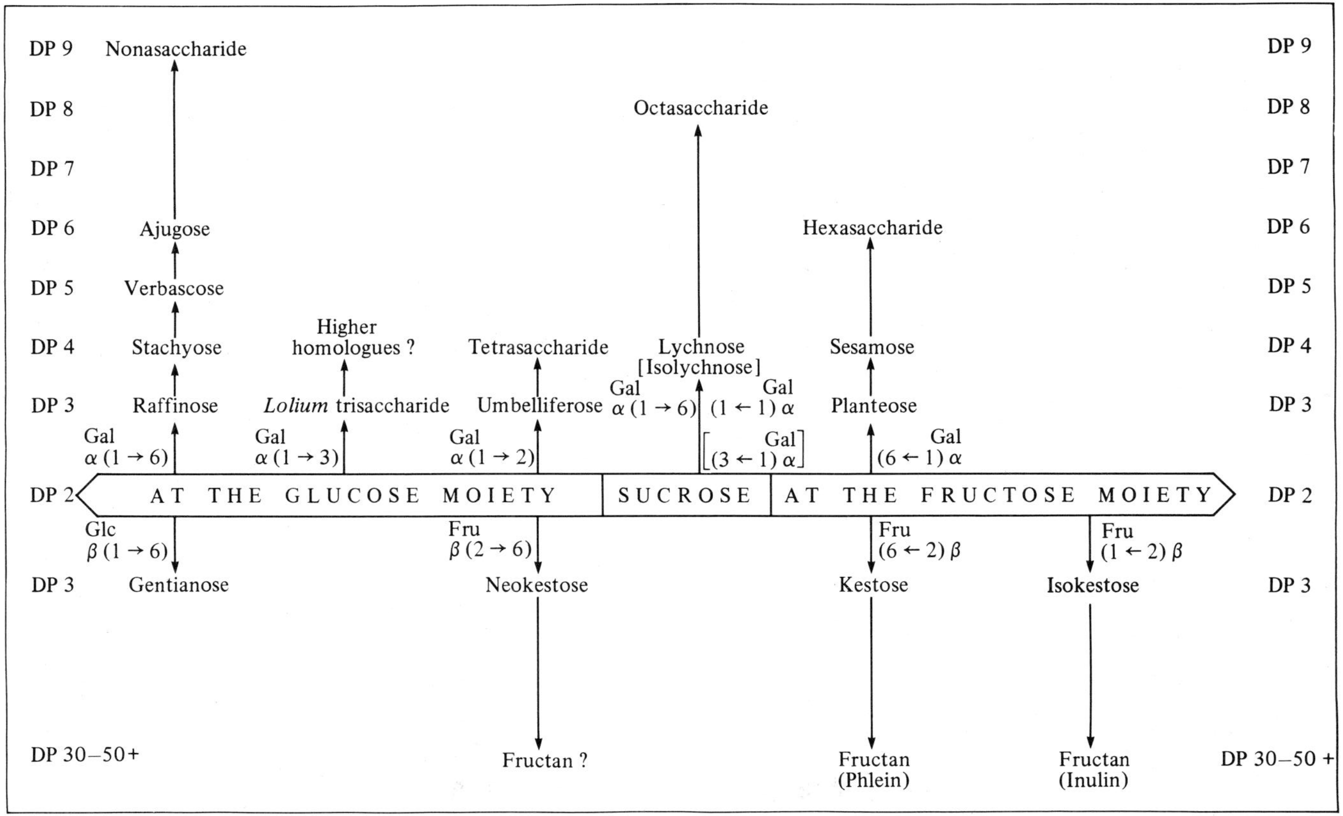

Fig. 1.5 The structural relationships of sucrosyl-fructans, -galactans and -glucoside to sucrose. Sugars in the upper part of the figure consist of galactose units α-linked to the glucose moiety (left) or fructose moiety (right) of sucrose. Both units of sucrose are involved in the lychnose/isolychnose series. Formulae of representative sugars are shown in Fig. 1.6. Sugars in the lower part of the figure consist of glucose or fructose units β-linked to the glucose moiety (left) or fructose moiety (right) of sucrose. Formulae of representative sugars are shown in Figs 1.4 and 1.6.

They are, for example, important constituents of many Leguminosae (Courtois & Percheron, 1971) and are translocated in plants as taxonomically diverse as the herbaceous Cucurbitaceae (squashes, melons, etc.) and woody Oleaceae (ash, lilac, etc.). Tables XII, XIV, XVI, and XVII of Staněk *et al.* (1965) record a large number of species from which this series of oligosaccharides has been isolated.

As Table 1.4 indicates, the other four kinds of sucrosyl-galactan are of much more restricted distribution. This has been reviewed by Kandler & Hopf (1980). Umbelliferose (Fig. 1.6) is confined to the related families Apiaceae (= Umbelliferae) and Araliaceae and occurs in both vegetative and reproductive parts of the plants. By contrast, the *Lolium*-trisaccharide and planteose (Fig. 1.6) and, sesamose, the higher homologue of the latter, are almost confined to seeds. Planteose is restricted to a group of related orders of vascular plants within the Asteridae as conceived by Stebbins (1974) – Gentianales, Polemoniales, Lamiales, Plantaginales and Scrophulariales. In

Table 1.4 *Distribution of sucrosyl-oligosaccharides other than the raffinose series in vascular plants (see Kandler & Hopf, 1980)*

Oligosaccharide	Occurrence restricted to
Gentianose	Gentianaceae (*Gentiana*, *Swertia*)
Kestose series	Asterales; Campanulales; Boranginaceae; Iridales; Poales
Lolium trisaccharide	Poaceae (= Gramineae)
Lychnose series	Caryophyllaceae
Planteose series	Polemoniales; Scrophulariales; Lamiales *sensu* Takhtajan (1969)
Umbelliferose	Apiaceae (= Umbelliferae); Araliaceae

Table 1.5 *Distribution of sucrosyl-oligosaccharides in organs of plants*

Oligosaccharide	Leaves	Roots/Rhizomes	Seeds
Gentianose	+	+ +	–
Kestose series	+	+ +	+
Lolium trisaccharide	–	–	+ +
Lychnose series	+	+ +	–
Planteose series	–	–	+ +[a]
Raffinose series	+	+ +	+ +
Umbelliferose	+	+ +	+ +

+, present; + +, present in considerable amounts; –, absent

[a] The occurrence of planteose in vegetative organs has only been reported once (in rhizomes of *Teucrium canadense*)

Fig. 1.6 The formulae of (*a*) gentianose, (*b*) raffinose, (*c*) stachyose, (*d*) *Lolium*-trisaccharide, (*e*) umbelliferose, (*f*) lychnose, (*g*) isolychnose and (*h*) planteose. (Compare with Fig. 1.5.)

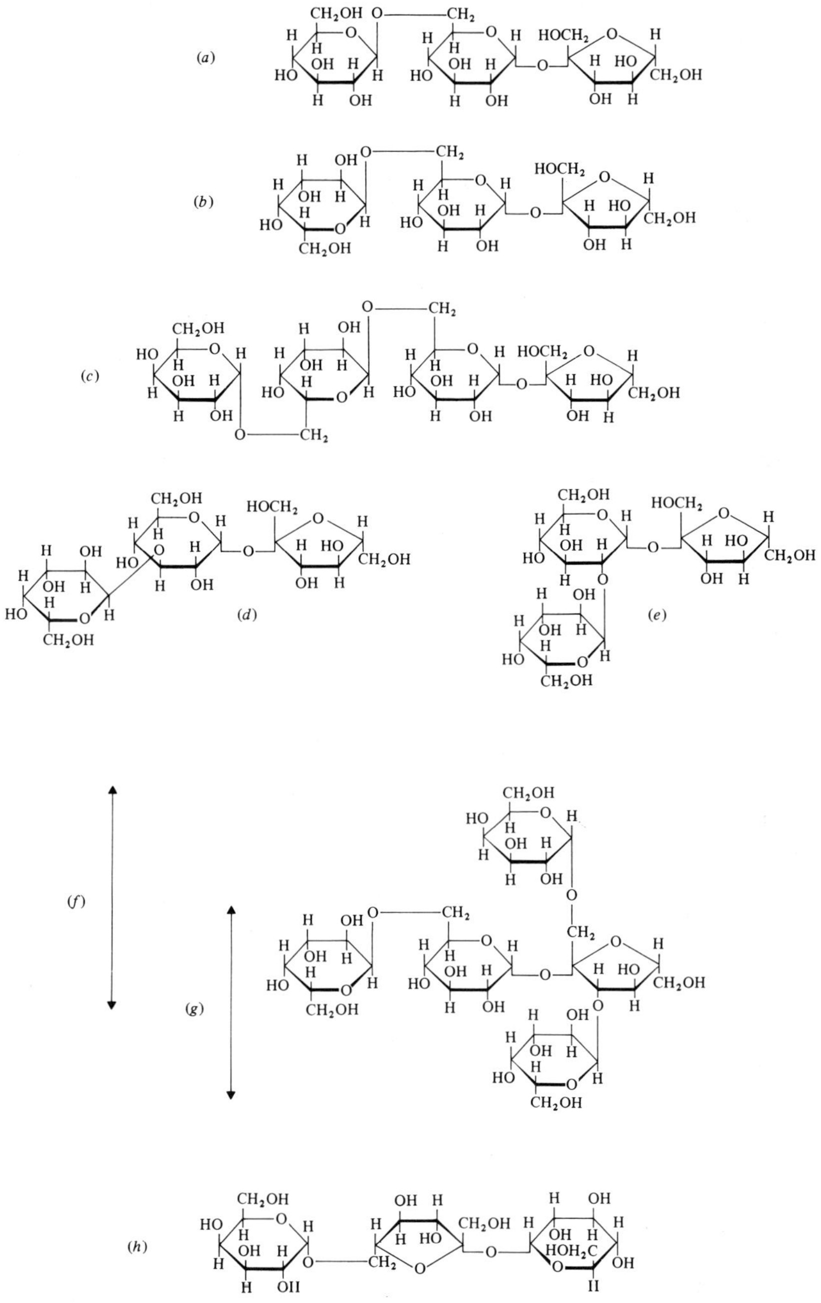

the scheme of Takhtajan (1969), it is found in fewer orders (Table 1.4). It has been reported only once from vegetative tissue – the rhizomes of *Teucrium canadense* (French, Wild, Young & James, 1953) but these were extracted with water, possibly allowing some non-physiological transgalactosylation of sucrose before sugars were precipitated with methanol. The single report from a species not within this taxonomic complex, cacao, turns out to be a post-harvest phenomenon during fermentation since planteose does not occur in freshly harvested beans (Kandler & Hopf, 1980). Again, by contrast, the two series based on the tetrasaccharides, lychnose and isolychnose (Fig. 1.6), are restricted to vegetative parts of species of the Caryophyllaceae where they co-exist with sugars of the raffinose series. The latter only is found in the seeds.

The only sucrosyl-glucoside yet isolated from plants, gentianose (Fig. 1.6), similarly is confined to the vegetative parts of species of *Gentiana* and *Swertia perennis* (also Gentianaceae). Chapter 5 considers the biosynthesis of this sugar and the sucrosyl-galactans.

The polysaccharides

Introduction

Two types of polysaccharide have a storage function in plants. These are starch and 'hemicelluloses' (see above for reservations about this term). The former, a homopolymer of glucose, is intracellular where it occurs as granules in membrane-bound organelles, the plastids (chloroplasts and amyloplasts). The latter are heteropolymers of a variety of sugars (see Table 1.6 below) and are extracellular, often components of cell walls. The literature on both types in both 'pure' and 'applied' science is extensive. Much is known of the chemistry and biology of starch (see below and Chapters 8–10), but, until recently, most studies of 'hemicelluloses' were by chemists and applied scientists (see below). However, there has recently been a welcome and fruitful upsurge of research into their biology (see Chapter 11).

Structure, occurrence and location

Starch and phytoglycogen. Details of the chemical and physical structure of starch may be found in many reviews and monographs such as those of Badenhuizen (1958, 1969), Whelan (1958), Whistler & Paschall (1965), Radley (1968), Greenwood (1970), Manners (1973), Banks & Greenwood (1975) and French (1975). The following is but a brief summary.

It has been customary to distinguish 'assimilatory' from 'storage' starch. The former is characteristically rapidly synthesized in chloroplasts by day and at least partially degraded by night, whereas the latter is deposited in storage organs, principally fruits, seeds and tubers, over a longer period during one season and remobilized at another. As pointed out in Chapter 8, however, this distinction is blurred.

The starch grain itself varies in both size (1–100 μm in diameter) and shape (spherical, polygonal, elliptical, grooved, etc.) between species and with environmental conditions. Czaja (1978) describes 14 types and discusses their taxonomic distribution. The grains consist principally of two polymers, amylose and amylopectin, although material of an intermediate nature also occurs, especially in starches rich in amylose. Although little studied compared to the two main polymers, this material may be important in the synthesis of the granules themselves (Banks & Greenwood, 1975). Amylose is an essentially linear, helically structured α-D-(1→4) glucan with a DP of several thousand. A few α-D-(1→6) linked branches are present (Banks & Greenwood, 1975). This is the component of starch which binds iodine and stains blue. The iodine occupies the cavity of the helix in which there are six glucose residues per turn. Amylopectin, an even larger molecule (DP > 10^4), is a branched glucan consisting of α-D-(1→4)-linked chains with an average length of 19–26 units joined together by α-D-(1→6) linkages (about 4–5% of the total). This stains purple but does not bind iodine appreciably, probably because the large number of branch-points disrupts formation of the iodine-binding helix. The chains of both polymers may be regarded as repeated units of maltose with isomaltose as the disaccharide linking the chains in amylopectin. In this way, amylose essentially has one reducing and one non-reducing end, whereas amylopectin has one reducing but many non-reducing terminal glucosyl units (Fig. 1.7). The structure of amylopectin in this figure is, as yet, only a tentative one (see Banks & Greenwood, 1975, and French, 1975, for detailed discussions). There is now good evidence, also discussed by these authors, that starch, especially from roots and tubers, contains a small amount of phosphorus (c. 0.06% as glucose 6-phosphate), principally in the amylopectin fraction.

The proportions of the two polymers varies between species, between cultivars within species, between different parts of the plant and within leaves, during the diurnal cycle (see Whelan, 1958; Matheson & Wheatley, 1963; Table 1 of Greenwood, 1970; Kakie & Sugizaki, 1970; Chang, 1979). In most situations, amylopectin usually predominates to the extent of > 99% in *waxy* mutants of cereals. Normally, it accounts for 75–85%. However, starches with high proportions of amylose occur in the Easter lily (32%), in certain varieties of maize (77%) and in wrinkled peas (72%) (Whelan, 1958). Lower values for 'amylomaize' are given by Greenwood (1970) and Banks & Greenwood (1975) because much of the material previously reported as amylose is better regarded as the intermediate material mentioned above.

Two further branched α-D-glucans occur in fruits of sweet corn varieties of maize and in Müllerian bodies of *Cecropia peltata*, structures predated by the protective colonies of ants harboured by this species. The glucans have more branching and shorter chain lengths than amylopectins and so resemble glycogen, the typical storage glucan of animals and fungi. They are termed phytoglycogens (Manners, 1973; Marshall & Rickson, 1973).

Analyses made over a century ago (e.g. Boehm, 1856 – cited by Parkin, 1899) revealed that certain plants never contain starch. In 1885, Meyer remarked that 'dicotyledons store starch plentifully in their leaves whilst monocotyledons have little or none'. Meyer classified a wide range of

Fig. 1.7 Formulae of (*a*) maltose and (*b*) isomaltose, and structures of (*c*) amylose and (*d*) amylopectin. In (*c*) and (*d*), the open circles represent non-reducing glucose units and the closed circles reducing glucose units. In the tree formula of amylopectin (*d*), an A chain is linked to the rest of the molecule only through its reducing end group (i.e. it is *not* substituted at C6); a B chain, in addition to being linked as an A-chain, is also substituted via the hydroxyl group on C6 in one or more of its constituent glucose units; the C-chain, which is substituted on C6 of one or more of its glucose units by A or B chains, terminates with a reducing end group. The dotted line leading to this indicates a repetition of the branching pattern (see Banks & Greenwood, 1975 and French, 1975, for details).

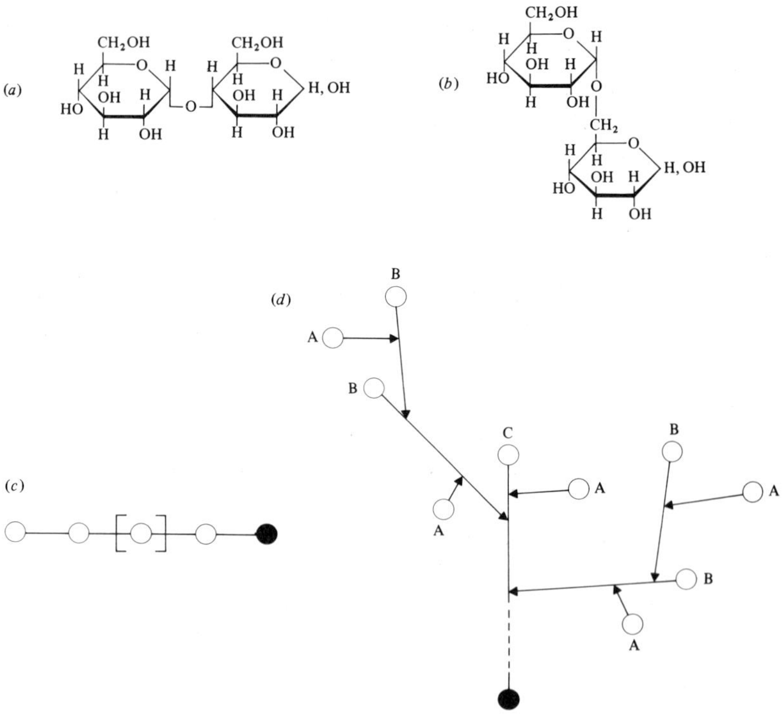

dicotyledons according to their starch content. For example, the Solanaceae and Leguminosae were grouped together since they contained more starch than any other families. Smith (1968) has discussed the distribution of starch in the monocotyledonous Gramineae. Parkin (1899) and Gates & Simpson (1968) have confirmed that production of starch is generally low in monocotyledons and absent in some species – notably the genus *Allium* (onion, garlic etc.) and *Yucca glauca* although Hart, Haldiman, Gries & Rogers (1960) indicate that synthetic enzymes are present but inhibited in garlic. While some, normally starch-free, species can be induced to accumulate this glucan by supplying exogenous sugars, this was not possible for various *Allium* spp. or *Y. glauca* (Boehm, 1883; Gates & Simpson, 1968). Visually starch appears the same whether synthesized during photosynthesis or from exogenously supplied sugar (Fig. 1.8) although its composition may differ (Badenhuizen, 1969).

Many endogenous and exogenous factors also markedly influence the amount of assimilatory starch accumulated. Characteristically, within leaves, the palisade cells contain more than the spongy mesophyll and, in C4 species, the bundle sheath cells contain more than the mesophyll. In leaves as a whole, content of starch varies with their age. In tobacco, for example, there is little starch during the period of expansion but the amount present rapidly increases in the full-sized leaf. Concentrations then drop as the leaves senesce (Matheson & Wheatley, 1963). The rate of consumption of photosynthetic products in other parts of the plant is also a major influence on accumulation of starch in leaves (see Herold, 1980, and Chapter 8). Among the exogenous factors which affect amounts of starch in leaves are temperature, irradiance, photosynthetic period, concentration of CO_2 and phosphorus nutrition (e.g. Brown & Escombe, 1902; Eaton, 1950, 1952; Madsden, 1968; Forde, Whitehead & Rowley, 1975; Nafziger & Koller, 1976; Chatterton & Silvius, 1979; Silvius, Chatterton & Kremer, 1979). Some aspects of these variations are considered in Chapters 8 and 9 but, in terms of their biochemical basis and biological rationale, the surface has only been scratched.

Two final points need to be stressed. Firstly, starch in the guard cells of stomata behaves quite differently from that in the rest of the leaf – characteristically it is synthesized at night and degraded by day, thereby acting as the source of balancing anions for the osmotically active potassium which controls the aperture of guard cells (Jarvis & Mansfield, 1981). Secondly, the assimilatory starch of leaves is, in quantitative terms, a minor component compared with the massive accumulation of starch in many seeds, particularly those of cereals, and vegetative storage organs. It is these that are of greatest economic importance (see Chapter 10).

Other storage polysaccharides. Many polysaccharides other than starch and fructan are used for storage by vascular plants. Some are hydrated mucilages but most are components of cell walls. They thus perform two functions, providing carbohydrate when the walls are degraded and contributing to the structure of the tissue beforehand. Polysaccharides of cell walls inevitably

Fig. 1.8 (*a*) Starch grains formed photosynthetically in young tobacco leaves. (*b*) Starch grains formed from exogenously supplied sucrose in young tobacco leaves. Leaf disks were floated on 0.5 M sucrose solution for 24 h in the dark.

(*a*)

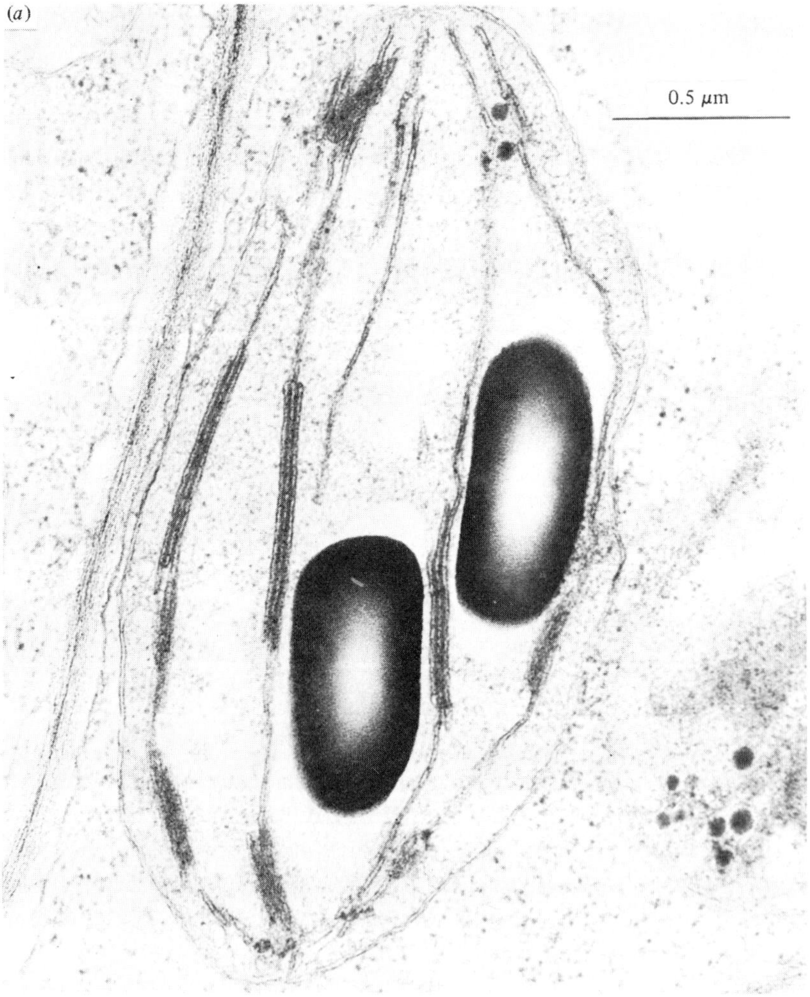

serve as a source of reserve carbohydrate when any specialized storage organ is consumed, e.g. endosperm, perisperm, megagametophyte or cotyledon of seeds, or tuber, corm, root or bulb. Cell walls in some of these organs are much thickened (1–5 μm). Indeed, in a few seeds, notably the galactomannan-storing legumes, the mannan-storing palms and the xyloglucan-storing tamarind, cell-wall polysaccharide accounts for 40–60% of organ dry weight. However, in most storage organs, where other reserve materials (starch, sugars, lipids, proteins, etc.) predominate, cell-wall polysaccharide content is low (e.g. $< 10\%$ dry weight). Cell walls in the otherwise not obviously storage roots and stems of certain perennial monocotyledons and deciduous trees

(*b*)

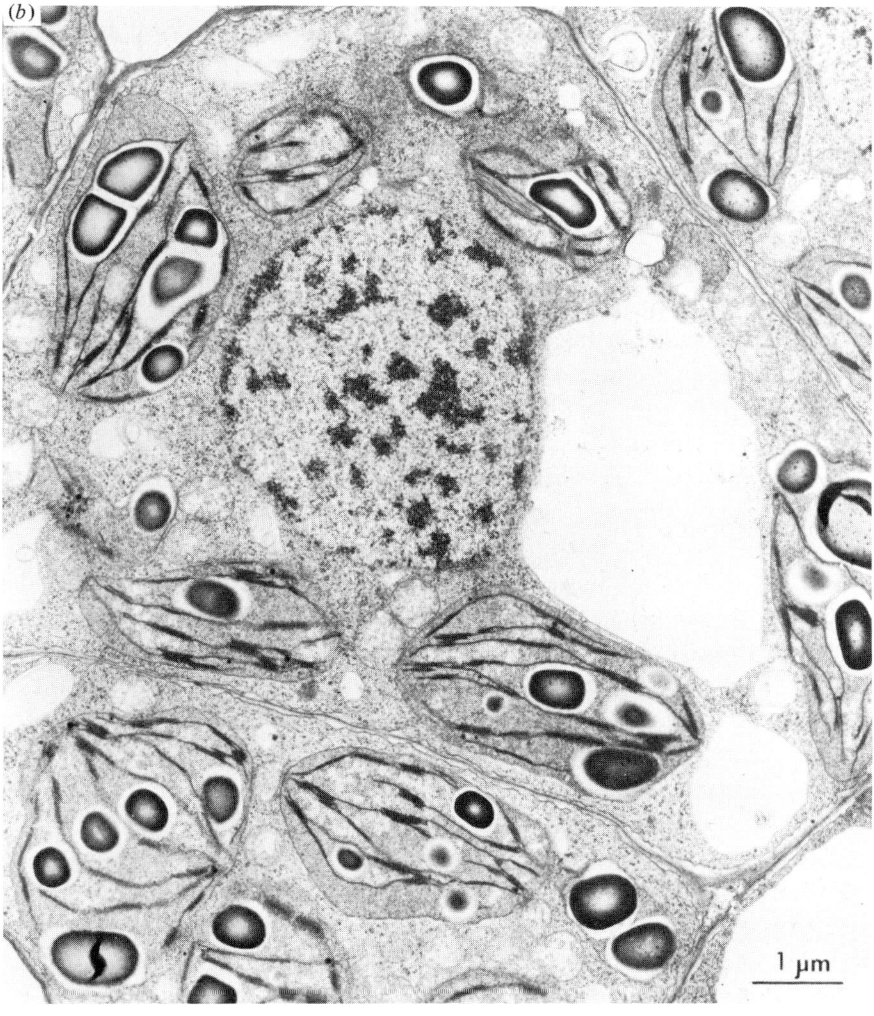

become temporarily thickened each autumn (see Murneek, 1929) and, then, the cell-wall polysaccharides supply a carbohydrate reserve when they are mobilized in the spring. Similarly, the large amounts of callose, a β-(1→3)-glucan deposited over winter in many phloem sieve tubes, are subsequently mobilized in species whose phloem is reactivated in the spring (Esau, 1969). Cell walls are also degraded, and their breakdown products consumed by neighbouring cells, when abscission layers, cavities and ducts are formed. These, however, are not normally regarded as instances of mobilization of reserve polysaccharides.

Cell-wall polysaccharides are commonly classified as 'pectins' or 'hemi-celluloses', and some as 'gums' and 'mucilages' also. Unfortunately these terms are not used with consistent meanings in the literature and are therefore liable to cause some confusion. Pectic (meaning 'gel-forming') polysaccharides, which are rich in galacturonan-rhamnans ('pectic acids'), and arabinans and arabinogalactans ('neutral pectins') are present in most cell walls, particularly in young growing cells and fleshy fruits (about 30–50% of wall carbohydrate). The 'hemicelluloses' (originally named because of their presumed chemical relationship to cellulose) are the other polysaccharides that, together with the pectins, constitute the amorphous matrix of the wall, in which the fibrillar polysaccharides (almost always cellulose in vascular plants) are embedded. Generally pectins are extractable in high yield by mild agents, such as cold or hot water with or without chelating agents, whereas 'hemicelluloses' in many plant tissues require warm alkali (e.g. 0.4–4.0 M NaOH or KOH, sometimes with added 0.5 M borate) for efficient extraction. It has thus become the practice to use these procedures to analyse cell walls in terms of pectic, 'hemicellulosic' and cellulosic content. In such cases, *operational* not *chemical* definitions are being used: respectively, 'water-extractable', 'alkali-extractable' and 'residual' polysaccharides. These simple extraction techniques, however, do not separate cleanly the polysaccharides of different types. Some hemicellulose-type polysaccharides, for instance, are commonly partly extracted by hot water whereas others resist the strongest alkali. Examples from the storage polysaccharides are the galacto-mannans of leguminous seeds which, although they can be almost completely extracted by hot water, are commonly classed as 'hemicelluloses' because of their structural resemblance to the alkali-extractable galacto-glucan-mannans of coniferous and angiospermous woods.

The reason for the great differences in the ease of extraction of similar polysaccharides from different tissues is that extractability is not merely related to polysaccharide size, chemical structure and solubility in water, but more to the nature of the covalent, ionic and hydrogen bonds between polysaccharide chains in the cell wall. Indeed, many 'hemicelluloses' which are only extractable in alkali are water-soluble once extracted from cell walls. Also, secondarily thickened cell walls usually contain lignin, a complex phenolic polymer which cross-links polysaccharides by ether and ester bonds. Lignin must be removed (e.g. by hot chlorous acid) if 'hemicelluloses' are to be extracted in good yield. This is not a problem for the extraction of storage polysaccharides, however, as lignin is not readily catabolized by plants and is virtually absent from food-reserve cell walls.

Certain cell-wall storage polysaccharides (arabinans, galacto-mannans, xylo-glucans and some glucans) are also classed with plant gums and mucilages, because of similarities in structure, composition or physical properties. Indeed, cell-wall polysaccharides with mucilaginous properties may aid uptake or retention of water during drought (Reid & Bewley, 1980). It should be noted, though, that the 'true' intracellular mucilages in the outer seed-coat cells of certain species of plantain and members of the Cruciferae, Loranthaceae and other families, which are used for adhesion of the seed to the soil or host-plant, are not mobilized after germination. Like cell-wall polysaccharides of the 'seed coat' (i.e. in testa and in pericarps of dry, indehiscent fruits), they do not serve as food reserves.

It follows that a terminology based, firstly, upon chemical structure (see below) and, secondly, on function is more useful than one based on extractability in arbitrary solvents.

Table 1.6 lists 11 kinds of polysaccharide used for storage in vascular plants and sample references to the large literature on the subject. There are many species, however, where the nature of the storage polysaccharides is incompletely known or not known at all. The Table excludes cellulose, a $(1\rightarrow4)$-β-D-glucan, and the structurally complex galacturonan-rhamnans with arabinan and galactan side chains (pectins) which are primarily structural in function but which, by virtue of turnover in some tissues such as softening fleshy fruit, may be considered to have a role in storage also.

The storage polysaccharides are principally composed of various combinations of five sugars: the hexoses, D-Glc, D-Gal and D-Man, which are stereochemical isomers (epimers) of each other, and the pentoses, D-Xyl and L-Ara (Figs. 1.1 and 1.8). The methyl pentoses (6-deoxy-hexoses), L-Rha and L-Fuc, and the uronic acids, D-GlcA and D-GalA, which are each a set of epimers (Fig. 1.9) are found in other polysaccharides. Methyl pentoses, uronic acids and pentoses are chemically (and biosynthetically – see Chapter 11) related to hexoses by the reduction, oxidation and decarboxylation, respectively, of the C6 hydroxy-methyl group (Fig. 1.9). All these sugars are present in polysaccharides in the 6-membered(pyranose)-ring conformation, except for L-Ara which is more usually present in the 5-membered(furanose)-ring form (Fig. 1.9). Most polysaccharides consist of several different component

Table 1.6 *Structure and distribution of storage polysaccharides (other than starch and fructan) in vascular plants*

Polysaccharide (Main chain linkage; approx. molecular weight)	Major side chains	Location (% dry weight) and occurrence in plant families	Comments
(1→3)-β-D-glucan ('callose') dilute-alkali-extractable MW 1.6×10^4	None	Pollen Major component of intine in some e.g. pine	Other reserves: starch + 'pectins' in intine (up to 50%). Low-MW carbohydrates (up to 10%) (Stanley & Linskens, 1974)
		Phloem sieve tubes	Primarily a protective substance blocking sieve plates. 'Dormancy callus' in certain angiosperms (e.g. grapevine and perennial monocots), where phloem reactivated: removed after winter. 'Definitive callus' in most species removed more slowly (Esau, 1969)
(1→3), (1→4)-β-D-glucan water-soluble 'gums'; partly extracted by water, completely by alkali; linkage ratio 3:7 MW 10^6; aggregates to 10^7 (with protein)	None	Cereal endosperm e.g. barley (5.6–7.5%) (Ballance & Manners, 1978); also oat and ryegrass but not wheat	Other reserves in barley: starch (85–90%), protein (8–11%), arabino-xylan (1.5–2.5%). Also lipid and protein in scutellum. (In brewing, glucans affect malting efficiency and beer quality.)
(1→5)-α-L-arabinan readily cold- or hot-water-extractable MW 10^4	(1→3)-α-L-Ara-	Seed cotyledons e.g. mustard, an epigeal germinator (9%) and peanut, a hypogeal germinator (1.7%) (Rees & Richardson, 1966; Wankhede, Tharanathan & Rao, 1979)	Other reserves in cotyledons: lipid (40%), protein (30%) lipid (50%), protein (30%)

Arabino-(1 → 4)-β-D-galactan	(1 → 3)-α-L-Ara-; (1 → 5)-α-L-Ara-		
generally hot-water-extractable Ara : Gal = 1:2.3		Seed cotyledons e.g. soybean, an epigeal germinator: major wall component	Other reserves: lipid (17%), protein (37%); starch (26%) in cotyledons, galactomannan (2%) in endosperm
Ara:Gal = 1:1.5 MW 1.5×10^4		lupin, an epigeal germinator (8%) (Matheson & Saini, 1977)	Raffinose/stachyose (20%), protein
Ara:Gal = 1:16 MW 10^4		Tubers e.g. potato: major component of cell walls (c. 3%) (Ring & Selvendran, 1978)	Starch (60–75%), Suc/Glc/Fru (1–5%), arabino-(1 → 3)-galactan-protein (1%)
	as above; also some (1 → 6)-α-D-Gal-	Bulbs e.g. garlic: major cell-wall component (8%) (Das & Das, 1977)	Fructan (20%)
Xylo-(1 → 4)-β-D-glucan ('amyloid' – so-named because of blue-staining reactions with KI) alkali-extractable, giving water-soluble 'gums'	(1 → 6)-α-D-Xyl-; or (1 → 2)-β-D-Gal-(1 → 6)-α-D-Xyl-		Few structurally analysed
Gal:Xyl:Glc = 1:2:3 Gal:Xyl:Glc = 2:3:4		Seed cotyledons Some or all members of 10 dicotyledonous families, including: tamarind, *Tropaeolum majus* and balsam	

Table 1.6 (*cont.*)

Polysaccharide (Main chain linkage; approx. molecular weight)	Major side chains	Location (% dry weight) and occurrence in plant families	Comments
	as above; also (1 → 4) -D-Gal-(1 → 6) -D-Glc-(1 → 6) -D-Glc-	*Brassica campestris* (Koimann 1960; Gould, Rees & Wight, 1971; Siddiqui & Wood, 1971)	
Gal:Xyl:Glc = 1:3:4		Seed endosperms: species in 7 dicotyledonous families, including: *Cyclamen* and *Primula* spp., *Paeonia* spp. and *Annona* (Koimann, 1960)	Not known in monocotyledonous seeds. Most of known seed xylo-glucan distribution is based on staining reaction alone (Koimann, 1960)
Galacto- (1 → 4)-β-D glucan-(1 → 4)-β-D-mannan	(1 → 6)-α-D-Gal- (?)	Seed endosperms	
alkali-extractable, some water-soluble Gal:Glc:Man = c. 0.05:1:1–1.3		*Iris* spp., bluebell and *Lilium* spp. (Jakimow-Barras, 1973)	
Gal:Glc:Man = 1:1:3		coffee (50%) (Wolfrom, Laver & Patin, 1961)	
Gal:Glc:Man = 1:1:6 (structure not known)		Lettuce 50% of endosperm, 3.5% of seed (Halmer, Bewley & Thorpe, 1975)	Other reserves in cotyledons 35% lipid, 20% protein, 2.5% suc/raffinose
Gal:Glc:Man = 0.2:1:18–23		Caraway 14% of seed (Hopf & Kandler, 1977)	
		Minor component in Gramineae	

(1→4)-β-D-glucan-(1→4)-β-D-mannan water-extractable ('glucomannan') c. 5–10% residues *O*-acetylated Glc:Man = 1:3.3		<u>Tubers</u> *Orchis* spp. (25%), *Amorphophallus* and *Arum* spp. (Buchala, Franz & Meier, 1974; Achtardjiev & Koleva, 1973)	Mucilage globules in specialized cells
Glc:Man = 1:2		<u>Bulbs</u> *Lilium* spp. (Andrews, Hough & Jones, 1956)	
Galacto-(1→4)-β-D-mannan (1) readily extractable in hot water	(1→6)-α-D-Gal-		
Gal:Man = 1:1.2–5.0 MW 10^5–10^6 (generally) Some have up to 16% mannose residues linked (1→3)-β-	Tend to be grouped in clusters along main chain	<u>Seed endosperms</u> (generally 2–35%)	
Gal:Man = 1:1.5–2.0		50 spp. of Fabaceae (Leguminosae) e.g. guar (35%)	Examples cited are commercially important as source of 'gums' for paper, mining, food, petroleum and pharmaceutical industries
Gal:Man = 1:3.5–5.2		18 spp. of Caesalpiniaceae (Leguminosae) e.g. locust bean (38%) Also members of 5 other dicotyledonous families	Other reserves: lipid and protein in cotyledons

Table 1.6 (*cont.*)

Polysaccharide (Main chain linkage; approx. molecular weight)	Major side chains	Location (% dry weight) and occurrence in plant families	Comments
(2) alkali-extractable, water-insoluble Gal:Man = 1:10		Also 6 spp. of Palmae e.g. date	
Gal:Man = 1:50–90 (but high degree of galactosylation in young endosperms – Gal:Man = 1:2.5) (Koimann, 1972)		ivory nut (Dea & Morrison, 1975)	
Arabino-(1 → 4)-β-D-xylan partly water-extractable, partly alkali-extractable, water-soluble 'gums'	(1 → 3)-α-L-Ara-; or (1 → 2)-α-L-Ara-: clustered on 2–4 adjacent Xyl residues	Cereal endosperms	
Ara:Xyl = 1:2 MW 5×10^4–5×10^6		e.g. wheat (0.9–1.8%) and	Other reserves: starch (68%), protein (8%)
Ara:Xyl = 1:1 Cross-linked by diferulic acid *O*-esters		barley (1.6–2.5%). Minor wall component but major in aleurone layer (Wilkie, 1979)	See (1 → 3), (1 → 4)-β-D-glucan (above) for other reserves
Arabino-galacto-(1 → 4)-β-D-xylan	(1 →)-α-L-Ara-; (1 → 3)-α-D-Gal-	Corms 'Internal mucilage' in *Watsonia* spp. (Shaw & Stephen, 1965)	
Ara:Gal:Xyl = 3.5:1:1.4	(1 →)-α-L-Ara-; (1 → 3)-α-D-Gal- (1 → 2)-α-L-Ara-		

	(1 →)-α-L-Ara-; attached to both C2 and C3 of xylose main-chain residues; 19 side chains per 10 main-chain residues		
Arabino-(1 → 3)-β-D-galactan	(1) Various combinations of e.g. (a) (1 → 6)-β-D-Gal- (1 → 6)-β-D-Gal-; (b) (1 → 3)-α-L-Ara- (1 → 6)-α-L-Ara-; (c) (1 → 6)-α-L-Ara-; (d) (1 → 6)-β-D-Gal-; and (e) (1 → 6)-β-D-GlcA-	Stemwood Larches: in nonfunctional tracheids and ray-cells of heartwood (5–30%): amounts highly variable between spp. and individuals (Coté, Day, Simson & Timell, 1966)	Function unknown but levels are lower in summer suggesting a food-storage role
Ara:Gal = 1:3		Stylar canals e.g. *Lilium longiflorum*	Major food reserve for pollen-tube growth in those species with stylar canals
Ara:Gal = 1:6.5 contains 3% protein		*Gladiolus gandavensis* (Clarke, Anderson & Stone, 1979)	
	(2) Oligo-galactosyl side chains, substituted with (1 → 6)-α-L-Ara-; and (1 → 3)-α-L-Ara-		
Ara:Gal = 1:1.7 contains 7% protein MW 2–3 × 10^5 Ara:Gal = 1:1.2 contains 6–8% protein		Seed endosperms and cotyledons e.g. Italian rye-grass in vesicles between starch grains Wheat (0.1%)	'All-β-Lectins'; widespread distribution but function unknown

sugars (hetero-polysaccharides), typically no more than five and often with two or three in small or trace amounts only. Some polysaccharides (homo-polysaccharides, e.g. the glucan callose and arabinan) consist of only one sugar.

As Table 1.6 shows, polysaccharides are readily classified into families which have common features of composition and linkage. Within such families, however, there are great differences in detailed structure which can only be outlined here. In general structural terms, all polysaccharides are constructed of a linear backbone (or main) chain containing one or two sugars and bond-linkage types joined to form polymers with as few as about 30, or as many as several thousand residues. Along the main chain may be attached one or more types of side chains, usually short (1–3 sugar residues) and

Fig. 1.9 The formulae and chemical inter-relationships of the sugars and uronic acids found in storage polysaccharides of vascular plants. All occur in the pyranose ring form (*p*) except arabinose, which is most often found in the furanose ring form (*f*) as shown. Structurally, L-fucose and L-rhamnose are derived from D-mannose and D-galactose, respectively, by reduction at C6 and epimerizations at C3 and C5. D-Glucuronic and D-galacturonic acids are derived from their corresponding aldoses by oxidation at C6 and L-arabinose from D-galacturonic acid by decarboxylation at C6.

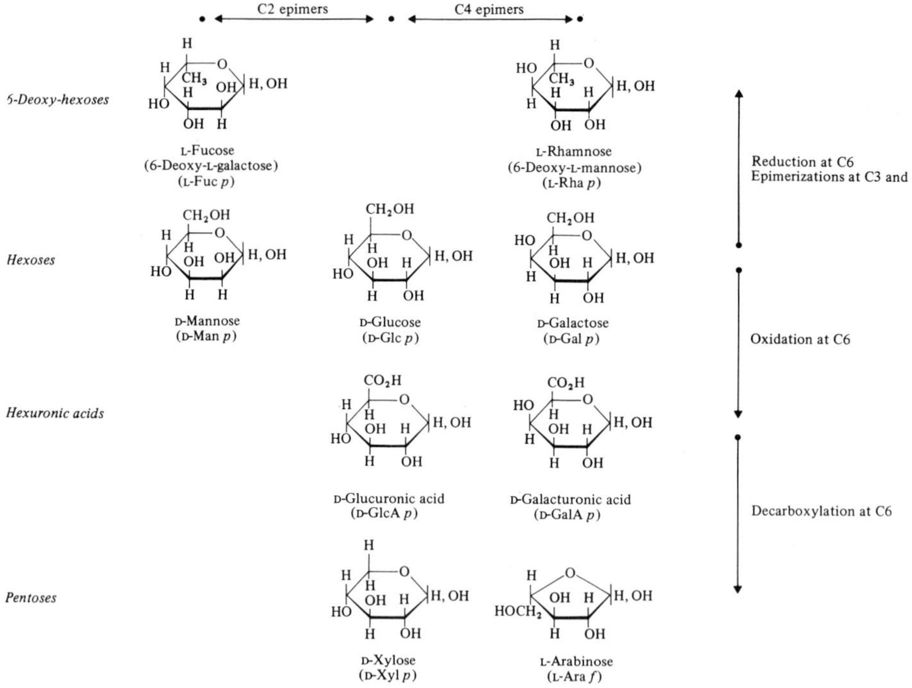

containing one or more different sugars, which are also generally different from the sugars comprising the main chain (Table 1.6). (Polysaccharides are systematically named here as follows: principal side-chain sugar (suffix '-o'); plus main-chain sugars (suffix '-an'), in alphabetical order; preceded by their linkage-bond(s). For convenience, names are also abbreviated, e.g. D-galacto-(1→4)-β-D-glucan-(1→4)-β-D-mannan becomes galacto-(1→4)-β-glucan-(1→4)-β-mannan or galacto-glucan-mannan.) In some cases (e.g. arabino-(1→3)-β-galactans of larch), the side chains are themselves branched. Side chains may be sometimes much larger than three residues, e.g. in the pectinaceous galacturonan-rhamnans, some of which have extended side chains containing Ara and Gal, and can be regarded as several different polysaccharides joined together. In architectural terms, then, the majority of polysaccharides have a comb-like structure, in which the length of the 'comb' and the numbers, lengths and types of 'teeth' are all variable. In free solution, the whole chain may have a twisted or kinked conformation, dictated chiefly by the structure of the main chain (Preston, 1979).

Most polysaccharides do not contain strictly repeating structural units. (Although, for convenience, the results of structural analyses are frequently published in the form of 'representative' structural units, these are not meant to be construed as showing reiterated building blocks.) However, a larger, variable, structural unit is found in many polysaccharides, such as endosperm arabino-xylans and galacto-mannans. In these polysaccharides, side chains are clustered together, resulting in variable substituted and unsubstituted regions along the main chain (Dea & Morrison, 1975; Wilkie, 1979).

A certain degree of variability in structure and composition is found in polysaccharides extracted from each plant source. Heterogeneity of size is very common and may be considerable. The relative molecular weights of cereal endosperm arabino-xylans, for example, are reported to extend over a hundred-fold range of magnitude (8×10^4–5×10^6 daltons – Mares & Stone, 1973). Size and structural heterogeneity in polysaccharides may reflect the presence of different cell types in the plant tissue or changes in the nature of polysaccharides deposited at different stages of cell development. In the case of cereal arabino-xylans, the presence of phenolic-ester cross-links involving ferulic acid may be another factor contributing to variation in size. A certain degree of variability is also an inevitable consequence of the fact that polysaccharide size is not specified directly by a template during biosynthesis (see Chapter 11), as is polymer length for proteins and nucleic acids. In addition, some variability may be an artefact introduced during extraction and fractionation (see Wilkie, 1979, for discussion of this). There are also substantial differences between related species in the detailed structural features of reserve polysaccharides – in the main-chain length, and in the distribution, composition and number of side chains.

Cyclitols

Structure and nomenclature

The cyclitols are polyhydroxy derivatives of cyclohexane. There are eight isomeric forms of cyclohexanehexol, or inositol, and one of these is optically active. The configurations of all nine forms are set out in Fig. 1.10, although only five are known to occur in plants. Nomenclature of the inositols and other cyclitols is complex since they possess stereochemical features not clearly displayed by standard stereochemical notations. The IUPAC–IUB Tentative Cyclitol Nomenclature Rules (1968) set out the problems and present a unified solution (see Cosgrove, 1980). It is generally accepted, however, that the hexols are depicted with the greater number of hydroxyl groups in the *cis* (superior) position and carbon atoms are numbered with C1 at the start of the longest consecutive series of *cis*-hydroxyls, running clockwise. Thus *myo*-inositol, the most widely distributed and most studied cyclitol, is 1,2,3,5/4,6 hexahydroxycyclohexane. It is isomeric with the hexoses (see Fig. 7.1 in Chapter 7).

Cyclitols with only five or four hydroxyl groups also occur in nature and may be considered cyclohexane pentols and tetrols or alternatively as mono- and di-deoxy-inositols; the former nomenclature is normally used. To date only four pentols and four tetrols have been found, all in plants, and their configurations are also set out in Fig. 1.10. That of betitol is very tentative, whilst leucanthemitol and conduritol are in fact cyclohexene tetrols. Conventionally the CH_2 group of pentols is considered as C6. The tetrols present more nomenclatural difficulties; here C1 must be the first of the longest series of similarly orientated, preferably *cis*, hydroxyls in a clockwise direction.

The diversity of naturally occurring cyclitols is completed by a number of mono- and di-methyl derivatives of hexols, and certain phosphorylated derivatives of *myo*-inositol. The methyl group may be esterified to a hydroxyl group (*O*-methyl-ethers) or combined with one of the ring carbon atoms (*C*-methyl cyclitols). The only plants known to contain the latter compounds are algae. Figure 1.11 shows the structures of naturally occurring methyl cyclitols. There are 11 such compounds and different conventions for numerical designation of the methylated position(s) have led to confusing and uninformative descriptions. The IUPAC–IUB Rules should be consulted; they set out, in addition to the Recommended Rules, a Standardized Traditional Procedure which is helpful in interpreting the literature.

That the trivial names of so many cyclitols are derived from the generic names of the plants from which they were initially isolated emphasizes that vascular plants are a rich source of these compounds. However, it must not be supposed that such names indicate that they are found only in these genera. For a detailed discussion of their distribution, the following reviews should be consulted: Plouvier (1963), Posternak (1965) and Alston & Turner (1963).

Two types of phosphorylated cyclitol derivatives are found in plants, and also animals. These are the mono- to hexaphosphates of *myo*-inositol and the phosphoinositides which are phospholipids containing *myo*-inositol. The latter will not be considered further since, as they are components of cell

Fig. 1.10 The formulae* and occurrence of all possible cyclohexane hexols (inositols) and of the naturally occurring pentols and tetrols. Vertical bars show the configuration of hydroxyl groups.

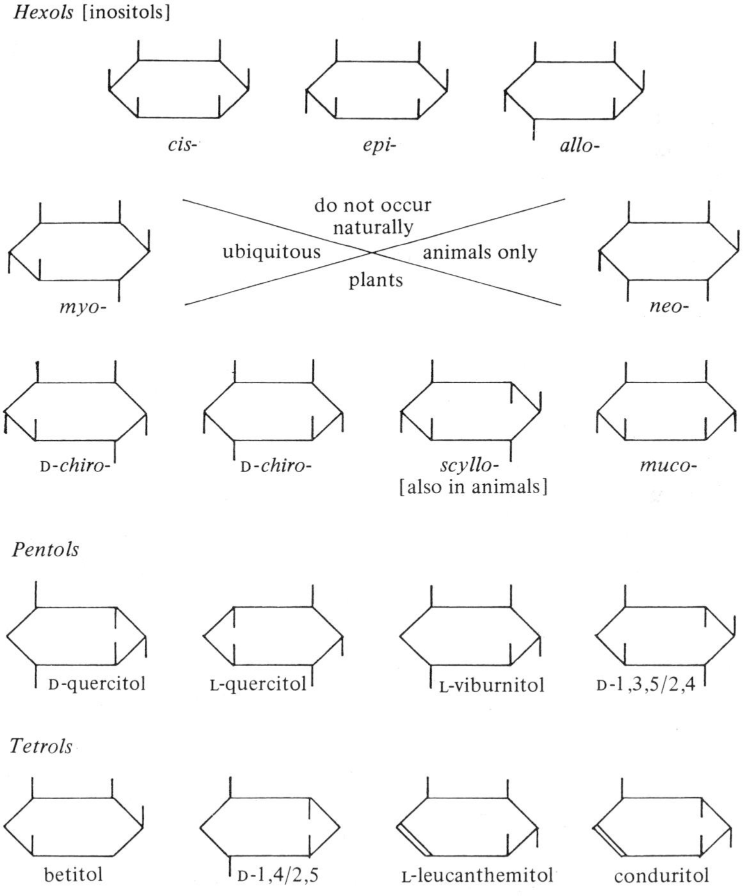

* There is inconsistency in the literature concerning use of upper or lower case prefixes for the inositols (D- or *d*-; L-or *l*-). The former are used in this chapter; the latter in Chapter 6.

membranes, they are outside the context of this book. However, the *myo*-inositol phosphates, and especially *myo*-inositol hexaphosphate* or phytic acid, are accumulated in many plants, often serving as phosphate stores in seeds which, when mobilized, release *myo*-inositol as well as inorganic phosphate. Figure 1.11 shows the structures of phytic acid and a typical plant phosphoinositide.

Fig. 1.11 The formulae of the methylated cyclitols of plants and of the major naturally occurring phosphorylated derivatives of *myo*-inositol. Vertical bars show the configuration of hydroxyl groups.

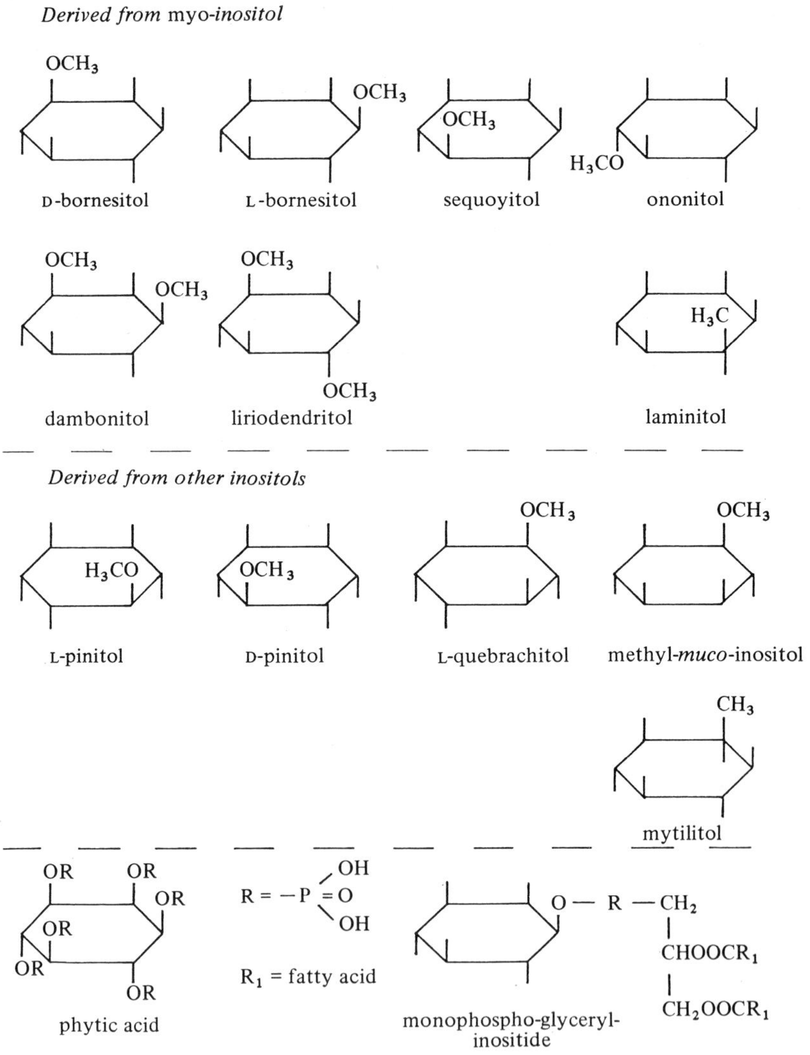

* Now called *myo*-inositol *hexakis* (phosphate) (Cosgrove, 1980).

Occurrence and location

Cyclitols and their methylated derivatives occur in gymnosperms and angiosperms (Table 1.7). No reports of these compounds in pteridophytes are available but this is probably because they have not been sought.

Positive identification of such compounds in a plant depends, of course, on the sensitivity and specificity of the methods used. Most of the work on vascular plants summarized in Table 1.7, by workers such as Plouvier, used gross chemical procedures such as crystallization from plant extracts. However, by extracting considerable quantities of plant material, concentrations down to about 0.01% dry weight were detected and identified. This is more than adequate sensitivity when considering the storage role of such compounds. However, the application of radiochromatography using ^{3}H and ^{14}C tracers has allowed detection of minute quantities of several cyclitols in plants where

Table 1.7 *Distribution of cyclitols and their methyl derivatives in vascular plants*

	Gymnosperms	Angiosperms	
		Dicotyledons	Monocotyledons
Tetrols			
Betitol	−	+	−
Conduritol	−	+	−
L-Leucanthemitol	+	+	−
Pentols			
D-Quercitol	−	+	+
L-Quercitol	−	+	−
L-Viburnitol	−	+	−
Hexols			
myo-Inositol	+	+	+
D-*chiro*-Inositol	+	−	−
L-*chiro*-Inositol	−	+	+
scyllo-Inositol	−	+	+
muco-Inositol	−	−	+
Mono-methyl ethers			
D-Bornesitol	−	+	−
L-Bornesitol	−	+	−
Sequoyitol	+	+	−
D-Ononitol	−	+	−
D-Pinitol	+	+	−
L-Pinitol	−	+	−
L-Quebrachitol	−	+	−
Methyl-*muco*-Inositol	+	+	+
Di-methyl ethers			
Dambonitol	−	+	−
Liriodendritol	−	+	−

they might not previously have been expected, as in the studies of Hoffmann-Ostenhof and his team, whilst gas–liquid chromatography now allows rapid separation, identification and quantitation of cyclitols and their derivatives in extracts from small amounts of plant material.

Details of these and other studies are given in Chapter 6. It appears that the cyclitols, when present at all, are widespread within a given plant although, when relatively high concentrations are encountered, there may be accumulation in specific organs, usually leaves. The occurrence and role of cyclitol glycosides is discussed in Chapter 5.

Alditols

Structure, occurrence and location

The alditols are chemically derived by the reduction of the carbonyl group of aldoses or ketoses. Each aldose produces only one alditol and this relationship determines the strict nomenclature of the alditols, i.e. D-glucose→D-glucitol, D-mannose→D-mannitol, etc. Although many have unrelated trivial names, again often reflecting the plants from which they were first isolated (Table 1.8), these will generally be avoided in this book for

Table 1.8 *The alditols of vascular plants*

Name	Synonyms	Formula
Erythritol	Erythrol, Erythrite Phycitol	Fig. 1.1[a]
2C-Methyl-erythritol	–	Fig. 1.12
Ribitol	Adonitol, Adonite	Fig. 1.1[a]
2C-Hydroxymethyl-ribitol	Hamamelitol	Fig. 1.12
D-Arabinitol (= D-lyxitol)	D-Arabitol	Fig. 1.1[a]
Xylitol	–	Fig. 1.1[a]
Allitol	Allodulcitol	Fig. 1.1[a]
D-Glucitol (= L-gulitol)	Sorbitol, Sorbite	Fig. 1.1[a, b]
1–5 Anhydro-D-glucitol	Polygalitol	Fig. 1.12
D-Mannitol	Mannite	Fig. 1.1[a, b]
1–5 Anhydro-D-mannitol	Styracitol	Fig. 1.12
Galactitol	Dulcitol, Dulcite Melampyrin, Melampyrite	Fig. 1.1[a, b]
L-Iditol	Sobierite	Fig. 1.12
D-*glycero*-D-*gluco*-Heptitol	β-Sedoheptitol	Table 1.1[a, b]
D-*glycero*-D-*manno*-Heptitol	Volemitol, α-Sedoheptitol β-Mannoheptitol	Table 1.1[a, b]
D-*glycero*-D-*galacto*-Heptitol	Perseitol, α-Mannoheptitol	Table 1.1[a, b]
D-*erythro*-D-*galacto*-Octitol	–	Table 1.1[b]

[a] Formula of corresponding aldose shown
[b] Formula of corresponding ketose shown

alditols with six or fewer carbon atoms. Since two configurations about carbon atom 2 are possible when ketoses are reduced, two different alditols are related to each ketose. These relationships are evident from Fig. 1.1, e.g. D-glucitol and D-mannitol, derived from D-glucose and D-mannose, respectively, may both be obtained by reduction of D-fructose. Table 1.8 lists the one tetritol, three pentitols, four hexitols, three heptitols and one octitol of the D-series that have been isolated from vascular plants (see Fig. 1.1 and Table 1.1 for their corresponding reducing sugars). Table 1.8 also lists the one L-hexitol, two anhydro-hexitols and two branched-chain alditols that are known to occur in vascular plants. Figure 1.12 records the formulae of these uncommon compounds.

The distribution of alditols has been most extensively reviewed by Plouvier (1963), Lewis & Smith (1967) and Brimacombe & Webber (1972). Many are so far only known from a few species, often from single genera or even sections within genera (see, for example, the distribution of hamamelitol and its galactoside, clusianose, in the genus *Primula* (Sellmair *et al.* 1977) and

Fig. 1.12 The formulae of the uncommon alditols of plants.
(*a*) 2C-Methyl-erythritol, (*b*) 2C-Hydroxymethyl-ribitol (Hamamelitol), (*c*) L-Iditol, (*d*) 1–5 Anhydro-D-glucitol (Polygalitol) and (*e*) 1–5 Anhydro-D-mannitol (Styracitol).

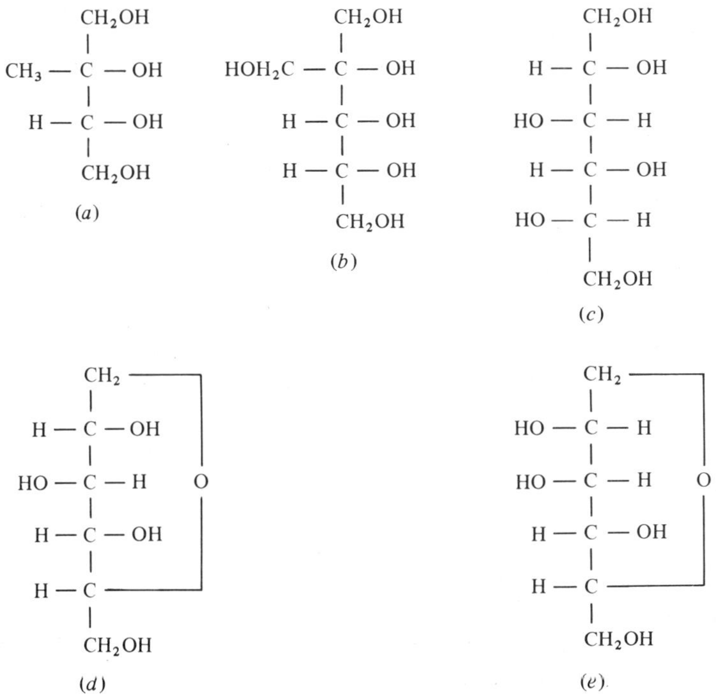

of 2C-methyl-erythritol in *Convolvulus glomeratus* (Anthonsen, Hagen, Kazi, Shah & Tagar, 1976)). However, three – D-glucitol, D-mannitol and galactitol – are widespread. D-glucitol is found mainly in the Rosaceae (Wallaart, 1980) and plantains (Plantaginaceae) (Galkowski, Gauvin & Santurri, 1966; Ahmed, Lahrer & Stewart 1979; Jeffries, Rudmic & Dillon, 1979; Lambeers, Blacquière & Stuiver, 1981), whereas D-mannitol is characteristic of the Oleaceae and Umbelliferae (Apiaceae). Galactitol occurs mainly in the Celastraceae, Scrophulariaceae and Orobanchaceae; in the last two related families, it frequently co-exists with D-mannitol (Plouvier, 1963). Smith, Muscatine & Lewis (1969) commented on the frequent presence of alditols in angiospermous parasites, often in large concentrations, e.g. 20% in the broom rape, *Orobanche cumana* (Kiesel, 1923).

The distribution of alditols within plants varies with species. They are always found in vegetative parts but may be absent from seeds, e.g. D-mannitol from those of Oleaceae and D-glucitol from those of Plantaginaceae (Jukes & Lewis, 1974; Kandler & Hopf, 1980). D-glucitol is, however, present in seeds of at least some rosaceous species, e.g. apple (Whetter & Taper, 1966). Within both vegetative parts and fruit, concentrations vary both diurnally and with season (see Lewis & Smith, 1967). These observations suggest that some alditols at least can be rapidly metabolized but others, although rapidly synthesized, are turned over very slowly (see Chapter 7, which also discusses sites of synthesis and degradation).

Chemotaxonomic significance of the distribution of carbohydrates

The taxonomic distribution of storage carbohydrates has been frequently mentioned in previous sections. To the papers cited above should be added substantial compilations of chemotaxonomists such as Hegnauer (1962–1977) and Gibbs (1974). In such listings, the evidence for the accuracy of observations is not always assessed and original papers should be consulted wherever possible, especially when anomalies occur.

It is clear that the presence or absence of many soluble carbohydrates and starch have taxonomic significance at the ordinal, familial, generic and even subgeneric levels, but a comprehensive phylogenetic scheme to accommodate all these data has yet to be attempted. Figure 1.13 shows the distribution of oligosaccharides and sucrosyl-fructans other than the widespread raffinose series of sucrosyl-galactans. Their restriction to groups of families at the periphery of this phylogenetic scheme of Dahlgren (1975) is striking. They are predominant in families considered to be evolutionarily most advanced (Edelman & Jefford, 1968; Hopf & Kandler, 1980).

The occurrence or structural features of storage polysaccharides other than starch and sucrosyl-fructans might well also be of taxonomic value. However, very little use has so far been made of them and work has hitherto only

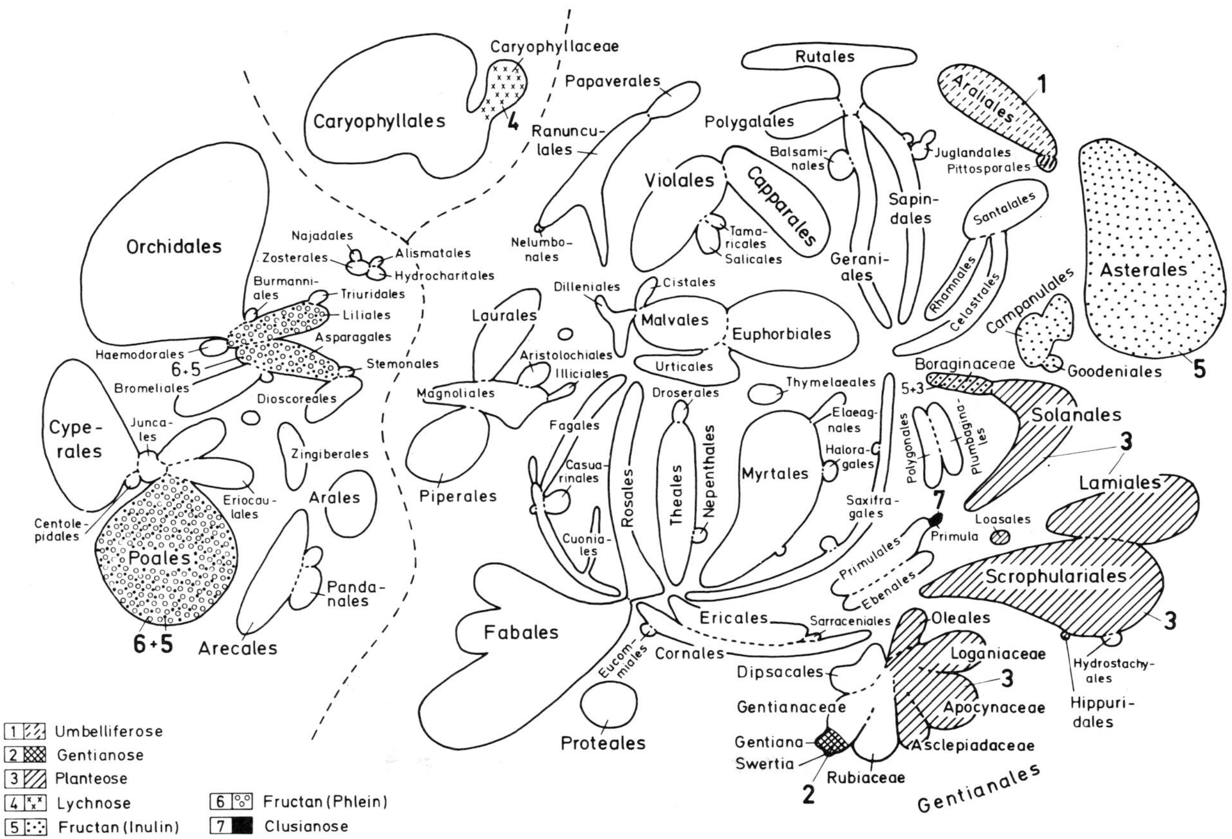

Fig. 1.13 The distribution of sucrosyl-fructans, sucrosyl-galactans and clusianose (hamamelitol galactoside) in vascular plants superimposed on the phylogenetic tree in transection of Dahlgren (1975). (Diagram from a slide presented at the Second International Congress of Systematic and Evolutionary Botany by Hopf & Kandler, 1980.)

produced observations confirming agreed groupings of species. It is obvious, for example, that some polysaccharides are predominant in certain families or orders, e.g. the glucan-mannans found variously in the cell walls and mucilages of endosperms, bulbs, tubers, leaves and other organs of many of the Arales, Iridales, Liliales, and Orchidales. In galacto-mannans of leguminous seeds, the Gal:Man ratio is highly conserved within tribes of the Caesalpiniaceae (-oideae). There are differences also between tribes in whether starch, xylo-glucan or galacto-mannan is the main storage polysaccharide of seeds (Koimann, 1960, 1972). From histochemical evidence alone and in need of verification, it is said that paeonies are the only members of the Ranunculaceae that store xylo-glucan in their endosperm, evidence which supports placing them in a separate family (Koimann, 1960). Generally though, the composition of storage polysaccharides (or non-storage ones for that matter) is not known for most members of any genus, let alone family. The fairly comprehensive information about galacto-mannans (Dea & Morrison, 1975) is due chiefly to commercial interest in these industrially important polysaccharides. The development in recent years of somewhat-simplified techniques for detailed structural analysis (e.g. methylation analysis using gas–liquid chromatography/mass spectrometry) may now make polysaccharides a feasible taxonomic tool. Because of the inherent variability in structure and composition within polysaccharides extracted from any plant source, it will obviously be very important to standardize the procedures used to extract and fractionate them.

References

Achtardjiev, C. Z. & Koleva, M. (1973). A glucomannan from the tubers of *Arum orientale*. *Phytochemistry*, **12**, 2897–900.

Ahmed, I., Lahrer, F. & Stewart, G. R. (1979). Sorbitol, a compatible osmotic solute in *Plantago maritima*. *New Phytologist*, **82**, 671–8.

Akazawa, T. (1965). Starch, inulin and other reserve polysaccharides. In *Plant Biochemistry*, ed. J. Bonner & J. E. Varner, pp. 259–97. New York & London: Academic Press.

Alston, R. E. & Turner, B. L. (1963). *Biochemical Systematics*. Englewood Cliffs, New Jersey: Prentice-Hall.

Andrews, P., Hough, L. & Jones, J. K. N. (1956). Mannose-containing polysaccharides. Part IV. The glucomannans of lily bulbs. *Journal of the Chemical Society*, 181–6.

Anselmino, O. & Gilg, E. (1913). Trehalose in *Selaginella*. *Bericht der Deutschen pharmazeutischen Gesellschaft*, **23**, 326–7.

Anthonsen, T., Hagen, S., Kazi, M. A., Shah, S. W. & Tagar, S. (1976). 2-C-Methyl-erythritol, a new branched alditol from *Convolvulus glomeratus*. *Acta Chemica Scandinavica*, **B30**, 91–3.

Archbold, H. K. (1940). Fructosans in the monocotyledons, a review. *New Phytologist*, **39**, 185–219.

Arni, P. C. & Percival, E. G. V. (1951). Studies on fructans. Part II. Triticin from the rhizomes of couch grass (*Triticum repens*). *Journal of the Chemical Society*, 1822–30.

Arnold, W. N. (1968). The selection of sucrose as the translocate of higher plants. *Journal of Theoretical Biology*, **21**, 13–20.

Aspinall, G. O. & Das Gupta, P. C. (1959). The structure of the fructosan from *Agave vera-cruz* Mill. *Journal of the Chemical Society*, 718–22.

Axelrod, B. (1965). Mono- and oligosaccharides. In *Plant Biochemistry*, ed. J. Bonner & J. E. Varner, pp. 231–57. New York & London: Academic Press.

Bacon, J. S. D. & Edelman, J. (1951). The carbohydrates of the Jerusalem artichoke and other Compositae. *Biochemical Journal*, **48**, 114–26.

Badenhuizen, N. P. (1958). Structure, properties and growth of starch granules. In *Encyclopedia of Plant Physiology*, vol. VI, ed. W. Ruhland, pp. 137–53. Berlin: Springer-Verlag.

Badenhuizen, N. P. (1969). *The Biogenesis of Starch Granules in Higher Plants*. New York: Appleton-Century-Crofts.

Ballance, G. M. & Manners, D. J. (1978). Structural analysis and enzymic solubilization of barley endosperm cell walls. *Carbohydrate Research*, **61**, 107–18.

Banks, W. & Greenwood, C. T. (1975). *Starch and its Components*. Edinburgh: Edinburgh University Press.

Bell, D. J. & Palmer, A. (1952). Structural studies on inulin from *Inula helenium* and on levan from *Dactylis glomerata* and *Lolium italicum*. *Journal of the Chemical Society*, 763–70.

Boehm, J. (1883). Über Stärkebildung aus Zucker. *Botanische Zeitung*, **41**, 33–54.

Brimacombe, J. S. & Webber, J. M. (1972). Alditols and derivatives. In *The Carbohydrates: Chemistry and Biochemistry*, 2nd edn, vol. IA, ed. W. Pigman & D. Horton, pp. 479–518. New York & London: Academic Press.

Brown, H. T. & Escombe, F. (1902). The influence of varying amounts of carbon dioxide in the air on the photosynthetic process of leaves and the mode of growth of plants. *Proceedings of the Royal Society of London*, **70**, 397–413.

Buchala, J. A., Franz, G. & Meier, H. (1974). A glucomannan from the tubers of *Orchis morio*. *Phytochemistry*, **13**, 163–6.

Chandorkar, K. R. & Collins, F. W. (1972). *De novo* synthesis of fructo-oligosaccharides in leaf disks of certain Asteraceae. *Canadian Journal of Botany*, **50**, 295–303.

Chang, C. W. (1979). Starch and its component ratio in developing cotton leaves. *Plant Physiology*, **63**, 973–7.

Chatterton, N. J. & Silvius, J. E. (1979). Photosynthetic partitioning into starch in soybean leaves. I. Effects of photoperiod *versus* photosynthetic period duration. *Plant Physiology*, **64**, 749–53.

Clarke, A. E., Anderson, R. L. & Stone, B. A. (1979). Form and function of arabinogalactans and arabinogalactan-proteins. *Phytochemistry*, **18**, 521–40.

Cosgrove, D. J. (1980). *Inositol Phosphates: Their Chemistry, Biochemistry and Physiology*. Amsterdam: Elsevier Scientific Publishing Company.

Coté, W. A., Day, A. C., Simson, B. W. & Timell, T. E. (1966). Studies on larch arabinogalactan. I. Distribution of arabinogalactan in larch wood. *Holzforschung*, **20**, 178–92.

Courtois, J. E. & Percheron, F. (1971). Distribution des monosaccharides, oligosaccharides et polyols. In *Chemotaxonomy of the Leguminosae*, ed. J. B. Harborne, D. Boulter & B. L. Turner, pp. 207–29. London & New York: Academic Press.

Czaja, A. T. (1978). Structure of starch grains and the classification of vascular plant families. *Taxon*, **27**, 463–70.

Dahlgren, R. (1975). A system of classification of the angiosperms to be used to demonstrate the distribution of characters. *Botaniska Notiser*, **128**, 119–47.

Darbyshire, B. & Henry, R. J. (1978). The distribution of fructan in onions. *New Phytologist*, **81**, 29–34.

Das, N. N. & Das, A. (1977). Structure of the D-galactan isolated from garlic (*Allium sativum*) bulbs. *Carbohydrate Research*, **56**, 337–49.

Davies, D. D., Giovanelli, J. & ap Rees, T. (1964). *Plant Biochemistry*. Oxford: Blackwell Scientific Publications.

Dea, I. C. M. & Morrison, A. (1975). Chemistry and interactions of seed galactomannans. *Advances in Carbohydrate Chemistry and Biochemistry*, **31**, 241–312.

Dey, P. M. (1980). Biochemistry of α-D-galactosidic linkages in the plant kingdom. *Advances in Carbohydrate Chemistry and Biochemistry*, **37**, 283–372.

Eaton, S. V. (1950). Effect of phosphorus deficiency on growth and metabolism of soybean. *Botanical Gazette*, **111**, 426–36.

Eaton, S. V. (1952). Effect of phosphorus deficiency on growth and metabolism of black mustard. *Botanical Gazette*, **113**, 301–9.

Edelman, J. & Jefford, T. G. (1968). The mechanism of fructosan metabolism in higher plants as exemplified in *Helianthus tuberosus*. *New Phytologist*, **67**, 517–31.

Esau, K. (1969). *The Phloem. Encyclopedia of Plant Anatomy*, vol. 5. New York, Heidelberg & Berlin: Springer-Verlag.

Fischer, M. & Kandler, O. (1975). Identifizierung von Selaginose und deren verbreitung in der Gattung *Selanginella*. *Phytochemistry*, **14**, 2629–33.

Forde, B. J., Whitehead, H. C. M. & Rowley, J. A. (1975). Effect of light intensity and temperature on photosynthetic rate, leaf starch content and ultrastructure of *Paspalum notatum*. *Australian Journal of Plant Physiology*, **2**, 185–95.

Franz, G. & Meier, H. (1972). Untersuchungen über Vorkommen und Physiologie des Trisaccharids Gentianose. *Zeitschrift für Pflanzenphysiologie*, **66**, 433–9.

French, D. (1975). Chemistry and biochemistry of starch. *MTP International Review of Science: Biochemistry*, Series I, ed. W. J. Whelan, pp. 267–375. London: Butterworth.

French, D. & Wild, G. M. (1953). Correlation of carbohydrate structure with papergram mobility. *Journal of the American Chemical Society*, **75**, 2612–16.

French, D., Wild, G. M., Young, B. & James, W. J. (1953). Constitution of planteose. *Journal of the American Society*, **75**, 709–12.

Galkowski, T. T., Gauvin, P. N. & Santurri, P. R. (1966). Sorbitol from common plantain. *Journal of Agricultural and Food Chemistry*, **14**, 324–5.

Gates, J. W. & Simpson, G. M. (1968). The presence of starch and alpha-amylase in the leaves of plants. *Canadian Journal of Botany*, **46**, 1459–62.

Gibbs, R. D. (1974). *Chemotaxonomy of Flowering Plants*, vols 1–4. Montreal: McGill–Queen's University Press.

Gould, S. E. B., Rees, D. A. & Wight, N. J. (1971). Polysaccharides in germination. Xyloglucans (Amyloids) from the cotyledons of white mustard. *Biochemical Journal*, **124**, 47–53.

Greenwood, C. T. (1970). Starch and glycogen. In *The Carbohydrates: Chemistry and Biochemistry*, 2nd edn, vol. IIB, ed. W. Pigman & D. Horton, pp. 471–513. New York & London: Academic Press.

Guthrie, R. D. & Honeyman, J. (1974). *An Introduction to the Chemistry of Carbohydrates*, 4th edn. Oxford: Clarendon Press.

Hall, J. L., Flowers, T. J. & Roberts, R. M. (1974). *Plant Cell Structure and Metabolism*. London: Longman.

Halmer, P., Bewley, J. D. & Thorpe, T. A. (1975). Enzyme to break down lettuce endosperm cell wall during gibberellin- and light-induced germination. *Nature*, **258**, 716–18.

Hanesslan, S. & Haskell, T. H. (1970). Antibiotics containing sugars. In *The Carbohydrates: Chemistry and Biochemistry*. 2nd edn, vol. IIA, ed. W. Pigman & D. Horton, pp. 139–211. New York & London: Academic Press.

Harrison, F. C., Tarr, H. L. A. & Hibbert, H. (1930). Studies on reactions relating to carbohydrates and polysaccharides XXXIII. The synthesis of polysaccharides by bacteria and enzymes. *Canadian Journal of Research*, **3**, 449–63.

Hart, R. D., Haldiman, E., Gries, G. A. & Rogers, B. J. (1960). Inhibition of phosphoglucomutase by an extract from garlic. *Botanical Gazette*, **122**, 148–50.

Hegnauer, R. (1962–1973). *Chemotaxonomie der Pflanzen*, vols 1–6. Basel: Birkhauser.

Herold, A. (1980). Regulation of photosynthetisis by sink activitity – the missing link. *New Phytologist*, **86**, 131–44.

Herold, A., McGee, E. E. M. & Lewis, D. H. (1980). The effect of orthophosphate concentration and exogenously supplied sugars on the distribution of newly fixed carbon in sugar beet leaf discs. *New Phytologist*, **85**, 1–13.

Hirst, E. L. (1957). Some aspects of the chemistry of the fructosans. *Proceedings of the Chemical Society*, 193–204.

Holligan, P. M. (1971). Routine analysis by gas–liquid chromatography of soluble carbohydrates in extracts of plant tissues. I. A review of techniques used for the separation, identification and estimation of carbohydrates by gas–liquid chromatography. *New Phytologist*, **70**, 239–69.

Holligan, P. M., Chen, C. & Lewis, D. H. (1973). Changes in the carbohydrate composition of leaves of *Tussilago farfara* during infection by *Puccinia poarum*. *New Phytologist*, **72**, 947–55.

Holligan, P. M. & Drew, E. A. (1971). Routine analysis by gas–liquid chromatography of soluble carbohydrates in extracts of plant tissues. II. Quantitative analysis of standard carbohydrates, and the separation and estimation of soluble sugars and polyols from a variety of plant tissues. *New Phytologist*, **70**, 271–97.

Hopf, H. & Kandler, O. (1976). Physiologie der Umbelliferose. *Biochemie and Physiologie der Pflanzen*, **169**, 5–36.

Hopf, H. & Kandler, O. (1977). Characterisation of the 'reserve cellulose' of the endosperm of *Carum carvi* as a $\beta(1 \rightarrow 4)$-mannan. *Phytochemistry*, **16**, 1715–17.

Hopf, H. & Kandler, O. (1980). Oligosaccharides as taxonomic and phylogenetic markers in angiosperms. *Abstracts of the second International Congress of Systematic and Evolutionary Biology*, Vancouver, Canada, p. 237.

IUPAC–IUB Commission on Biochemical Nomenclature (1968). The nomenclature of cyclitols: tentative rules. *European Journal of Biochemistry*, **5**, 1–12. (Also *Biochimica et Biophysica Acta*, **165**, 1–20.)

Jakimow-Barras, N. (1973). Les polysaccharides des graines de quelques Liliacées et Iridacées. *Phytochemistry*, **12**, 1331–9.

Jarvis, P. G. & Mansfield, T. A. (eds) (1981). *Stomatal Physiology*. Cambridge: Cambridge University Press.

Jeffries, R. L., Rudmic, T. & Dillon, E. M. (1979). Response of halophytes to high salinities and low water potentials. *Plant Physiology*, **64**, 989–94.

Jukes, C. & Lewis, D. H. (1974). Planteose, the major soluble carbohydrate of seeds of *Fraxinus excelsior*. *Phytochemistry*, **13**, 1519–21.

Kakie, T. & Sugizaki, Y. (1970). Diurnal changes in starch and sugars of tobacco leaves. *Soil Science and Plant Nutrition*, **16**, 201–3.

Kandler, O. & Hopf, H. (1980). Occurrence, metabolism and function of oligosaccharides. In *The Biochemistry of Plants*, vol. 3, ed. J. Preiss, pp. 220–70. New York and London: Academic Press.

Kiesel, A. (1923). Mannit aus *Orobanche cumana*. *Hoppe-Seyler's Zeitschrift für physiologische Chemie*, **126**, 257–60.

Koimann, P. (1960). On the occurrence of amyloids in plant seeds. *Acta Botanica Neerlandica*, **9**, 208–19.

Koimann, P. (1972). The relation between monosaccharide composition and specific optical rotation of galactomannans from plant seeds. *Carbohydrate Research*, **25**, 1–9.

Kühbauch, W. (1974). Fructosangehalt, -polymerisationgrad und-Struktur in verschiendenen Pflanzenteilen von Lieschgras. *Zeitschrift für Pflanzenphysiologie*, **74**, 121–9.

Laidlaw, R. A. & Reid, S. G. (1951). Studies on fructosans III. A fructosan from *Lolium perenne*. *Journal of the Chemical Society*, 1830–4.

Lambeers, H., Blacquière, T. & Stuiver, B. (1981). Interactions between osmoregulation and the alternative respiratory pathway in *Plantago coronopus* as affected by salinity. *Physiologia Plantarum*, **51**, 63–8.

Lewis, D. H. & Smith, D. C. (1967). Sugar alcohols (polyols) in fungi and green plants. I. Distribution, physiology and metabolism. *New Phytologist*, **66**, 143–84.

Loewus, F. A. & Labarca, C. (1973). Pistil secretion product and pollen tube wall formation. In *Biogenesis of Plant Cell Wall Polysaccharides*, ed. F. A. Loewus, pp. 175–93. New York & London: Academic Press.

Loewus, F. A. & Tanner, W. (eds) (1982). *Plant Carbohydrates I. Intracellular Carbohydrates. Encyclopedia of Plant Physiology*, New Series, vol. 13A. Berlin: Springer-Verlag.

Maas, W. S. G. & Craigie, J. S. (1964). Examination of some soluble constituents of *Sphagnum* gametophytes. *Canadian Journal of Botany*, **42**, 805–13.

Madsden, E. (1968). Effect of CO_2-concentration on the accumulation of starch and sugar in tomato plants. *Physiologia Plantarum*, **21**, 168–75.

Manners, D. J. (1973). Starch and inulin. In *Phytochemistry*, vol. 1, ed. L. P. Miller, pp. 176–97. New York: Van Nostrand Reinhold Company.

Mares, D. J. & Stone, B. A. (1973). Studies on wheat endosperm. II. Properties of the wall components and studies on their organization in the wall. *Australian Journal of Biological Sciences*, **26**, 813–30.

Marshall, J. J. & Rickson, F. R. (1973). Characterization of α-D-glucan from the plastids of *Cercropia peltata* as a glycogen-type polysaccharide. *Carbohydrate Research*, **28**, 31–7.

Matheson, N. K. & Saini, H. S. (1977). Polysaccharide and oligosaccharide changes in germinating lupin cotyledons. *Phytochemistry*, **16**, 59–66.

Matheson, N. K. & Wheatley, J. M. (1963). Diurnal-nocturnal changes in the starch of tobacco leaves. *Australian Journal of Biological Sciences*, **16**, 70–6.

McDonald, E. J. (1946). The polyfructosans and difructose anhydrides. *Advances in Carbohydrate Chemistry*, **2**, 253–77.

Meeuse, B. J. D. (1963). Inulin in the green alga *Batophora oerstedi* J. Ag. (Dasycladales). *Acta Botanica Neerlandica*, **12**, 315–18.

Meredith, P. & Jenkins, L. D. (1973). Saccharides and saccharidases of developing wheat grains. *New Zealand Journal of Science*, **16**, 599–616.

Meyer, A. (1885). Über die Assimilations producte der Laubblatter angiospermer Pflanzen. *Botanische Zeitung*, **43**, 417–512.

Miller, L. P. (ed.) (1973). *Phytochemistry*, vol. 1. *The Process and Products of Photosynthesis*. New York: Van Nostrand Reinhold Company.

Murneek, A. E. (1929). Hemicellulose as a storage carbohydrate in woody plants, with special reference to the apple. *Plant Physiology*, **4**, 251–64.

Nafziger, E. D. & Koller, H. R. (1976). Influence of leaf starch concentration on CO_2 assimilation in soybean. *Plant Physiology*, **57**, 560–3.

Parkin, J. (1899). Starch synthesis from endogenous sugars. II. Contributions to our knowledge of the formation, storage and depletion of carbohydrates in monocotyledons. *Philosophical Transactions of the Royal Society of London*, **B191**, 35–79.

Pigman, W. & Horton, D. (eds) (1970–72). *The Carbohydrates: Chemistry and Biochemistry*, vols. IA, IIA, IIB. New York & London: Academic Press

Plouvier, V. (1963). Distribution of aliphatic polyols and cyclitols. In *Chemical Plant Taxonomy*, ed. T. Swain, pp. 313–36. New York & London: Academic Press.

Pollock, C. J. (1979). Pathway of fructosan synthesis in leaf bases of *Dactylis glomerata*. *Phytochemistry*, **18**, 777–9.

Pollock, C. J., Hall, M. A. & Roberts, D. P. (1979). Structural analysis of fructose polymers by gas–liquid chromatography and gel filtration. *Journal of Chromatography*, **171**, 411–15.

Pollock, C. J. & Jones, T. (1979). Seasonal patterns of fructan metabolism in forage grasses. *New Phytologist*, **83**, 9–15.

Pontis, H. G. (1977). Riddle of sucrose. In *MTP International Review of Science. Biochemistry, Series II*, vol. 13, *Plant Biochemistry*, ed. D. H. Northcote, pp. 79–117. London: Butterworth.

Posternak, T. (1965). *The Cyclitols*. San Francisco: Holden-Day.

Preiss, J. (ed.) (1980). *The Biochemistry of Plants*, vol. 3, *Carbohydrates: Structure and Function*. New York & London: Academic Press.

Preston, R. D. (1979). Polysaccharide conformation and cell wall function. *Annual Review of Plant Physiology*, **30**, 55–78.

Quillet, M. (1957). Séparation des glucofructosanes et des fructosanes par la chromatographie bidimensionelle. Comparaison de l'extrait fructosidique naturel du tuburcule de Topinambour et de l'hydrolysat d'inuline par l'acide dilué. *Comptes Rendus hebdomadaires des Séances de l'Académie des Sciences*, **244**, 2177–80.

Quillet, M. & Bourdon, D. (1956). Sur le métabolisme glucidique du soja (*S. hispida* var. *Vilnensis*). Le maltose, fraction importante du stock glucidique extrait des petioles des tiges et des racines. *Comptes Rendus hebdomadaires des Séances de l'Académie des Sciences*, **242**, 1054–6.

Quillet, M. & Soulet, M. (1964). Sur l'accumulation concomitante du saccharose et du tréhalose chex plusieurs espèces de Sélaginelles indigènes et exotiques. *Comptes Rendus hebdomadaires des Séances de l'Académie des Sciences*, **259**, 635–7.

Radley, J. A. (1968). *Starch and its Derivatives*. London: Chapman & Hall.

Rees, D. A. & Richardson, N. G. (1966). Polysaccharides in germination. Occurrence, fine structure, and possible biological role of the pectic araban in white mustard cotyledons. *Biochemistry*, **5**, 3099–107.

Reid, J. S. G. & Bewley, J. D. (1980). A dual role for the endosperm and its galactomannan reserves in the germ in active physiology of fenugreek (*Trigonella foenum-graecum* L.), an endospermic leguminous seed. *Planta*, **147**, 145–50.

Ring, S. G. & Selvendran, R. R. (1978). Purification and methylation analysis of cell wall material from *Solanum tuberosum*. *Phytochemistry*, **17**, 745–52.

Rutherford, P. P., Griffiths, C. M. & Wain, R. L. (1966). Studies on plant growth regulating substances XXXIII. Water uptake by discs from inulin-storing roots and tubers in response to growth substance treatment. *Annals of Applied Biology*, **58**, 467–76.

Schaffer, R. (1972). Occurrence, properties, and preparation of naturally occurring monosaccharides (including 6-deoxy sugars). In *The Carbo-*

hydrates: Chemistry and Biochemistry, 2nd edn, vol. IA, ed. W. Pigman & D. Horton, pp. 69–111. New York & London: Academic Press.
Scott, R. W. (1968). Transfructosylation in higher plants containing fructose polymers. Ph.D. thesis, University of London.
Sellmair, J., Beck, E., Kandler, O & Kress, A. (1977). Hamamelose and its derivatives as chemotaxonomic markers in the genus *Primula*. *Phytochemistry*, **16**, 1201–4.
Shaw, D. H. & Stephen, A. M. (1965). Structure of a polysaccharide found in the corm sacs of *Watsonia pyrimidata* (Andr.) Stapf. *Carbohydrate Research*, **1**, 400–13.
Shiomi, N. (1981). Two novel hexasaccharides from the roots of *Asparagus officinalis*. *Phytochemistry*, **20**, 2581–3.
Siddiqui, I. R. & Wood, P. J. (1971). Structural investigation of water soluble rapeseed (*Brassica campestris*) polysaccharides. I. Rape seed amyloid. *Carbohydrate Research*, **17**, 97–108.
Silvius, J. E., Chatterton, N. J. & Kremer, D. F. (1979). Photosynthate partitioning in soybean leaves at two irradiance levels. Comparative responses of acclimated and unacclimated leaves. *Plant Physiology*, **64**, 872–5.
Smith, D. (1968). Classification of several native North American grasses as starch or fructosan accumulators in relation to taxonomy. *Journal of the British Grassland Society*, **23**, 306–9.
Smith, D., Muscatine, L. & Lewis, D. (1969). Carbohydrate movement from autotrophs to heterotrophs in parasitic and mutualistic symbiosis. *Biological Reviews*, **44**, 17–90.
Staněk, J., Černý, M., Kocourek, J. & Pacák, J. (1963). *The Monosaccharides*. New York & London: Academic Press.
Staněk, J., Černý, M. & Pacák, J. (1965). *The Oligosaccharides*. New York & London: Academic Press.
Stanley, R. G. & Linskens, J. F. (1974). *Pollen*. New York, Heidelberg & Berlin: Springer-Verlag.
Stebbins, G. L. (1974). *Flowering Plants: Evolution above the Species Level*. London: Arnold.
Stoddart, J. F. (1971). *Stereochemistry of Carbohydrates*. New York: Wiley–Interscience.
Takhtajan, A. (1969). *Flowering Plants, Origin and Dispersal*. Edinburgh: Oliver & Boyd.
Tetley, R. M. & Krivak, B. M. (1976). Carbohydrate metabolism in oat leaves I. Photosynthetic accumulation of fructans in detached leaf segments. *Plant Physiology*, **57**(Suppl.), 48.
Tomašić, J., Jennings, H. J. & Glaudemans, C. P. J. (1978). Evidence for a single type of linkage in a fructofuranan from *Lolium perenne*. *Carbohydrate Research*, **62**, 127–33.
Wallaart, R. A. M. (1980). Distribution of sorbitol in Rosaceae. *Phytochemistry*, **19**, 2603–10.
Wankhede, D. B., Tharanathan, R. N. & Rao, M. R. R. R. (1979). Structural investigations on two hemicellulosic polysaccharides from ground nut (*Arachis hypogea*) seed endosperm. *Carbohydrate Research*, **74**, 207–15.

Watson, R. R. & Orenstein, N. S. (1975). Chemistry and biochemistry of apiose. *Advances in Chemistry and Biochemistry*, **31**, 135–84.

Whelan, W. J. (1958). Starch and similar polysaccharides. In *Encyclopedia of Plant Physiology*, vol. VI., ed. W. Ruhland, pp. 154–240. Berlin: Springer-Verlag.

Whetter, J. M. & Taper, C. D. (1966). Occurrence of sorbitol (D-glucitol) and certain related sugars in germinating seeds and developing seedlings of *Malus*. *Canadian Journal of Botany*, **44**, 51–5.

Whistler, R. L. & Paschall, E. F. (1965). *Starch: Chemistry and Technology. Fundamental Aspects*, vol. 1. New York & London: Academic Press.

Whiting, G. C. (1970). Sugars. In *The Biochemistry of Fruits and their Products*, vol. 1, ed. A. C. Hulme, pp. 1–31. New York & London: Academic Press.

Wilkie, K. C. B. (1979). The hemicelluloses of grasses and cereals. *Advances in Carbohydrate Chemistry and Biochemistry*, **36**, 215–64.

Wolfrom, M. L., Laver, M. L. & Patin, D. L. (1961). Carbohydrates of the coffee bean. II. Isolation and characterisation of a mannan. *Journal of Organic Chemistry*, **26**, 4533–5.

T. ap REES

2 Sucrose metabolism

Introduction

The purpose of this essay is to present the principal features of sucrose metabolism by vascular plants.* The central importance of sucrose is demonstrated and this is followed by an account of the mechanisms responsible for the synthesis, breakdown and storage of sucrose. Full understanding of sucrose metabolism requires that the above topics be considered in relation to the physiology of the plant as a whole, particularly the transport of sucrose. Thus, this chapter should be read in conjunction with the next.

Importance of sucrose

Sucrose plays three vital and inter-related roles in plant metabolism. It is a major product of photosynthesis, the principal form of translocated carbon and the main storage sugar. Any one of these functions is enough to make sucrose very important. In combination, they ensure that sucrose is not merely a crucial sugar of vascular plants but is pre-eminently *the* sugar of vascular plants.

The role of sucrose as a primary product of photosynthesis is demonstrated by its accumulation during photosynthesis, and by its labelling during photosynthesis in $^{14}CO_2$. For example, over 50% of the CO_2 absorbed by sunflower plants in an hour's photosynthesis has been accounted for as sucrose (Smith, 1944). The kinetics of labelling of sucrose in illuminated leaves in $^{14}CO_2$, an initial lag followed by steady accumulation of ^{14}C, also demonstrate that sucrose is an important product and not an intermediate. Similar evidence is now available from a sufficient range of species, both C_3 and C_4, to show that sucrose is by far the dominant sugar formed in photosynthesis. Indeed, as a product of photosynthesis sucrose is rivalled only by starch. The relative amounts of sucrose and starch formed vary with the species and the conditions. Sucrose appears to be a universal product of vascular plant photosynthesis, but there is a range of species in which the contribution made by starch is negligible (see Chapters 1 and 8). Where

* This account was delivered to the volume editor in December 1980.

balance sheets are available, they show that generally the amount of sucrose formed exceeds the amount of starch (Walker & Herold, 1977). These observations suggest that sucrose may be regarded as the prime product of photosynthesis in vascular plants. This view is strengthened by evidence which suggests that, in some instances at least, starch synthesis is most marked under conditions that restrict sucrose synthesis (Herold, 1980).

The fact that sucrose is the major compound translocated from leaves can be demonstrated by analysis of the contents of sieve tubes (Ziegler, 1975) and by determination of the distribution of label in petiole and stem when a leaf is allowed to photosynthesize in $^{14}CO_2$. The latter approach generally reveals that most of the exported label is present, initially, as labelled sucrose in the phloem (Canny, 1973). A similar dominance of sucrose is found if we consider the translocation of material stored in non-photosynthetic tissues. During the germination of seeds, the mobilization of starch, fat, and, to some extent, even protein involves their conversion to sucrose for transport from the storage tissues to the growing points (ap Rees, Thomas, Fuller & Chapman, 1974).

The importance of sucrose as the dominant storage sugar is revealed by analyses of plant tissues. For example, sucrose constitutes up to 20% of the dry weight of leaves of snowdrop, and up to 19, 4.5 and 7.5% of the fresh weight of the storage tissue of sugar cane, carrot root and red beet, respectively. An additional aspect of sucrose accumulation in higher plants is that it is often enhanced by lowering the temperature to 0–10 °C. For example, potato tubers sweeten when stored in the cold and the accumulated sugar makes them unpalatable and unfit for making into crisps. The mechanisms involved, and the possibility that this low-temperature sweetening contributes to frost hardiness, have been discussed recently (ap Rees, Dixon, Pollock & Franks, 1981). We concluded that there was insufficient definitive evidence available to settle this question, but the rather meagre data that are available favour the view that low-temperature sweetening does contribute to hardiness in some plants.

One of the most striking features of sucrose metabolism is that not only does sucrose fulfil its three roles in higher plants in general but it is also by far the dominant compound to do so. In certain families, oligosaccharides of the raffinose series and, in other families, sugar alcohols are major products of photosynthesis and are important in transport and storage (see Chapters 5 and 7, respectively). However, even where such compounds are clearly of major importance, they do not appear to replace sucrose entirely (Grant & ap Rees, 1981). Possible reasons for this dominance of sucrose have been discussed by Arnold (1968) and by Pontis (1977) but no clear answer has emerged. It is probably pertinent to bear in mind that, in insects, trehalose, a compound closely comparable to sucrose (see Chapter 1), is the main form of transported carbon and that fungi use trehalose and sugar alcohols for transport and storage. In respect of their biological functions, all the above compounds and sucrose have three features in common. All may be present at high concentrations in the organism, all are non-reducing and all are separated from the respiratory sequence of enzymes by one or, at the most, two steps. The most likely of the explanations proposed for the natural selection of these compounds appears to be their biochemical and chemical inactivity.

Enzymes of sucrose metabolism

The first, but, but by no means sole, requirement for an understanding of sucrose metabolism is a knowledge of the properties of the relevant enzymes. The following account is based on two detailed reviews (ap Rees, 1974; Pontis, 1977), which give the authorities for the statements I make.

Invertase

This enzyme (β-D-fructofuranoside fructohydrolase: EC 3.2.1.26), the first known to act on sucrose, catalyses

$$\text{sucrose} + H_2O \rightarrow \text{glucose} + \text{fructose} \; (\Delta G' = -27.6 \text{ kJ at pH } 7.0)$$

Invertase activity in vascular plants is appreciable and widespread. It is clearly not due to a single protein but we do not have enough information to specify all the activities present in plants. It is permissible and useful to distinguish between those with pH optima in the range pH 3.5–5.1, acid invertases, and those with optima in the range of pH 7.0–7.8, neutral or alkaline invertases, referred to hereafter as alkaline invertases. The distinction between acid and alkaline invertases rests on the fact that they can be separated by fractionation with ammonium sulphate and that their activities vary independently during development.

Acid invertase. In most plant tissues, this is so much the dominant invertase that for many the term acid invertase is synonymous with plant invertase. Acid invertase has been partially purified from a wide range of tissues and recently extensively purified from tubers of potato (Anderson & Ewing, 1978) and fruit of banana (Sum, Rogers, Jenkins & Guthrie, 1980). The enzyme is a β-fructofuranosidase, acts typically on sucrose for which the K_m is in the range 2–50 mM, and shows Michaelis–Menten kinetics.

The low pH optimum suggests that acid invertase functions in the vacuole and, or, the free space. Careful fractionation has shown that most of the acid invertase in the storage tissue of red beet is in the vacuole (Leigh, ap Rees, Fuller & Banfield, 1979). The crucial evidence for this statement is the appearance of the enzyme in intact isolated vacuoles and the extremely close correlation between the distribution of acid invertase in the different cell fractions and that of betanin, a unique marker for vacuoles in this tissue (Fig. 2.1).

There is also substantial evidence that in many plant tissues some of the acid invertase is located in the free space. Appreciable proportions of the acid invertase in plant homogenates are recovered in the cell-wall fraction. This indicates but does not prove a cell-wall location for the enzyme. This caution is necessary because the appearance of acid invertase in the cell-wall fraction may be an artefact. In extracts of grapes it is likely that acid invertase sediments with the cell wall because of the formation of protein–tannin–cell-wall complexes during homogenization of the fruit. Studies with extracts of tubers of Jerusalem artichoke provide strong evidence that acid invertase can become reversibly attached to components of the cell-wall fraction via salt linkages formed during homogenization. Supplementary evidence that some of the acid invertase is located in the free space is provided by the fact that in sugar cane the properties of the cell-wall enzyme differ from those of the soluble enzyme. Further support is provided by the observation that in

Fig. 2.1 Use of a correlative method to demonstrate the location of acid invertase in the vacuole of storage tissue of red beet. Homogenates were centrifuged at 1300 **g** and the resulting pellet was divided into five fractions by density-gradient centrifugation. The amounts of betanin and protein and the activity of acid invertase were determined in the unfractionated homogenate, the 1300 **g** pellet, the 1300 **g** supernatant and the five fractions from the gradient. For each of the above fractions the specific content of betanin (μmol mg^{-1} protein) is plotted against that of acid invertase (μmol h^{-1} mg^{-1} protein) (Leigh *et al.* 1979).

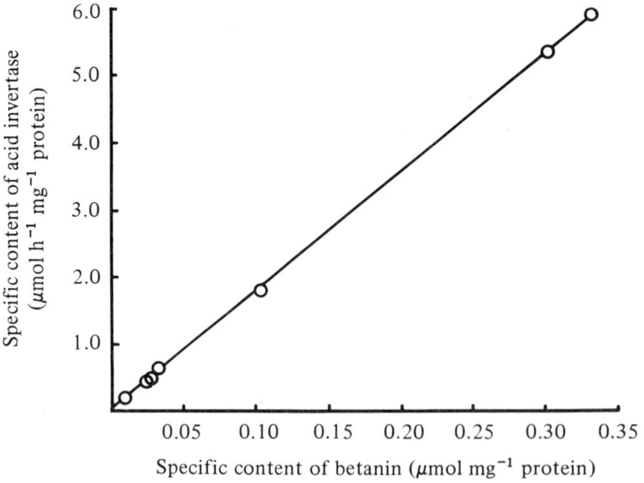

some, but by no means all, tissues uptake of sucrose is preceded by extracellular hydrolysis. The available evidence suggests that acid invertase has a dual location in plant tissues as a whole and that the relative proportions in the vacuole and free space vary with the tissue and its state of differentiation. A good deal more work will be needed to establish whether this suggestion is correct.

Two recent studies of acid invertase from vascular plants have provided appreciable evidence that it is a glycoprotein (Anderson & Ewing, 1978; Sum *et al.* 1980). This is consistent with what we know of the location of the enzyme. By analogy with animal cells, it would be expected that glycosylation of cytoplasmically synthesized acid invertase would be necessary for its transport across either the tonoplast or the cell membrane.

Finally, a particularly important aspect of acid invertases of vascular plants is the discovery that plants contain a proteinaceous inhibitor which specifically inactivates acid invertase from plants. The inhibitor binds to the enzyme to form an inactive complex which is normally undissociable. This inhibitor could contribute to the control of the maximum catalytic activity of invertase in plants (Anderson & Ewing, 1978).

Alkaline invertase. Invertase activity which is optimum over the range pH 7.0–7.8 has been demonstrated in enough plants to suggest that it is general. As a rule this activity does not reach the high levels characteristic of acid invertase; this probably accounts for the neglect of alkaline invertase. Although partially purified preparations have been obtained from a number of species, the enzyme has yet to be fully characterized. The K_m for sucrose is in the range 8–25 mM and the enzyme appears to be a β-fructofuranosidase. Its intracellular location has not been studied thoroughly. To date, it has been recovered in the soluble fraction of tissue homogenates. This observation, the pH optimum, and the fact that there is a constant ratio between alkaline-invertase activity and total protein in extracts of different regions of pea roots (Lyne & ap Rees, 1971) suggest that the enzyme may be located in the cytoplasm.

Sucrose synthetase

This enzyme (UDP-glucose: D-fructose-2-α-D-glucosyltransferase: EC 2.4.1.13) catalyses

$$\text{UDP-glucose} + \text{fructose} \rightleftharpoons \text{sucrose} + \text{UDP}$$
$$(\Delta G' = -4.18 \text{ kJ at pH 7.4 and 37 °C})$$

Sucrose synthetase has been found in so many different plants that it is almost certainly characteristic of all living cells of vascular plants. It has been extensively purified from tubers of Jerusalem artichoke, rice and mung bean, and partially purified from a range of other plants. It is a tetramer of identical subunits and has no significant non-protein component. An earlier suggestion that sucrose synthetase contained lipid has not been confirmed.

The reaction catalysed by sucrose synthetase is readily reversible and this distinguishes the enzyme from all other known sugar nucleotide transglucosylases. A second distinction is the very wide specificity towards the nucleoside base. UDP-, ADP-, GDP-, TDP- and CDP-glucose can all act as glucosyl donors, and the corresponding nucleoside diphosphates are substrates for the reverse reactions. Specificity towards sucrose and fructose is very high. K_m for the sugar nucleotide ranges from 0.2 to 5.3 mM. With the purified enzyme from mung bean, K_m for UDPG was almost a tenth of that for ADPG but such a large difference has not been reported for preparations from many other plants. Estimates of K_m for the nucleoside diphosphates range from 0.1 to 0.4 mM and there is no evidence that K_m for UDP is much different from that for the other nucleotides, although maximum rates of sucrose breakdown are generally observed with UDP. With preparations from a number of tissues, it has been shown that if UDP and ADP are both available, then the former is used preferentially. Estimates of K_m for fructose and sucrose range, respectively, from 1 to 40 mM and 10 to 400 mM.

Sucrose synthetase has a broad pH optimum for synthesis, pH 7.5–8.5, but a sharp and quite different optimum, pH 6.0–6.5, for sucrose cleavage. The enzyme shows no absolute need for a metal. UDP inhibits competitively with respect to UDPG, and *vice versa.* The effects of a range of sugar phosphates on the enzyme have been investigated but no clear picture has emerged yet. In studies of the location of sucrose synthetase, the enzyme was found to be almost entirely confined to the soluble fraction of the homogenates (Dick & ap Rees, 1976; Nishimura & Beevers, 1979). In the second of these studies, the fractionation procedure was so good that very high yields of organelles as fragile as proplastids were obtained. It thus seems clear that sucrose synthetase is located in the soluble phase of the cytoplasm.

Sucrose phosphate synthetase

This is UDP-glucose:D-fructose 6-phosphate 2-α-D-glucosyltransferase (EC 2.4.1.14) and it catalyses

$$\text{UDP-glucose} + \text{fructose 6-phosphate} \rightarrow \text{sucrose 6-phosphate} + \text{UDP}$$
$$(\Delta G' = -18.7 \text{ kJ at pH 7.5 and 38 °C})$$

Sucrose phosphate synthetase is as widely distributed as sucrose synthetase and it has been purified from a number of tissues. Perhaps the most definitive study is that of Salerno & Pontis (1978). Sucrose phosphate synthetase is very specific: neither UDPG nor fructose 6-phosphate can be effectively replaced. Estimates of K_m range from 2 to 8 mM for UDPG and from 0.6 to 4 mM for fructose 6-phosphate. Thus the enzyme needs relatively high concentrations of its substrates for maximum activity. Some preparations of sucrose phosphate synthetase have shown co-operativity towards the substrates but this was not very marked and has not been shown to be a general feature of the enzyme. As yet no obvious regulatory property of the enzyme has been demonstrated to be of general occurrence. This may be due to the absence of such properties in the native enzyme, or to their loss during extraction. Competent fractionations of carefully prepared extracts of photosynthetic (Bird, Cornelius, Keys & Whittingham, 1974) and non-photosynthetic (Nishimura & Beevers, 1979) tissues leave little doubt that sucrose phosphate synthetase is in the soluble phase of the cytoplasm and not in the plastids.

Sucrose phosphate phosphatase

This phosphatase (sucrose-6^F-phosphate phosphohydrolase, EC 3.1.3.24) catalyses

$$\text{sucrose 6-phosphate} + H_2O \rightarrow \text{sucrose} + \text{orthophosphate}$$

Despite its importance we know little about this enzyme. The difficulty of preparing an adequate supply of sucrose phosphate explains but does not excuse our neglect. Sufficient work has been done to show that the enzyme is a specific phosphatase. Preparations from carrot and sugar cane showed little or no activity towards the commonly occurring hexose phosphates. In the latter preparations, the K_m was 0.13–0.17 mM but the enzymes from castor bean and broad beans had appreciably lower values, 0.05–0.07 mM. The pH optimum is between pH 6.4 and 6.7. The enzyme needs Mg^{2+} and is inhibited by EDTA, but perhaps its most interesting property is its inhibition by sucrose (K_i = 10 mM). Its intracellular location is not yet known.

Sucrose synthesis

Pathway

An ability to synthesize sucrose is a general characteristic of both the photosynthetic and non-photosynthetic cells of higher plants. Any suggestion as to how sucrose is made must be consistent with the fact that rapid synthesis can occur in tissues that have accumulated large amounts of sucrose. The pathway of sucrose synthesis is revealed by the combined results of three different types of investigation, thermodynamic data, feeding experiments, and measurements of enzyme activities.

The large change in free energy and the dominance of the equilibrium position by water rule out invertases as a means of significant synthesis of sucrose in plants. Thermodynamic considerations suggest synthesis via sucrose phosphate synthetase and sucrose phosphate phosphatase. The overall change in free energy of these coupled reactions strongly favours synthesis even in the presence of considerable accumulation of sucrose. The reversibility of sucrose synthetase argues against a role for it in sucrose synthesis. This argument is not conclusive because the enzyme could contribute to sucrose synthesis if, once formed, the sucrose were transported to another cellular compartment.

Two aspects of the kinetics of labelling of hexose phosphates, UDPG, sucrose phosphate and sucrose during photosynthesis in $^{14}CO_2$ indicate the pathway of sucrose synthesis. First these kinetics are entirely consistent with the synthesis of sucrose via sucrose phosphate. Second, sucrose is labelled before free fructose. A similar view of sucrose synthesis is obtained if [^{14}C]glucose is fed to leaves: fructose 6-phosphate is labelled before fructose, label appears in sucrose phosphate, and the fructosyl group of sucrose is labelled before fructose (Whittingham, Keys & Bird, 1979). In all the above

Table 2.1 *Activities of sucrose phosphate synthetase and sucrose synthetase in extracts of plant tissues*

Tissue	Activity (nmol min^{-1} g^{-1} fresh wt) Sucrose phosphate synthetase	Sucrose synthetase	Reference
Sucrose synthesis predominating			
Shoot of pea	317	75	Bird *et al.* (1974)
Leaves of spinach	388	4	
Leaves of broad bean	430	55	
Cold-stored tubers of potato	115	8	Pollock & ap Rees (1975)
Scutellum of germinating corn (maize)	500	105	Hawker (1971)
Polysaccharide synthesis predominating			
Developing endosperm of corn (maize)	200	683	Hawker (1971)
Developing cotyledons of broad bean	150	260	
Seed of rice	8	178	
Stele of root of pea	77	460	Lyne & ap Rees (1972)

experiments the appearance of label in sucrose before free hexoses argues strongly against synthesis by sucrose synthetase.

Table 2.1 lists estimates of the maximum catalytic activities of sucrose phosphate synthetase and sucrose synthetase in a range of tissues. Neither enzyme is easy to measure in crude extracts, and some values may be underestimates. Nonetheless, in tissues in which sucrose synthesis is marked, the values for sucrose phosphate synthetase are generally higher. The values for potatoes were compared with estimates of the rate of sucrose synthesis *in vivo* (Pollock & ap Rees, 1975). This showed that there was more than enough sucrose phosphate synthetase to mediate sucrose synthesis but insufficient sucrose synthetase.

If we take all the above investigations together they form compelling evidence that both photosynthetic and non-photosynthetic cells of plants synthesize sucrose predominantly, if not entirely, via sucrose phosphate synthetase.

Location

Although it has been suggested that sucrose is made in the chloroplast, the evidence now overwhelmingly favours an exclusively cytoplasmic location. Claims for the chloroplast were based on relatively rare demonstrations that chloroplast preparations showed sucrose phosphate synthetase activity and fixed $^{14}CO_2$ into sucrose. With the benefit of hindsight, it may now be argued that in none of the above demonstrations was the ability to synthesize sucrose shown to be within the chloroplast and it is probable that it was due to cytoplasmic contamination of the chloroplast preparations. Positive evidence

Fig. 2.2 Pathway of sucrose synthesis in a photosynthetic cell. The enzymes are numbered as follows: (1) fructose-1,6-bisphosphate aldolase; (2) fructose-1,6-bisphosphatase; (3) glucosephosphate isomerase; (4) phosphoglucomutase; (5) UDP-glucose pyrophosphorylase; (6) sucrose phosphate synthetase; (7) sucrose phosphate phosphatase.

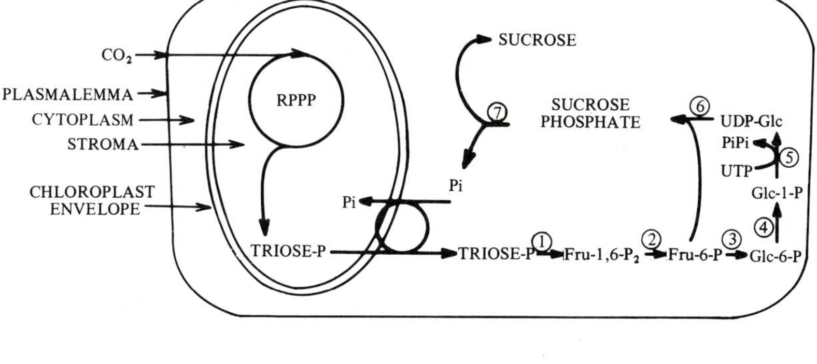

that sucrose is made in the cytoplasm in photosynthetic cells is provided by the cytoplasmic location of sucrose phosphate synthetase, the fact that sucrose does not readily cross the chloroplast envelope, and the knowledge that the products of photosynthesis move out of the chloroplast mainly as triose phosphates. The presence of a cytoplasmic fructose-1,6-bisphosphatase permits the conversion of the exported triose phosphates to sucrose as shown in Fig. 2.2 (ap Rees, 1980).

Sucrose is also made in the cytoplasm in gluconeogenic tissues. Nishimura & Beevers (1979) have shown that not only sucrose phosphate synthetase but also the whole sequence from oxaloacetate to sucrose is located in the cytoplasm in castor-bean endosperm. There have been no definitive studies of the location of sucrose synthesis in cells that are neither photosynthetic nor gluconeogenic. However, it will be surprising if sucrose synthesis in such cells does not also prove to be cytoplasmic.

Regulation

Changes in intermediates when the rate of sucrose synthesis is varied suggest that in leaves synthesis is regulated, at least in part, by sucrose phosphate synthetase or sucrose phosphate phosphatase or both (Stitt, Wirtz & Heldt, 1980). Our knowledge of the properties of these enzymes is still not good enough to reveal any mechanism of control. The most important feature of the available data is the inhibition of sucrose phosphate phosphatase by sucrose. Herold, Lewis and Walker (Herold, 1980) have suggested that if sucrose synthesis exceeds transport from the site of synthesis, the consequent accumulation of sucrose restricts further synthesis through inhibition of sucrose phosphate phosphatase. They also suggested that the attendant accumulation of sugar phosphates in the cytoplasm could reduce cytoplasmic inorganic phosphate with the following effects. The availability of phosphate in the chloroplast would decline and reduce photosynthesis. Lack of cytoplasmic phosphate would curtail the activity of the phosphate translocator by which chloroplastic triose phosphate is exchanged for inorganic phosphate (see Chapters 8 and 9). The net effect of these changes would be to increase the ratio of 3-phosphoglycerate to orthophosphate in the chloroplast and stimulate starch synthesis (again see Chapters 8 and 9). This hypothesis would mean that if photosynthesis exceeds the rates at which sucrose can be transported from, stored in, or metabolized by the leaf then carbon assimilation is not stopped but is merely reduced and diverted to starch, which can be converted to sucrose in the dark when the level of sucrose falls. There is indirect evidence in favour of this attractive hypothesis but as yet there has been no direct investigation of control of sucrose synthesis. We particularly need to know whether accumulation of sucrose in the cytoplasm does inhibit sucrose phosphate phosphatase, and whether this causes a sufficient accumulation of sugar phosphates to reduce cytoplasmic orthophosphate significantly.

There has been little definitive study of the control of sucrose synthesis in non-photosynthetic cells. There is evidence that, during cold-induced sweetening of potatoes, control may be exerted mainly through availability of substrate (ap Rees *et al.* 1981). It is possible that, as a prime storage compound, synthesis of sucrose is regulated in part by availability of substrate and in part by feedback inhibition as suggested for leaves.

Sucrose breakdown

A capacity to break down sucrose is almost certainly a universal feature of the living cells of vascular plants. In general this breakdown serves the two main purposes of providing carbon for polysaccharide synthesis and for respiration. Plants contain three enzymes, sucrose synthetase, acid invertase and alkaline invertase, that could catalyse this breakdown. Thus, the central problem is to discover the roles of these enzymes. The similarity of the enzymes' products and the shortness of the pathways involved, make it very difficult to provide a definitive answer. Perhaps the best way to proceed is to examine the working hypothesis that conversion to polysaccharide is mainly via sucrose synthetase whilst conversion to respiratory substrate is mainly via invertases.

Conversion to polysaccharide

Except for fructans (see Chapter 4), the major polysaccharides of vascular plants appear to be made via sugar nucleotides. It is likely that, in the main, structural polysaccharides are made via UDPG whilst the principal storage polysaccharide, starch, is made via ADPG (see Chapter 8–10). Both UDPG and ADPG could be made directly from sucrose by sucrose synthetase and the resulting fructose metabolized via a specific fructokinase as follows.

$$\text{Sucrose synthetase: sucrose} + \begin{matrix}\text{UDP}\\\text{ADP}\end{matrix} \rightleftharpoons \begin{matrix}\text{UDPG}\\\text{ADPG}\end{matrix} + \text{fructose}$$

$$\text{Fructokinase: fructose} + \text{ATP} \rightarrow \text{fructose 6-phosphate} + \text{ADP}$$

$$\text{Phosphohexoisomerase: fructose 6-phosphate} \rightleftharpoons \text{glucose 6-phosphate}$$

$$\text{Phosphoglucomutase: glucose 6-phosphate} \rightleftharpoons \text{glucose 1-phosphate}$$

$$\text{Pyrophosphorylase: glucose 1-phosphate} + \begin{matrix}\text{UTP}\\\text{ATP}\end{matrix} \rightleftharpoons \begin{matrix}\text{UDPG}\\\text{ADPG}\end{matrix} + \text{PPi}$$

$$\text{Pyrophosphatase: PPi} + H_2O \rightarrow \text{Pi} + \text{Pi}$$

The evidence for this view is:

(1) Conversion to polysaccharide via sucrose synthetase conserves much of the energy of the glycosidic bond of sucrose. Conversion via invertase loses

it all. This argument, though persuasive, is essentially teleological and thus not decisive.

(2) Cell-free systems that contain sucrose synthetase and the appropriate glucosyl transferase transfer the glucosyl group of sucrose to β-1,4 glucan via UDP (Rollit & Maclachlan, 1974), and to α-1,4 glucan via UDP (Murata, Sugiyama & Akazawa, 1964) and via ADP (de Fekete & Cardini, 1964). This is important evidence in favour of the hypothesis but the demonstration of a pathway *in vitro* is never proof of its operation *in vivo*.

(3) When [glucosyl-^{14}C]sucrose was supplied to rice pannicles, both ADPG and UDPG became labelled (Murata & Akazawa, 1966). This is consistent with the scheme but does not prove it because a similar result could be obtained if the [^{14}C] sucrose were hydrolysed by invertase.

(4) There is considerable evidence (Dick & ap Rees, 1975) that the cells of pea roots receive their carbon for polysaccharide synthesis as sucrose via the symplasm. Thus the transport system appears to deliver sucrose directly to sucrose synthetase.

(5) Tissues known to be particularly active in converting sucrose to polysaccharide often have a high maximum catalytic activity of sucrose synthetase (Table 2.1). This correlation has been demonstrated for both the activity and the amount of the enzyme. Su & Priess (1978) showed that, in

Table 2.2 *Estimates of the maximum catalytic activities of sucrose synthetase and invertase in tissues synthesizing polysaccharides. Data from Hawker (1971) except those for barley which are from Baxter & Duffus (1973)*

	Enzyme activity			
Tissue	Sucrose synthetase	Acid invertase	Alkaline invertase	Units
Developing cotyledons of broad bean	260	3	70	nmol min^{-1} g^{-1} fresh wt
Developing endosperm of corn (maize)	9500	170	170	nmol min^{-1} mg^{-1} protein
Developing grains of barley				
8 days after anthesis	20	15	15	nmol min^{-1} per grain
28 days after anthesis	620	15	15	nmol min^{-1} per grain

developing kernels of maize, sucrose synthetase is one of the major proteins and accounts for 2.8% of the total protein. The most telling evidence of this type is from tissues where invertase activity appears to be very low (Table 2.2). Claims for low activities of enzymes in plant extracts should always be treated with caution. However, invertase is a particularly robust enzyme whereas sucrose synthetase is not. Thus, it is unlikely that the values in Table 2.2 were due to preferential loss of invertase during extraction.

Collectively, the above observations provide substantial support for the view that conversion of sucrose to polysaccharide proceeds mainly via sucrose synthetase. However, except possibly in the instances in Table 2.2, the evidence is not conclusive and we cannot rule out significant contributions from either or both the invertases. Such contributions could vary according to the polysaccharide or the tissue. Two aspects of the proposed role for sucrose synthetase require particular attention. The first is the intracellular location of the steps from sucrose to polysaccharide. Sucrose synthetase is in the cytoplasm. It is very probable that the major transglycosylations from sugar nucleotides that result in synthesis of structural polysaccharides occur either on membranes or within membrane-bound organelles (Hanke, 1977). The major storage polysaccharide, starch, is made entirely within plastids. The movement of UDPG from cytoplasm to endoplasmic reticulum presents no obvious problem, but movement into Golgi bodies, and transport of ADPG into plastids, would probably need specific transport mechanisms. The problem is most marked in respect of starch synthesis in non-photosynthetic tissues. Until we know what crosses the amyloplast envelope we must remain cautious about the route from sucrose to starch.

A further reason for caution is our ignorance of the mechanism of the formation of ADPG for starch synthesis. Although sucrose synthetase functions with ADP, it does so less effectively than with UDP. Further, even quite low concentrations of UDP reduce activity with ADP. Estimates of the amounts of UDP and ADP in plant cells suggest that there is enough UDP to interfere seriously with ADPG synthesis by sucrose synthetase. It has been suggested (Turner & Turner, 1975) that this problem is solved as follows:

$$\text{sucrose} + \text{UDP} \rightleftharpoons \text{UDP-glucose} + \text{fructose}$$

$$\text{UDP-glucose} + \text{PPi} \rightleftharpoons \text{glucose 1-phosphate} + \text{UTP}$$

$$\text{glucose 1-phosphate} + \text{ATP} \rightleftharpoons \text{ADP-glucose} + \text{PPi}$$

This would conserve energy but requires pyrophosphate to be present in sufficient amounts to reverse UDPG pyrophosphorylase. This seems unlikely in view of the high activity of pyrophosphatase in plants (Simmons & Butler, 1969) and of the importance of the immediate hydrolysis of pyrophosphate for polymer synthesis in general. The operation of the above scheme appears to require some form of compartmentation for which, as yet, we have no evidence.

Provision of respiratory substrate

There is appreciable evidence that tissues with a high demand for respiratory substrate have very high maximum catalytic activities of acid invertase (Table 2.3). Such high activities correlate closely with rapid expansion growth in internodes of sugar cane, epicotyls of lentil and pea, petals of morning glory, and roots of carrot, pea (ap Rees, 1974) and sugar beet (Giaquinta, 1979). The rapid rise in respiration which occurs when thin slices of mature tissue are incubated under physiological conditions, is accompanied by a marked and parallel increase in the maximum catalytic activity of acid invertase (ap Rees, 1974). In red beet (Leigh *et al.* 1979) and sugar cane, pea root and carrot root (ap Rees, 1974), there is reasonable evidence that most of this acid invertase is intracellular. There is also evidence

Table 2.3 *Estimates of sucrose content and maximum catalytic activities of invertases during development of sugar-cane internode (Hatch & Glasziou, 1963), carrot root (Ricardo & ap Rees, 1970) and sugar beet (Giaquinta, 1979)*

Tissue	Sucrose (mg g^{-1} fresh wt)	Invertase (nmol sucrose min^{-1} g^{-1} fresh wt) Acid	Alkaline
Sugar-cane internode			
Immature	14	1310	3
Maturing	25	12	15
Mature	35	3	90
Carrot roots			
20 days old	0.5	366	9
40 days old	3.4	139	12
55 days old	15.8	81	46
Maturing	23	44	81
Mature	30	12	32
Sugar-beet roots	(mg g^{-1} dry wt)	(nmol min^{-1} mg^{-1} protein)	
9 days old	–	500	
12 days old	–	375	
22 days old	30	–	
28 days old	–	10	
42 days old	250	4	
58 days old	320	2	

that in rapidly respiring tissues the activity of acid invertase greatly exceeds those of alkaline invertase and sucrose synthetase (ap Rees, 1974). These observations suggest that in rapidly respiring tissues, conversion of sucrose to the hexose required by the respiratory pathways is mediated, at least in part, by acid invertase. This view is not invalidated by observations that high total acid-invertase activity does not always correlate with rapid respiration. Many of the exceptions are tissue cultures in which the bulk of the acid invertase appears to be associated with the cell wall. Further, high acid-invertase activity may serve other purposes as well as provision of respiratory substrate.

Different tissues of plants vary widely in their activity of acid invertase. This activity generally falls during development and is often very low in mature tissues. Those mature tissues that have been studied often have appreciable activities of alkaline invertase (Table 2.3; ap Rees, 1974). This information does not reveal the role of alkaline invertase but it does suggest that alkaline invertase could meet the demands of respiration in tissues with low acid-invertase activity.

The available data are consistent with the suggestion that conversion of sucrose to respiratory substrate is mediated largely by the invertases. However, once again the evidence is not conclusive. We cannot rule out an important role in this process for sucrose synthetase. The cytoplasmic location of the latter suggests that fructose formed by it would be available for respiration as well as for polysaccharide synthesis. It is also conceivable that sugar nucleotides formed by sucrose synthetase could be respired. Both photosynthetic and non-photosynthetic tissues of plants (Feingold & Avigad, 1980) contain ADPG phosphorylase (EC 2.7.7.36) which catalyses

$$\text{ADP-glucose} + \text{Pi} \rightarrow \text{ADP} + \text{glucose 1-phosphate}$$

This could supply hexose phosphate for respiration without loss of all the energy of the glycosidic bond of sucrose. UDPG could be metabolized via a phosphodiesterase, which has also been demonstrated in plants (Feingold & Avigad, 1980) and which catalyses

$$\text{UDP-glucose} \rightarrow \text{UMP} + \text{glucose 1-phosphate}$$

The view that conversion of sucrose to polysaccharides proceeds mainly via sucrose synthetase, and to respiratory substrate mainly via invertase, has just enough support to make it a valuable working hypothesis in urgent need of decisive testing.

Sucrose storage

For sucrose in the plant as a whole, storage is the alternative fate to breakdown. The amounts of sucrose stored are often so high that it is generally assumed that storage is in the vacuole. Proof that this assumption is correct is now available for castor-bean endosperm (Nishimura & Beevers, 1979) and the storage tissue of red beet (Leigh *et al.* 1979). Two major questions must be answered before much progress can be made towards understanding the mechanism of sucrose storage. These are: How does sucrose get into the vacuole?; and What keeps it there? The recent development of techniques for isolating vacuoles offers a means of direct study of these questions.

Recently, Wyse (1979) showed that sucrose uptake by disks of sugar-beet storage tissue does not involve appreciable hydrolysis and follows non-saturating kinetics over the range 1–500 mM. This led to the proposal that a non-saturating carrier was involved in the transport of sucrose into the vacuole. We do not know whether sucrose itself or some derivative of it crosses the tonoplast into the vacuole. Glasziou & Gayler (1972) have argued persuasively that sucrose phosphate is the direct precursor of the sucrose in the storage compartment of sugar cane. This is an attractive hypothesis as some of the energy of the sucrose phosphate could be used to drive sucrose accumulation. As yet, reported studies with isolated vacuoles have been confined to investigation of the uptake of free sugars. Doll, Rodier & Willenbrink (1979) have presented evidence that vacuoles isolated from red beet will absorb sucrose and that a relatively small fraction (25–30%) of the absorption is dependent upon ATP. Further work is needed before we can assess the significance of these observations and decide whether they reflect the normal mechanism of transport of sucrose into the vacuole. We particularly need to determine the intracellular location of sucrose phosphate phosphatase and to compare uptake into isolated vacuoles from sucrose with that from sucrose phosphate.

We have equally little understanding of the machinery that keeps sucrose in the vacuole. Perhaps the most pertinent observation (Table 2.3) is the striking inverse relationship between sucrose content and the maximum catalytic activity of acid invertase in sugar-cane internodes, carrot storage tissue, pea roots (ap Rees, 1974), slices of red beet (Bacon, MacDonald & Knight, 1965) and sugar beet (Giaquinta, 1979). In all but the latter instance there is clear evidence that most of the acid invertase is intracellular. For red beet the inverse relationship has been demonstrated in vacuoles (Fig. 2.3). It seems clear that appreciable storage of sucrose does not occur in the presence of high vacuolar acid invertase. In the instances quoted, sucrose storage occurs only when the high activities of acid invertase, characteristic of young tissues, decline during development. A particularly clear illustration of this point is that high concentrations of sucrose in sugar cane are only found in those tissues or varieties that show the marked decline of acid invertase during development (Hatch & Glasziou, 1963).

The above observations strongly suggest that vacuolar acid invertase plays a dominant role in regulating sucrose storage. The following hypothesis (Ricardo & ap Rees, 1970) is consistent with the main experimental observations. High activity of vacuolar acid invertase ensures that sucrose stored in or entering the vacuole is rapidly converted to hexoses. These move into the cytoplasm where they are used in respiration. Thus, in rapidly growing tissues, the available sucrose is directed into the pathways required for growth. As growth declines so does the maximum catalytic activity of vacuolar acid invertase and the demand for hexose. Under these conditions sucrose storage can occur. If the decline in acid invertase is complete then the tissue could be left without hexose for respiration. This need could be met by the development of alkaline invertase in the cytoplasm. The separation of the alkaline invertase from the sucrose would permit storage of the latter and, provided there was some movement of sucrose from the vacuole into the cytoplasm, would not deprive the cell of respiratory substrate.

Fig. 2.3 Relationship between sucrose content and acid invertase activity of vacuoles isolated from storage tissue of red beet. Storage tissue in different physiological conditions was obtained by incubating slices in aerated water at 25 °C (Leigh *et al.* 1979).

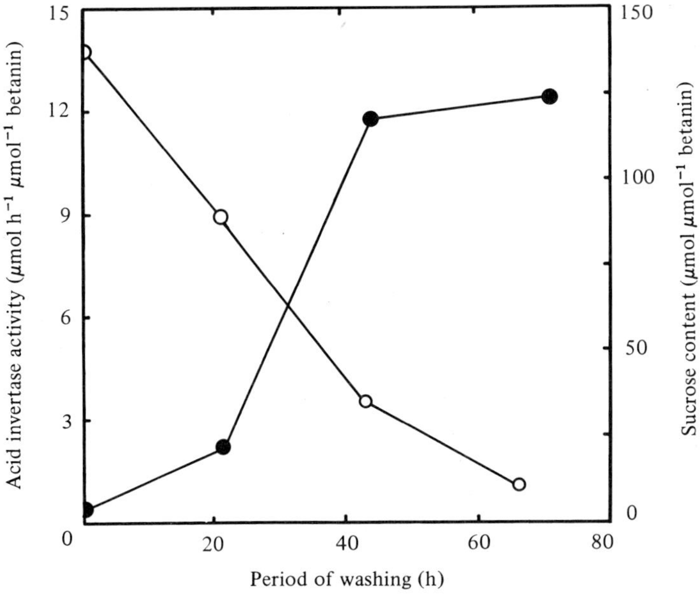

The above hypothesis is supported by comparison of the maximum catalytic activities of the enzymes of sucrose metabolism in different regions of the pea root (Table 2.4) with the way in which these regions metabolized [^{14}C]sucrose (Table 2.5). The labelled sucrose was delivered to the root apex via the translocation stream and its fate was determined in a series of pulse and chase experiments (Dick & ap Rees, 1976). The results show that [^{14}C]sucrose that arrived in regions of the root that had high activity of acid invertase, the apical 6 mm and the cortex, was very rapidly metabolized. Much of the label entered the respiratory pathways and hardly any remained as [^{14}C]sucrose. In the stele where acid invertase activity was much lower, far less of the [^{14}C]sucrose was respired and there was a corresponding increase

Table 2.4 *Maximum catalytic activities of enzymes of sucrose metabolism in different regions of the roots of 5-day-old seedlings of pea (from Dick & ap Rees, 1976)*

	Activity (nmol min^{-1} mg^{-1} protein)		
Region of root	Sucrose synthetase	Acid invertase	Alkaline invertase
Apical 6 mm	19	100	20
Stele 6–24 mm from apex	34	11	26
Cortex 6–24 mm from apex	17	91	19

Table 2.5 *Metabolism of [^{14}C]sucrose by roots of 5-day-old seedlings of pea*

		Per cent ^{14}C recovered from each sample found in:		
Tissue		Sucrose	Products of respiration	Polysaccharides
Apical 6 mm	Pulse	17	53	17
	Chase	1	57	23
Stele 6–24 mm from apex	Pulse	65	11	11
	Chase	38	29	17

Cortex was removed from the region of the roots 36–46 mm from the apex and [U-^{14}C]sucrose was supplied directly and specifically to the bared stele for 2 h. Then the apical 26 mm of the roots were excised, dissected into apical 6 mm, stele, and cortex. Each type of tissue was divided into duplicate samples: one was killed and analysed at once, (the pulse); the other was incubated in 0.02 M KH_2PO_4 (pH 5.2) for 2 h and then killed and analysed (the chase). Products of respiration were calculated by summing ^{14}C in CO_2, acidic and basic components of the soluble substances, and the difference between the labelling of the insoluble material and that of polysaccharides. Cortex and apex behaved in the same way (Dick & ap Rees, 1976).

in the proportion stored. The availability of isolated vacuoles should permit a more direct test of the hypothesis.

I thank Dr N. J. Kruger for his helpful criticisms.

References

Anderson, R. S. & Ewing, E. E. (1978). Partial purification of potato tuber invertase and its proteinaceous inhibitor. *Phytochemistry*, **17**, 1007–81.

ap Rees, T. (1974). Pathways of carbohydrate breakdown in plants. In *MTP International Review of Science, Biochemistry*, Series 1, vol. XI, *Plant Biochemistry*, ed. D. H. Northcote, pp. 89–127. London: Butterworth.

ap Rees, T. (1980). Integration of pathways of synthesis and degradation of hexose phosphates. In *The Biochemistry of Plants*, vol. III, *Carbohydrates: Structure and Function*, ed. J. Preiss, pp. 1–42. London and New York: Academic Press.

ap Rees, T. Thomas, S. M., Fuller, W. A. & Chapman, B. (1974). The relationship between gluconeogenesis and carbohydrate oxidation in higher plants. In *Plant Carbohydrate Biochemistry*, ed. J. B. Pridham, pp. 27–46. London and New York: Academic Press.

ap Rees, T., Dixon, W. L., Pollock, C. J. & Franks, F. (1981). Low temperature sweetening of higher plants. In *Recent Advances in the Biochemistry of Fruits and Vegetables*, ed. J. Friend & M. C. Rhodes, pp. 41–61. London and New York: Academic Press.

Arnold, W. N. (1968). The selection of sucrose as the translocate of higher plants. *Journal of Theoretical Biology*, **21**, 13–20.

Bacon, J. S. D., MacDonald, I. R. & Knight, A. H. (1965). The development of invertase activity in slices of the roots of *Beta vulgaris* L. washed under aseptic conditions. *Biochemical Journal*, **94**, 175–82.

Baxter, E. D. & Duffus, C. M. (1973). Enzymes of carbohydrate metabolism in developing *Hordeum distichum* grain. *Phytochemistry*, **12**, 1923–8.

Bird, I. F., Cornelius, M. J., Keys, A. J. & Whittingham, C. P. (1974). Intracellular site of sucrose synthesis. *Phytochemistry*, **13**, 59–64.

Canny, M. J. (1973). *Phloem Translocation.* London: Cambridge University Press.

Dick, P. S. & ap Rees, T. (1975). The pathway of sugar transport in roots of *Pisum sativum. Journal of Experimental Botany*, **26**, 305–14.

Dick, P. S. & ap Rees, T. (1976). Sucrose metabolism by roots of *Pisum sativum. Phytochemistry*, **15**, 255–9.

Doll, S., Rodier, F. & Willenbrink, J. (1979). Accumulation of sucrose in vacuoles isolated from red beet tissue. *Planta*, **144**, 407–11.

de Fekete, M. A. R. & Cardini, C. E. (1964). Mechanism of glucose transfer from sucrose into the starch granule of sweet corn. *Archives of Biochemistry and Biophysics*, **104**, 173–84.

Feingold, D. S. & Avigad, G. (1980). Sugar nucleotide transformations in plants. In *Biochemistry of Plants*, vol. III, *Carbohydrates: Structure and Function*, ed. J. Preiss, pp. 101–70. London and New York: Academic Press.

Giaquinta, R. T. (1979). Sucrose translocation and storage in the sugar beet. *Plant Physiology*, **63**, 828–32.
Glasziou, K. T. & Gayler, K. R. (1972). Storage of sugars in stalks of sugar cane. *Botanical Review*, **38**, 471–90.
Grant, C. R. & ap Rees, T. (1981). Sorbitol metabolism by apple seedlings. *Phytochemistry*, **20**, 1505–12.
Hanke, D. E. (1977). The endomembrane system and the integration of cellular activities. In *The Molecular Biology of Plant Cells*, ed. H. Smith, pp. 182–207. Oxford: Blackwell.
Hatch, M. D. & Glasziou, K. T. (1963). Sugar accumulation cycle in sugar cane. *Plant Physiology*, **38**, 344–8.
Hawker, J. S. (1971). Enzymes concerned with sucrose synthesis and transformation in seeds of maize, broad bean and castor bean. *Phytochemistry*, **10**, 2312–22.
Herold, A. (1980). Regulation of photosynthesis by sink activity – the missing link. *New Phytologist*, **86**, 131–44.
Leigh, R. A., ap Rees, T., Fuller, W. A. & Banfield, J. (1979). The location of acid invertase activity and sucrose in the vacuoles of storage roots of beetroot (*Beta vulgaris*). *Biochemical Journal*, **178**, 539–47.
Lyne, R. L. & ap Rees, T. (1971). Invertase and sugar content during differentiation of roots of *Pisum sativum*. *Phytochemistry*, **10**, 2593–9.
Lyne, R. L. & ap Rees, T. (1972). Sucrose metabolism in stele and cortex isolated from roots of *Pisum sativum*. *Phytochemistry*, **11**, 2171–6.
Murata, T. & Akazawa, T. (1966). Enzymic mechanism of starch synthesis in ripening starch grains. IV. Starch synthesis in glutinous rice grains. *Archives of Biochemistry and Biophysics*, **114**, 76–87.
Murata, T., Sugiyama, T. & Akazawa, T. (1964). Enzymic mechanism of starch synthesis in ripening starch grains. II. Adenosine diphosphate glucose pathway. *Archives of Biochemistry and Biophysics*, **107**, 92–101.
Nishimura, M. & Beevers, H. (1979). Subcellular distribution of gluconeogenetic enzymes in germinating castor bean endosperm. *Plant Physiology*, **64**, 31–7.
Pollock, C. J. & ap Rees, T. (1975). Activities of enzymes of sugar metabolism in cold-stored tubers of *Solanum tuberosum*. *Phytochemistry*, **14**, 613–17.
Pontis, H. G. (1977). Riddle of sucrose. In *MTP International Review of Science, Biochemistry*, Series II, vol. XIII. *Plant Biochemistry*, ed. D. H. Northcote, pp. 79–117. London: Butterworth.
Ricardo, C. P. P. & ap Rees, T. (1970). Invertase activity during the development of carrot roots. *Phytochemistry*, **9**, 239–47.
Rollit, J. & MacLachlan, G. A. (1974). Synthesis of wall glucan from sucrose by enzyme preparations from *Pisum sativum*. *Phytochemistry*, **13**, 367–74.
Salerno, G. L. & Pontis, H. G. (1978). Sucrose phosphate synthetase. Separation from sucrose synthetase and a study of its properties. *Planta*, **142**, 41–8.
Simmons, S. & Butler, L. G. (1969). Alkaline inorganic pyrophosphatase of maize leaves. *Biochimica et Biophysica Acta*, **172**, 150–7.
Smith, J. H. C. (1944). Concurrency of carbohydrate formation and carbon

dioxide absorption during photosynthesis in sunflower leaves. *Plant Physiology*, **19**, 394–403.

Stitt, M., Wirtz, W. & Heldt, H. W. (1980). Metabolite levels during induction in the chloroplast and extrachloroplast compartments of spinach protoplasts. *Biochimica et Biophysica Acta*, **593**, 85–102.

Su, J. C. & Preiss, J. (1978). Purification and properties of sucrose synthase from maize. *Plant Physiology*, **61**, 389–93.

Sum, W. F., Rogers, P. J., Jenkins, I. D. & Guthrie, R. D. (1980). Isolation of invertase from banana fruit (*Musa cavendishii*). *Phytochemistry*, **19**, 399–401.

Turner, J. F. & Turner, D. H. (1975). The regulation of carbohydrate metabolism. *Annual Review of Plant Physiology*, **26**, 159–86.

Walker, D. A. & Herold, A. (1977). Can the chloroplast support photosynthesis unaided? In *Photosynthetic Organelles*, ed. S. Miyachi, S. Katoh, Y. Fujita, & K. Shibata, pp. 295–310. Japan: Japanese Society of Plant Physiologists Centre for Academic Publications.

Wyse, R. (1979). Sucrose uptake by sugar beet tap root tissue. *Plant Physiology*, **64**, 837–41.

Whittingham, C. P., Keys, A. J. & Bird, I. F. (1979). The enzymology of sucrose synthesis in leaves. In *Encyclopedia of Plant Physiology*, New Series, vol. VI, *Photosynthesis II*, ed. M. Gibbs & E. Latzko, pp. 313–36. Berlin, Heidelberg and New York: Springer-Verlag.

Ziegler, H. (1975). Nature of transported substances. In *Encyclopedia of Plant Physiology*, New Series, vol. 1, *Transport in Plants*, ed. M. H. Zimmerman & J. A. Milburn, pp. 59–100. Berlin, Heidelberg and New York: Springer-Verlag.

J. H. THORNE and R. T. GIAQUINTA

3 Pathways and mechanisms associated with carbohydrate translocation in plants

Introduction

In terms of the carbon economy of crops, yield is to a large extent influenced by the conversion of atmospheric CO_2 into carbohydrate by photosynthesis and the subsequent partitioning of the products of photosynthesis to the harvestable versus non-harvestable portion of the crop. The potential for manipulating both the production and translocation of assimilates by genetic means or by use of plant growth regulators offers substantial opportunities for increased agricultural productivity. Unfortunately, our knowledge and understanding of the events which govern the manner in which assimilates are translocated from their sites of synthesis in photosynthesizing leaves to their sites of utilization in the harvested region is presently too meagre to allow us to identify, let alone control, the biochemical and physiological processes which govern partitioning of assimilates and thus yield.

This Chapter attempts to present an overview of the structural features and physiological processes operating in source and sink regions of plants which relate to the pattern and rate of distribution of assimilate. Firstly, in the source leaves, attention will focus on the cellular events which control: (1) the partitioning of photosynthetic carbon at the level of the chloroplast; (2) synthesis of sucrose; (3) compartmentation of assimilate; and (4) transfer and entry of assimilate into the phloem. Collectively, these processes govern the availability of assimilates for export to the rest of the plant. Secondly, in sinks, where comparatively little information is available on the pathways and mechanisms associated with unloading and utilization of assimilate, particular and detailed attention will be paid to structure/function relationships of unloading in developing seeds – an area which has not been the subject of previous reviews.

It is important to stress at the outset that, as distribution of assimilate in an intact plant is a well-integrated and -orchestrated process, it cannot be described adequately by dealing with various cellular processes in isolation.

With this *caveat* in mind, an attempt is made at least to identify the various levels at which control of translocation may be exerted.

Physiology of export – factors controlling the availability of sucrose for translocation

Partitioning of assimilate between starch and sucrose

The reactions governing the fixation and distribution of the initial products of photosynthesis between starch, a chloroplastic non-transportable carbohydrate, and cytoplasmic transportable carbohydrates is the first stage in determining the proportion of fixed carbon which will be available for export. Sucrose is the major transported carbohydrate in most plants and the chapter will be concerned exclusively with this disaccharide. However, oligosaccharides based on sucrose as well as sugar alcohols are loaded into the phloem, translocated and unloaded in many plants (see Giaquinta, 1980*a* and Chapters 5 and 7).

The factors which control the partitioning of assimilate between chloroplastic starch and extra-chloroplastic sucrose, and which control the synthesis and degradation of these two carbohydrates in leaves are discussed in Chapters 2, 8 and 9. Particularly important in the regulation of the metabolism of starch and of the transport of carbon, as triose phosphate (TP), across the chloroplast membrane are the concentrations of inorganic phosphate inside and outside the chloroplast (see also Preiss & Levi, 1980).

Biosynthesis of sucrose in the cytoplasm

Although the rates of photosynthesis and translocation are closely coupled, the relationship is indirect in that the principal determinants of export are the concentration of sucrose and its availability for export. The concentration of sucrose in leaves is essentially regulated by the activities and spatial distribution of four enzymes involved in its metabolism: (1) sucrose phosphate synthetase (SPS); (2) sucrose phosphate phosphatase (SPP); (3) invertase; and (4) sucrose synthetase. Collectively these enzymes can adjust the sucrose concentration of the leaf in response to metabolic needs and export status. From the TP exported from chloroplasts, formation of sucrose in the mesophyll cytoplasm by SPS in conjunction with SPP is clearly the major biosynthetic route for sucrose in photosynthesizing leaves (see Chapter 2).

The liberation of inorganic phosphate (Pi) in the cytoplasm during formation of sucrose favours continued export of TP from the chloroplast via the phosphate translocator. Therefore, synthesis of sucrose plays an important role in photosynthetic carbon metabolism, both as the major transport species and as a mechanism for recycling Pi for continued photosynthesis. The enzymes responsible for biosynthesis of sucrose have been proposed to figure prominently in the regulation of its translocation in response to sink demand. For example, when translocation is reduced by a decrease in sink demand, the resulting increase in concentration of sucrose in the source leaf has been proposed to inhibit SPP, which in turn sequesters intracellular phosphate in the form of sugar phosphates. The lower cytoplasmic [Pi] promotes the retention of TP within the chloroplast for synthesis of starch. Alternatively, an increase in sink demand, and thus export, should favour the synthesis of sucrose through liberation of Pi into the cytoplasm which catalyses the further export of TP into the cytoplasm for additional synthesis of sucrose (Herold & Lewis, 1977). These proposals require experimental testing (see Chapter 2). That SPS also plays an important role in export is indicated by the temporal correlation between the appearance of its activity and the onset of the capacity to export in developing leaves (Giaquinta, 1978; Silvius, Kremer & Lee, 1978). Interestingly, Silvius, Chatterton & Kremer (1979) later found that, even though soybeans acclimated to relatively high irradiance (950 $\mu E\ m^{-2}\ s^{-1}$) had similar photosynthetic rates compared to plants acclimated at moderate irradiance (600 $\mu E\ m^{-2}\ s^{-1}$), plants acclimated at high irradiance had lower rates of starch accumulation and faster rates of translocation. The increased rates of export in these plants was accompanied by a two-fold increase in the specific activity of SPS on a protein basis. These authors suggested that the activity of SPS, and thus the rate of synthesis of sucrose, plays an important role in partitioning of carbon and sucrose export. This proposal is supported by the recent studies of Huber (1981*a*, *b*) on inter- and intra-species differences in formation of sucrose and starch in isolated leaf protoplasts. Those derived from high-starch-forming plants, such as tobacco, peanut and soybean, partitioned approximately equal amounts (32–39%) of fixed carbon between starch and sucrose compared to protoplasts from low-starch-forming plants (wheat, barley) which partitioned 56–65 and 11–12% of the fixed carbon into sucrose and starch, respectively. He attributed this difference in partitioning to the presence and regulatory properties of SPS, since the protoplasts from high-starch-forming species had less SPS activity per cell compared to the high sucrose formers. These studies also showed that the SPS activity from species having a high potential for sucrose formation was not inhibited by the end product, sucrose, whereas the SPS activity from low sucrose formers (i.e. high starch formers) was inhibited by exogenous sucrose. This proposal that SPS is regulated by sucrose as well as by intermediates of carbon metabolism (Whittingham, Keys & Bird, 1979) and that this regulation plays an important role in translocation, in terms of sucrose availability, certainly warrants further attention.

Compartmentation of sucrose within the mesophyll

Once sucrose is formed in the mesophyll cytoplasm via SPS, several lines of evidence indicate that prior to export, it undergoes both a temporal and spatial compartmentation, both within the mesophyll cell itself and among the various cell types within the source leaf. Using $^{14}CO_2$-pulse–chase studies and cell-dissection techniques, Outlaw & Fisher (1975) and Outlaw, Fisher & Christy (1975) found that all the incorporated ^{14}C was initially located in the palisade mesophyll cells but moved rapidly to the vascular tissue where it accumulated to its greatest extent. Based on determinations of the specific activity of sucrose, these authors identified kinetically distinct pools of sucrose among each of the various cell types within the leaf – an 'histological compartmentation' of sucrose. Subsequently, Fisher & Outlaw (1979) also showed that, within a given mesophyll cell, the cytoplasmic pool of sucrose was three times more concentrated than that in the vacuole, even though the vacuolar fraction contained a larger amount of sucrose owing to its greater volume. Similarly, in sugar beet leaves, Fondy & Geiger (1980) calculated that, although 60% of the sucrose was in the mesophyll cytoplasm versus 40% in the vacuole, both pools were turning over. The rate of turnover of the vacuolar pool, however, was $\frac{1}{6}$ the rate of the transport or cytoplasmic pool, whose kinetics ($t_{\frac{1}{2}} = 22$ min) matched those of export as determined by steady-state labelling experiments with $^{14}CO_2$. These data therefore indicate that sucrose is partitioned into a 'transport' pool located in the cytoplasm which is rapidly turning over, as one would expect if it were being evacuated toward the loading sites, and a vacuolar pool. This latter pool may be a temporary site of storage which may contribute sucrose for export in response to changes in the export status of the leaf (Geiger & Giaquinta, 1981).

Movement of sucrose to the phloem

Because of the symplastic continuity between the various photosynthetic cells in the leaf and the uniformity of concentration of solute within the mesophyll, assimilates are envisioned to move within the symplasm via plasmodesmatal connections to the region of the sieve element–companion cell complex of the phloem. However, the results of several kinds of experiments suggest that assimilates then enter the apoplast prior to their active accumulation into the phloem. These, which have been reviewed by Geiger (1975) and Giaquinta (1980*a*) include: (1) the ability of exogenous sucrose to be loaded into the phloem and translocated in sugar beet leaves with disrupted symplastic connections at rates comparable to those found for $^{14}CO_2$-derived assimilates in intact leaves; (2) the ability isotopically to trap photosynthetically-derived sucrose from the apoplast during its transit to the phloem; (3) the ability of non-permeating chemical modifiers, such as *p*-chloromercuribenzene sulphonate, to inhibit the loading and translocation of exogenous- and $^{14}CO_2$-derived sucrose; and (4) the lack of abundant plasmodesmatal connections between the mesophyll and companion cells at the mesophyll-phloem interface.

Although assimilates appear to enter the apoplast in species such as sugar beet and corn (maize) (Heyser, 1980), the paucity of detailed and quantitative information on the pathways of loading of sucrose in any one species or among different species prevents us from stating with confidence that all loading into the phloem occurs exclusively from the apoplast. In fact, the presence of symplastic continuity from the mesophyll to the companion cells via the phloem parenchyma may indicate that some, but certainly not all, loading could occur entirely within the symplasm. Since at least some sucrose does enter the apoplast prior to loading into the phloem, the mechanism of its efflux from the mesophyll into the apoplast is an important control point for export. Very little information exists on the mechanism, specificity, energetics and possible control of this efflux, except that it appears to be stimulated by increases in photosynthesis and when exogenous ATP, EDTA and Na^+ and K^+ ions are added to the free space of translocating sugar beet leaves (Doman & Geiger, 1979; Geiger & Giaquinta, 1981). Recently, Huber (1981*c*), using isolated mesophyll protoplasts, has suggested that leakage of sucrose may be coupled to a K^+ cotransport system. This proposed mechanism, particularly its possible tandem operation with the proton/sucrose cotransport mechanism involved in phloem loading (see below), deserves further study.

Loading of sucrose into the phloem

Phloem loading is the process whereby the major translocated sugars are selectively and actively accumulated into the sieve tubes prior to their export from the leaves. It is generally accepted that this loading generates the driving force for the mass flow of assimilates from source to sink by decreasing the osmotic potential of the sieve tubes which in turn generates a water potential gradient and water influx into the phloem (Giaquinta, 1980*a*).

Freeze-substitution electron microscopy coupled with studies involving incipient plasmolysis shows that the companion cell–sieve element complex of the phloem contains a much higher concentration of solute than the surrounding cell types in sugar beet source leaves. This observation and the physiological studies alluded to in the previous section indicate that sucrose enters the free space near the phloem region and is then actively and selectively accumulated by a carrier-mediated process at the phloem membranes (see Giaquinta, 1980*b*). Although the coupling mechanism between the putative sucrosyl carrier and metabolic energy is far from certain, there is a growing body of literature which is consistent with the view that loading of sucrose into the phloem is coupled to a proton cotransport system (Baker, 1978; Giaquinta, 1980*a*). In such a system, the driving force for the loading of sucrose is an electrochemical potential gradient of protons across the phloem membranes, which is established by a proton-extruding mechanism, presumably a membrane ATPase (Giaquinta, 1979). Sucrose is thought to enter the apoplast near the phloem, interact with a sucrose-specific carrier on the phloem membrane, and cross the membrane in response to the proton motive force established by the higher proton concentration of the apoplast relative to the interior of the phloem. In this manner, the energetically downhill movement of protons into the phloem is coupled to the secondary active loading of sucrose. Protons are continuously extruded to maintain the proton gradient by an active proton pump on the phloem plasma membrane. Recent studies by Komor, Rotter, Waldhauser, Martin & Cho (1980) show that the transient influx of sucrose accompanying proton uptake is also coupled to a charge-compensating efflux of K^+ during the depolarization of the membrane potential caused by the influx of protons. The studies of Komor *et al.* (1980), Heyser (1980), Baker (1978) and Giaquinta (1980*b*) present several lines of evidence which support a sucrose-proton cotransport system in phloem loading. Future emphasis needs to focus on possible feedback and feedforward control mechanisms which may operate at the loading sites in response to changes in the translocation status of the plant. The components of the loading mechanism which warrant further study in this regard include the carrier protein, turgor changes, ion fluxes and the resulting transmembrane electrical potential during loading.

Physiology of import – unloading and utilization of sucrose

Definition and occurrence

Unloading involves the cellular events associated with the exit of solutes and water at various locations along the phloem in maturing tissues and at terminal regions of phloem in developing tissues. The duration and intensity of unloading (an intrinsic aspect of sink strength) influences patterns of the distribution of solutes in the plant. Within a given source–sink configuration, transport is influenced by the steepness of concentration gradients in the files of sieve tubes linking sources and sinks. The localized depletion of translocate that occurs within sink tissues as a result of utilization, transformation or compartmentation apparently maintains gradients favouring transport. Under natural growing conditions, these gradients are undoubtedly very dynamic and respond to the developmental changes of the sinks and the environment acting upon them (Walker & Ho, 1977*a*, *b*). This potential for rapid, reversible change in the direction and magnitude of transport is readily demonstrated by manipulations of sources and sinks

(Swanson, Hoddinott & Sij, 1976; Wardlaw & Moncur, 1976; Pickard, Minchin & Troughton, 1978; Fellows, Egli & Leggett, 1979; Fondy & Geiger, 1980). Competition for translocate among sinks of similar vascular connection is often determined by their relative proximity to a source and their capacity or metabolic activity (Wyse & Saftner, 1982). The importance of direct vascular connection in determining primary source–sink routes is readily demonstrated, for example, by the transport of assimilates from leaves of a given orthostichy to a particular file of seeds in the developing head of sunflower (Prokofyev, Zhadanova & Sobolev, 1957). (Orthostichy is the name given to a directly vertical arrangement of leaves and other organs, interconnected by common vascular bundles.)

Unloading of phloem is often spatially and temporally distinct from the processes involved in the utilization of assimilates. In these systems, regulatory effects that the consumer cells would have on efflux from the phloem would necessarily be remote. For example, within developing seeds, the embryonic tissues lack direct vascular connection with the unloading sites in maternal tissues (Thorne, 1980, 1981; Felker & Shannon, 1980). Furthermore, primary roots store assimilates, notably starch, in layers of cortical parenchyma often considerably distant from the sites of unloading in the stele (Esau, 1965). In storage sinks, however, there is often a close association between sites of unloading in vascular tissue and the consumer cells of the storage parenchyma (Esau, 1965). Here, the consumer cells may more directly regulate efflux of solute from the phloem.

The wide assortment of sinks found within a plant and among plant species suggests that several mechanisms and structural specializations regulate unloading. However, regardless of type of sink, the import of assimilates appears to be dependent on energy generated by the sink tissues, for inhibitors, anoxia or low temperatures generally reduce import (Geiger & Sovonick, 1975). While not enough is known of the tissues and processes involved to specify whether unloading or subsequent metabolic events within sink tissues act as regulatory steps, possible sites of expenditure of energy include: (1) unloading of solutes from the sieve tubes; (2) active accumulation of solutes across the plasmalemma of consumer cells; (3) catabolic metabolism or biosynthesis of secondary products for growth or storage; and (4) intracellular compartmentation as within vesicles, plastids, storage vacuoles, etc.

Structural aspects of unloading

Pathways and mechanisms of unloading apparently vary with species, type of sink tissues involved, stage of development and physiological conditions (Giaquinta, 1980*a*). In the absence of biochemical or physiological data, many clues may be provided from morphological and ultrastructural studies of the vascularization of sinks. In the following discussion, an attempt is made to integrate available information from the standpoint of relations between structure and function.

In many sinks, the unloading of solutes from sieve tubes may involve tissues that are traditionally ascribed to phloem loading. For example, minor veins are produced by repeated vascular branching in developing sinks, just as at the sites of phloem loading in leaves. These are often flanked by one to three larger companion cells or transfer cells, dwarfing the diminutive sieve elements (Turgeon & Webb, 1976; Thorne, 1981; Kramer, Romheld, Landsberg & Marschner, 1980). In the limited number of studies available, companion cells are intimately associated with the phloem at the putative unloading sites. From studies of companion cells in other tissues, it is known that sieve elements and companion cells are sister cells which arise from the same phloem mother cell and eventually die at the same time (Behnke, 1975). Cell size may indicate involvement in loading and unloading of the phloem, for large companion cells dwarf the sieve elements of minor veins in both source (Geiger, 1975) and sink regions (Turgeon & Webb, 1976; Thorne, 1981). This relation is reversed in stem or petiolar tissue not involved in loading or unloading (Geiger, 1975). The cytoplasm of companion cells is rich in organelles and clearly distinguishable from that of surrounding phloem parenchyma (Behnke, 1975; Thorne, 1981). This density of organelles, particularly mitochondria, is consistent with concept that the high metabolic activity of these cells has functional significance in unloading (Thorne, 1981). The cell wall between sieve elements and companion cells is penetrated by numerous complex plasmodesmatal pores linking the symplasm into a sieve element–companion cell complex of similar osmotic potential (Geiger *et al.* 1973; Behnke, 1975).

Companion cells, phloem parenchyma cells and other cells through which a particularly rapid flux of solutes occurs become modified in some species to transfer cells (Gunning & Pate, 1969). These are strategically located and clearly specialized for enhancing transfer of solute across plasma membranes; from symplast to apoplast or *vice versa* (Pate & Gunning, 1972). In these cells, fingerlike wall ingrowths amplify the surface area of the associated plasmalemma up to 20-fold as compared to equivalent walls without ingrowths (Gunning, Pate & Green, 1970). Transfer cells are located at numerous places within the plant which are characterized by rapid flux of solute; structural differences indicate their possible function in each location. The wall ingrowths which develop relatively late in the life of the cell represent a specialized form of secondary wall, deposited on the inner surface of the original primary wall. There are companion transfer cells, xylem and phloem parenchyma transfer cells, and endosperm-, pericycle- and epidermal transfer cells, etc. (Pate & Gunning, 1972). Among transfer cells associated with minor veins of sinks, by far the most prevalent are modified companion cells. These are easily distinguished from modified phloem parenchyma, xylem parenchyma and bundle sheath cells by their extremely dense cytoplasm, the symmetric arrangement of wall ingrowths around the entire cell wall and the plasmodesmata linking them to contiguous sieve elements (Behnke, 1975; Pate, Kuo & Hocking, 1978; Kramer *et al.* 1980). The wall ingrowths develop just as intensive transport starts and become best developed on those faces of the cell most active in solute transport (Gunning & Pate, 1969).

In sink tissues, transfer cells are presumed to play important roles in both secretion from vascular tissues and subsequent absorption from internal extracytoplasmic compartments (Rost & Lersten, 1970; Gunning & Pate, 1969). The assumed transport function is based on the hypothesis that the amplification of plasmalemmal surface area increases the number of solute carriers that can be accommodated on the membrane (Pate & Gunning, 1972). Simple diffusion as the force for transport is thought to be unlikely since wall ingrowths considerably lengthen the diffusion paths of substances (Behnke, 1975). Instead, various solute pumps have been suggested as the driving forces of transport (Gunning & Pate, 1974). The localization of ATPases and acid phosphatases next to wall ingrowths lends support to this concept (Cranshaw, 1981).

Pathways and energetics of unloading in various sinks

Unloading *down* a concentration gradient may occur in many source–sink systems where unloading and transfer of solute to nearby consuming cells occurs without further expenditure of energy by sieve elements or their accessory cells (Geiger & Fondy, 1980). In these systems, unloading is thought to occur down a concentration gradient with utilization of solute by consumer cells as a driving force. Consumer cells are thought to exert a covert regulatory role in passive unloading by maintaining a low concentration of solute in the region of unloading. For example, Walker & Ho (1977*b*) demonstrated maximal import of photosynthate by small fruit whose rapid metabolism maintained a low internal concentration of sucrose. Unloading declined in fruit with higher sucrose concentrations and net *export* from the tomato fruit occurred when concentrations of sucrose exceeded 0.27% fresh weight or when fruits were cooled (Walker & Ho, 1977*a*, *b*). According to this hypothesis, the sensitivity of assimilate import to metabolic inhibitors such as low temperature, chemical inhibitors or anoxia is due to effects on the metabolic status of the sink consumer cells with only an indirect effect on the actual unloading process.

Transfer to the consumer cells following unloading from the phloem in such systems may be within the symplast or apoplast of surrounding cells but down a concentration gradient established by the consumer cells.

A symplastic route may be common to both developing organs such as young leaves, buds or root tips and to maturing sinks such as legume root nodules and some meristematic tissues. A symplastic pathway would seem to involve less expenditure of energy since two membrane transport steps are eliminated. Also, diffusion would occur in direct response to the lowering of the solute concentration by either metabolism or compartmentation, and would not be dependent upon specialized membrane carriers (Giaquinta, 1980). While a clear understanding of the actual unloading mechanism will await further work, some evidence is available. For example, differentiated sinks, at various stages of immaturity, often lack direct routes for import. Here solutes are thought to diffuse from mature phloem through immature phloem cells and into the consumer parenchyma of the sink (Geiger & Fondy, 1980). These workers suggest that, in young squash leaves undergoing sink-to-source transition, the bicollateral vascularization of this species permits symplastic diffusion from mature sieve elements to developing tissue via plasmodesmata. [The presence of phloem on both sides of the xylem makes the bundle bicollateral (Esau, 1965). The internal phloem is often primary but the external phloem often occurs as primary and secondary – perhaps linked radially by plasmodesmata.]

In other sink tissues, a symplastic route following unloading is not associated with immaturity of the sink tissue or the absence of functional phloem or companion cells. For example, the root nodules of legumes contain a vascular network which is peripheral to a central core of nitrogen-fixing bacterial tissue. Each vascular strand is bounded by an endodermis with well-developed casparian strips blocking an apoplastic route from phloem to bacteroids. Sugars unloaded from the phloem apparently pass to the bacteroids via a symplastic route (Pate, Gunning & Briarty, 1969). A symplastic transfer route is also indicated during the unloading of assimilates in pea roots (Dick & ap Rees, 1975). Asymmetrically labelled sucrose ([^{14}C]glucosyl-sucrose) was absorbed by the stele and transported without hydrolysis to the cortical and apical cells of the root even though the cell walls of these tissues contained considerable invertase activity. Since flooding the free space with unlabelled sugar did not reduce the amount of label transported to these tissues, it appears that a symplastic route was employed.

An apoplastic route following unloading has been clearly identified in several maturing sink systems and proposed for many others, but not enough is known of the actual unloading process or the tissues involved to specify regulatory sites. In many of these systems, the efflux of solutes from the phloem is spatially and temporally distinct from the subsequent, often well-characterized accumulation processes of the consumer cells in the sink. With few exceptions, neither the presence nor potential roles of companion cells or transfer cells in unloading have been investigated. As the consumer cells usually possess the capacity for active accumulation and rapid utilization (i.e. storage or metabolism) of unloaded solutes, efflux from the phloem possibly occurs in response to the resulting concentration gradient.

In storage parenchyma, such as the stems of sugar cane, sucrose is accumulated from the apoplast following unloading from the many vascular bundles regularly scattered throughout the parenchyma. These small bundles contain several sieve tubes and companion or transfer cells enclosed in a sheath of sclerenchyma (Esau, 1965). Sucrose is unloaded from these bundles into the apoplast of the stem parenchyma. From there it is accumulated by parenchyma cells at a rate which depends on the action of cell-wall acid invertase to hydrolyse sucrose (Glasziou & Gayler, 1972*a*; Gayler & Glasziou, 1972; Giaquinta, 1977*a*, *b*). These hexoses are actively accumulated without phosphorylation into the parenchyma cytoplasm and then converted to sucrose phosphate by sucrose phosphate synthetase. A sucrosyl carrier system incorporating a specific sucrose phosphate phosphatase is thought to regulate transfer of sucrose into the vacuole for storage (see Giaquinta, 1980*a*, *b*, for details).

Similarly, the storage parenchyma of tap roots of sugar beet apparently accumulate sucrose from the apoplast (Saftner & Wyse, 1980). Unlike sugar cane, in sugar beet sucrose uptake occurs by a non-saturable membrane carrier located in the tonoplast of the storage vacuole (Wyse, 1979; Saftner & Wyse, 1980). Also unlike sugar cane, accumulation of sucrose does not involve hydrolysis and resynthesis during storage (Giaquinta, 1977*b*). To account for this, Wyse (1979) proposed that accumulation occurs at sites where the plasmalemma and tonoplast are appressed and cytoplasm excluded, thus preventing sucrose metabolism. Recent studies (Saftner & Wyse, 1980) indicate the existence of an alkali cation (i.e. K^+)/sucrose cotransport mechanism in the tonoplast of the root sink of sugar beet.

In developing seeds, the import of assimilates for growth and formation of embryonic reserves involves the transfer of material between maternal and embryonic tissues that are not symplastically linked. As was the case with storage parenchyma, much less is known of the mechanisms and pathways involved in unloading of solutes from the phloem than of their subsequent transcellular transport and utilization by the consumer cells. While the mechanism is unknown, the region where unloading occurs has been identified in several seeds. Apparently, the large anatomical and biochemical diversity among species can be consolidated into fairly representative groups. Review of these groups may serve as examples of a majority of the known systems.

In corn (maize), and several other C_4 cereals, sucrose is unloaded from phloem terminals located in a small area of the pedicel at the very base of the fruit (Felker & Shannon, 1980). The veins divide and repeatedly anastamose throughout the pedicel, where they terminate. The pedicel is part of a thin maternal pericarp that surrounds the entire seed, enclosing the large starchy endosperm and the monocotyledonous embryo. Since there is no symplastic connection between maternal, endospermic and embryonic tissues, sucrose from the phloem must enter the apoplast prior to being accumulated by the endosperm. There appears to be some question as to whether sucrose is unloaded directly from the phloem into the maternal apoplast or whether there is a short passage through maternal symplasm *en route* to the free space between the generations. Hydrolysis of sucrose by cell-wall invertase appears to occur prior to uptake by the endosperm (Shannon & Dougherty, 1972; Shannon, 1972; Felker & Shannon, 1980), suggesting a cell-wall route through the funiculus–chalazal tissues. Furthermore, after labelling a plant with $^{14}CO_2$, much of the ^{14}C-photosynthate in the pedicel of the kernel can be removed by leaching the tissue in ice-cold water (Felker & Shannon, 1980).

Passage of the monosaccharides from the free space into the symplasm of the apposing endospermic cells is apparently facilitated by the conversion of some of these cells to a pad of endospermic transfer cells in the millets and in corn (maize) (Kiesselbach & Walker, 1952). There is some question as to whether diffusion alone is responsible for the subsequent distribution of sucrose to all areas of the endosperm for starch synthesis, a distance of up to 10 mm (Jenner, 1980). The rate of growth per kernel can be up to five times greater in corn (maize) than in many temperate cereals, suggesting that intercellular transfer of assimilates within the endosperm may be more than diffusion alone. These translocatory problems have not yet been studied.

Seeds of several other cereals are apparently very similar in the anatomy and manner in which assimilates are imported. These include pearl millet (Fussell & Dwarte, 1980), sorghum (Eastin, Hultquist & Sullivan, 1973) and yellow foxtail (Rost & Lersten, 1970).

Unloading of assimilate in C_3 cereal grains (wheat, rye, and barley) is not restricted to a small area at the base of the kernel as in C_4 grains, but occurs along the entire length of a large vascular bundle which extends from one end of the grain to the other in a longitudinal crease of maternal tissues. There is no symplastic connection, however, between this vascular cylinder and the surrounding consumer cells of the embryo and its endosperm (Frazier & Appalanaidu, 1965; Zee & O'Brien, 1971). Although the xylem is discontinuous at the base of the kernel, there appears to be no anatomical discontinuity of phloem sieve tubes between the kernel base and the vascular bundle in the longitudinal crease. Radioactivity was detected in the vascular tissue of the crease 1 h after supply of $^{14}CO_2$ to the flag leaf (Sakri & Shannon, 1975). The transfer cells observed by Zee & O'Brien (1971) at the point of attachment of the grain were associated with both the xylem and phloem, suggesting that materials released there from the xylem may be loaded into the phloem for transport into the kernel. Since import is not limited to a small region at the base of the grain as in corn (maize), the pathlength from the site of unloading to the endospermic consumer cells in wheat is often less than 2 mm. Also not as in corn, sucrose is apparently unloaded and transferred without hydrolysis through the apoplast of several layers of maternal cells and into the endosperm. That hydrolysis is not a prerequisite for import was demonstrated by Jenner (1974, 1976), who found that Tris buffer markedly inhibited activity of cell-wall invertase without inhibiting the rate of uptake of sucrose or deposition of starch in the endosperm. Also, the accumulation of symmetrically labelled sucrose in the presence of unlabelled hexoses did not alter the symmetry of the accumulated sucrose (Jenner, 1974, 1976). While the accumulation of asymmetrically labelled sucrose did not result in significant randomization of the label, Sakri & Shannon (1975) differ from Jenner and conclude that hydrolysis does occur during import in wheat. Continuing work by several groups may soon resolve this question.

There is evidence that the composition and concentration of assimilates in the developing grain is regulated within the crease. Thus *in vivo* and *in vitro*, the maximum concentration of sucrose available to the wheat endosperm is 58 mM, regardless of diurnal fluctuations in photosynthesis of the canopy (Jenner, 1976). Furthermore, the activities of starch-synthesizing enzymes in wheat endosperm are sensitive to irradiance during grain development and poor production of starch in shaded plants is not due to an insufficient supply of sucrose (Mengel & Judel, 1981). Under adverse growing conditions and limiting photosynthate, optimal assimilate levels for grain development are maintained by initiating the development of fewer primordia of spikelets (Scott, 1976).

Metabolic conversion of sucrose (primarily to starch) was thought to control the rate of accumulation of dry matter in wheat grains, for Jenner (1974) concluded that import to the endosperm occurs down a concentration gradient. However, recent studies indicate that active transport plays a major role in assimilation of sucrose by developing barley embryos (Cameron-Mills & Duffus, 1979). Below 50 mM sucrose, uptake was substrate-specific and sensitive to uncoupling agents. At higher concentrations, passive diffusion made an increasing contribution. It is unknown whether such a mechanism exists in the endosperm.

At maturation, the decline in import of sucrose and formation of starch in the grain is apparently due to a fall in the capacity of the endosperm to synthesize starch (Jenner & Rathjen, 1975) rather than the formation of a black layer as in the C_4 cereals such as corn, millet and sorghum. Formation of this layer in those species is associated with the cessation of import and the crushing of the basal endospermic transfer cells by the embryo (Fussell & Dwarte, 1980).

In dicotyledonous seeds, transport of assimilate has been studied primarily with commercially important non-endospermic legumes, particularly the garden pea, field pea and broad, field, or horse bean with fewer studies on the French, dwarf, or bush bean, and soybean. Additionally, several of the endospermic legumes and lipid-storing dicots have been examined. Most studies have focused on the biochemical processes associated with seed growth and the establishment of cotyledonary reserves of carbohydrates, proteins and lipids. These have been reviewed recently by Bewley & Black (1978).

Recent studies of translocation, however, have provided qualified support for the operation of a mechanism of unloading and import in leguminous seeds similar to that of the storage parenchyma of sugar beet. Translocation of nutrients to the developing cotyledons of legumes occurs from the phloem of the mother plant through the vascular bundles of the pod, which in turn enter the seeds via the funiculi. Upon entering the seeds, photosynthate is rapidly and evenly distributed throughout the seed coat within an anastamosing network of minor veins (Patrick & McDonald, 1980; Thorne, 1980, 1981). In soybeans, these minor veins are enclosed in a parenchymatous sheath and contain from two to ten sieve tubes with accompanying companion cells but no xylem tissue (Thorne, 1981). In garden pea, one major vascular bundle supplies the seed coat with photosynthate, apparently without the network of minor veins (Hardham, 1976). Indeed, there are numerous vascular patterns in leguminous seed coats (Corner, 1951). Very little information is available on the nature of the mechanism of transport which facilitates exit of sugar from the sieve tubes within the seed coat. Large companion cells are linked by plasmodesmata to the sieve-tube symplasm (Thorne, 1981) providing circumstantial evidence for a metabolic component in unloading. That unloading is energy-dependent is demonstrated by its sensitivity to anoxia in that ^{14}C-photosynthate is retained within the vascular bundles of the seed coat when the fruit is exposed to localized anoxia. The embryos are totally unlabelled under these conditions, although when they are removed from the seed coats in anoxia and incubated in [^{14}C]sucrose equilibrated with N_2, considerable amounts of label are absorbed (Thorne, 1982*a*). The basis for the energy-dependence in unloading is not apparent if unloading through the companion cells is passive, unless anoxia induces specific changes in the permeability of the membranes involved.

The accumulation by leguminous embryos of assimilates released from the seed coats appears to include both passive and active components. Lichtner & Spanswick (1981) have provided evidence that the active uptake of sucrose is by a sucrose/proton cotransport system in the plasmalemma of soybean cotyledonary cells. Sucrose uptake occurs without extracellular hydrolysis, for no randomization of the label occurs during active accumulation of [^{14}C](fructosyl)-sucrose and the symmetry of uniformly labelled sucrose is not altered when uptake is in the presence of unlabelled hexoses (Thorne, 1982*b*). Exogenous solutes have free access to these cells through the apoplast (Lichtner, 1980; Patrick & McDonald, 1980). However, in some species the accumulation of solutes may be amplified by the presence of epidermal transfer cells (Gunning & Pate, 1974). Such structural modifications would allow rapid uptake with a minimum of solvent, an important consideration in developing seeds which lack appreciable transpiration.

Unloading *against* a concentration gradient is probable in sugar-secreting glands in which extremely large rates of flux of sugars across membranes are a salient feature of their physiological activity. Best known are floral and extra-floral nectaries, although many types have been studied (Lüttge & Schnepf, 1976). The nectary glands are supplied by sieve tubes and the concentration of sugar in the nectar depends on the proportion of phloem and xylem elements in the gland parenchyma (Frey-Wyssling, 1955). A consideration of the sensitivity of nectar secretion to metabolic inhibitors and of the average concentration gradients between the phloem sap and the nectar itself suggests that in some cases, active unloading of phloem and/or subsequent symplastic transport of sugars occurs via the parenchyma and secretory cells of nectary glands into the nectar (Lüttge, 1977). Average concentrations of sugar in the phloem sap range from 18 to 30% while that of the nectar may exceed the limits of solubility in the nectar (e.g. 58–92%, Brewer, Collyard & Lott, 1974). Symplastic coupling within the system of sieve tubes, companion cells, parenchyma cells and secretory cells of nectary glands allows many possible sites for active transport: phloem unloading, passage through the stalk cells or elimination from the secretory cells. Although active transport occurs at one or more sites along the symplastic route, the location and driving forces involved are unknown.

Unloading may take the form of induced leakage from the phloem. For example, soon after the apical region of a searching 'hypha' of the parasitic dodder locates a functional sieve element of the host, a special contact cell is formed. Resembling a hand which grasps around the host sieve element with many fingers, this parasitic contact cell develops the extensive wall ingrowths of a transfer cell, thus enlarging the absorbing surface up to 20 times. The protoplast of the 'hypha' then differentiates in a manner characteristic of sieve-element development: enucleation, disintegration of the tonoplast, and occurrence of typical plasmatic filaments (Dörr, 1972). Jacob & Neumann (1968) concluded that the contact cells of the 'hypha' first cause the sieve-tube elements to leak and then absorb the solutes from the apoplast. Using ^{14}C-metabolites, it has been demonstrated that dodder is able to withdraw all assimilates which normally are transported from a photosynthesizing leaf to developing fruit of broad bean (Wolswinkel, 1974*a*). The enhanced unloading of assimilates to the free space was apparently under metabolic control since it could not be demonstrated at 0 °C nor after addition of dinitrophenol or sodium azide (Wolswinkel, 1974*b*). Similarly, the efflux of ^{14}C-labelled assimilates into the free space from the phloem was accompanied by an active efflux of Mg^{2+} and, to a lesser extent, K^{+} ions (Wolswinkel, 1975). Alterations to patterns of translocation induced by fungal pathogens and some possible mechanisms involved have been reviewed in this Seminar Series by Whipps & Lewis (1981).

Larvae of root-knot nematodes invade susceptible host plant roots and induce the formation of giant multinucleate transfer cells at the site where they become sedentary. The larvae obtain nutrients from the giant cells which enlarge as the nematodes grow. Extensive protuberances develop on walls of the giant cells adjacent to xylem vessels and, to a lesser extent, next to sieve elements. These protuberances are absent from walls adjacent to phloem parenchyma cells. Solutes are rapidly unloaded from the phloem into the apoplast in the region of the giant cells. The protuberances then concentrate selective solutes within the giant cells, upon which the larvae feed. Extremely rapid cellular metabolism is necessary in the giant transfer cells, not only to synthesize the wall protuberances and to provide the energy for massive accumulation of assimilates, but also to conduct these functions while simultaneously replacing cytoplasmic components and organelles consumed by the nematode during feeding (Jones & Northcote, 1972).

In all the examples cited, the mechanism of efflux from the phloem supplying assimilates to the sink consumer cells is unknown. In each case, however, it appears that efflux may occur in response to the solute concentration and energy gradient established by actively accumulating consumer cells. Inasmuch as the phloem is highly pressurized and generally resistant to leakage along its entire length (Wardlaw, 1965), the companion cells and, in some cases transfer cells, may facilitate unloading, perhaps by providing a variably permeable plasmalemma for efflux of solute. Possible effectors include plant hormones (Patrick, 1976; Tanner, 1980), specific ions (Fondy & Geiger, 1980), or, in the case of parasites, perhaps specific chemicals. The nature of such promotive mechanisms and their possible relation to the regulation of import of photosynthate deserve further study.

Note added in proof: The literature cited in this review was current as of 1981.

References

Baker, D. A. (1978). Proton co-transport of organic solutes by plant cells. *New Phytologist*, **81**, 485–97.

Baker, D. A., Malek, F. & Dehvar, F. D. (1980). Phloem loading of amino acids from the petioles of *Ricinus* leaves. *Berichte der Deutschen botanischen Gesellschaft*, **93**, 203–9.

Behnke, H.-D. (1975). Companion cells and transfer cells. In *Phloem Transport*, ed. S. Aronoff, J. Dainty, P. R. Gorham, L. M. Srivastava & C. A. Swanson, pp. 153–75. New York: Plenum Press.

Bewley, J. D. & Black, M. (1978). *Physiology and Biochemistry of Seeds in Relation to Germination*, vol. 1, *Development, Germination and Growth*. Berlin: Springer-Verlag.

Brewer, J. W., Collyard, K. J. & Lott, C. E. (1974). Analysis of sugars in dwarf mistletoe nectar. *Canadian Journal of Botany*, **52**, 2533–8.

Cameron-Mills, V. & Duffus, C. M. (1979). Sucrose transport in isolated immature barley embryos. *Annals of Botany*, **43**, 559–69.

Corner, E. J. H. (1951). The leguminous seed. *Phytomorphology*, **1**, 117–50.

Cranshaw, J. (1981). Phloem structure and function. *Annual Review of Plant Physiology*, **32**, 465–84.

Dick, P. S. & ap Rees, T. (1975). The pathway of sugar transport in roots of *Pisum sativum*. *Journal of Experimental Botany*, **26**, 305–14.

Doman, D. C. & Geiger, D. R. (1979). Effect of exogenously supplied foliar potassium on phloem loading in *Beta vulgaris* L. *Plant Physiology*, **64**, 528–33.

Dörr, I. (1972). The contact of *Cuscuta* hyphae with the sieve tubes of its host plants. *Protoplasma*, **75**, 167–84.

Eastin, J. K., Hultquist, J. H. & Sullivan, C. V. (1973). Physiologic maturity in grain sorghum. *Crop Science*, **13**, 175–8.

Esau, K. (1965). *Plant Anatomy*, 2nd edn, 767 pp. New York: John Wiley & Sons, Inc.

Felker, F. C. & Shannon, J. C. (1980). Movement of ^{14}C-labeled assimilates in kernels of *Zea mays* L. *Plant Physiology*, **65**, 864–70.

Fellows, R. J., Egli, D. B. & Leggett, J. E. (1979). Rapid changes in translocation patterns in soybeans following source–sink alterations. *Plant Physiology*, **64**, 652–5.

Fisher, D. B. & Outlaw Jr., W. H. (1979). Sucrose compartmentation in the palisade parenchyma of *Vicia fava* L. *Plant Physiology*, **64**, 481–3.

Fondy, B. R. & Geiger, D. R. (1980). Effect of rapid changes in sink–source ratio on export and distribution of products of photosynthesis in leaves of *Beta vulgaris* L. and *Phaseolus vulgaris* L. *Plant Physiology*, **66**, 945–9.

Frazier, J. C. & Appalanaidu, B. (1965). The wheat grain during development with reference to nature, location, and role of its translocation tissues. *American Journal of Botany*, **52**, 193–8.

Frey-Wyssling, A. (1955). The phloem supply to the nectaries. *Acta Botanica Neerlandica*, **4**, 358–69.

Fussell, L. K. & Dwarte, D. M. (1980). Structural changes of the grain associated with black region formation in *Pennisetum americanum*. *Journal of Experimental Botany*, **31**, 645–54.

Gayler, K. R. & Glasziou, K. T. (1972). Sugar accumulation in sugarcane. *Plant Physiology*, **49**, 563–8.

Geiger, D. R. (1975). Phloem loading. In *Encyclopedia of Plant Physiology*, vol. 1, ed. M. H. Zimmermann & J. A. Milburn, pp. 395–431. Berlin: Springer-Verlag.

Geiger, D. R. & Fondy, B. R. (1980). Phloem loading and unloading: pathways and mechanisms. *What's New in Plant Physiology*, **11**, 25–8.

Geiger, D. R. & Giaquinta, R. T. (1981). Translocation of photosynthates. Photosynthesis translocation, and crop yield. In *Photosynthesis: CO_2 Assimilation and Plant Productivity*, ed. Govindjee (in press).

Geiger, D. R., Giaquinta, R. T., Sovonick, S. A. & Fellows, R. J. (1973). Solute distribution in sugar beet leaves in relation to phloem loading and translocation. *Plant Physiology*, **52**, 585–9.

Geiger, D. R. & Sovonick, S. A. (1975). Effects of temperature, anoxia and other metabolic inhibitors on translocation. In *Encyclopedia of Plant Physiology*, vol. 1, ed. M. H. Zimmermann & J. A. Milburn, pp. 256–86. Berlin: Springer-Verlag.

Giaquinta, R. T. (1977*a*). Possible role of pH gradient and membrane ATPase in the loading of sucrose into the sieve tubes. *Nature*, **267**, 369–70.

Giaquinta, R. T. (1977*b*). Sucrose hydrolysis in relation to phloem translocation in *Beta vulgaris*. *Plant Physiology*, **60**, 339–43.

Giaquinta, R. T. (1978). Source and sink leaf metabolism in relation to phloem translocation. Carbon partitioning and enzymology. *Plant Physiology*, **61**, 380–5.

Giaquinta, R. T. (1979). Phloem loading of sucrose. Involvement of membrane ATPase and proton transport. *Plant Physiology*, **63**, 744–8.

Giaquinta, R. T. (1980*a*). Translocation of sucrose and oligosaccharides. In *Biochemistry of Plants*, vol. 3, ed. J. Preiss, pp. 271–319. New York: Academic Press.

Giaquinta, R. T. (1980*b*). Mechanism and control of phloem loading of sucrose. *Berichte der Deutschen botanischen Gesellschaft*, **93**, 187–201.

Glasziou, K. T. & Gayler, K. R. (1972). Storage of sugars in stalks of sugarcane. *Botanical Review*, **38**, 471–90.

Gunning, B. E. S. & Pate, J. S. (1969). Plant cells with wall ingrowths, specialized in relation to short distance transport of solutes – their occurrence, structure and development. *Protoplasma*, **68**, 107–33.

Gunning, B. E. S. & Pate, J. S. (1974). Transfer cells. In *Dynamic Aspects of Plant Ultrastructure*, ed. A. W. Robards, pp. 441–80. Maidenhead, UK: McGraw-Hill.

Gunning, B. E. S., Pate, J. S. & Green, L. W. (1970). Transfer cells in the vascular system of stems: taxonomy, association with nodes and structure. *Protoplasma*, **71**, 147–71.

Hardham, A. R. (1976). Structural aspects of the pathways of nutrient flow to the developing embryo and cotyledons of *Pisum sativum* L. *Australian Journal of Botany*, **24**, 711–21.

Hatch, M. D. (1964). Sugar accumulation by sugarcane storage tissue. The role of sucrose phosphate. *Biochemistry Journal*, **93**, 521–6.

Herold, A. & Lewis, D. H. (1977). Mannose and green plants: occurrence, physiology and metabolism, and use as a tool to study the role of orthophosphate. *New Phytologist*, **79**, 1–40.

Heyser, W. (1980). Phloem loading in the maize leaf. *Berichte der Deutschen botanischen Gesellschaft*, **93**, 221–8.

Huber, S. (1981*a*). Inter- and intra-specific variation in photosynthetic formation of starch and sucrose. *Zeitschrift für Pflanzenphysiologie*, **101**, 49–54.

Huber, S. (1981*b*). Interspecific variation in activity and regulation of leaf sucrose phosphate synthetase. *Zeitschrift für Pflanzenphysiologie*, **102**, 443–50.

Huber, S. (1981*c*). Cotransport of potassium and sugars across the plasmalemma of mesophyll protoplasts. *Plant Physiology*, **67**, 163–9.

Jacob, V. F. & Neumann, S. (1968). Studien an *Cuscuta reflexa* Roxb. 1. Zur Funktion der Haustorien bei der Aufnahme von Saccharose. *Flora*, **159**, 191–203.

Jenner, C. F. (1974). An investigation of the association between hydrolysis of sucrose and its absorption by grains of wheat. *Australian Journal of Plant Physiology*, **1**, 319–29.

Jenner, C. F. (1976). Physiological investigations on restrictions to transport of sucrose in ears of wheat. *Australian Journal of Plant Physiology*, **3**, 337–47.

Jenner, C. F. (1980). The conversion of sucrose to starch in developing fruits. *Berichte der Deutschen botanischen Gesellschaft*, **93**, 289–98.

Jenner, C. F. & Rathjen, A. J. (1975). Factors regulating the accumulation of starch in the ripening wheat grain. *Australian Journal of Plant Physiology*, **2**, 311–22.

Jones, M. G. K. & Northcote, D. H. (1972). Multinucleate transfer cells induced in *Coleus* roots by the root-knot nematode *Meloidogyne arenaria*. *Protoplasma*, **75**, 381–95.

Kiesselbach, T. A. & Walker, E. R. (1952). Structure of certain specialized tissues in the kernel of corn. *American Journal of Botany*, **39**, 561–9.

Komor, E., Rotter, M., Waldhauser, J., Martin, E. & Cho, B. H. (1980). Sucrose proton symport for phloem loading in the *Ricinus* seedling. *Berichte der Deutschen botanischen Gesellschaft*, **93**, 211–19.

Kramer, D., Romheld, V., Landsberg, E. & Marschner, H. (1980). Induction of transfer-cell formation by iron deficiency in the root epidermis of *Helianthus annuus* L. *Planta*, **147**, 335–9.

Lichtner, F. T. (1980). Sucrose/proton cotransport in developing soybean cotyledons. Ph.D. Thesis, Cornell University, Ithaca, New York.

Lichtner, F. T. & Spanswick, R. M. (1981). Electrogenic sucrose transport in developing soybean cotyledons. *Plant Physiology*, **67**, 869–74.

Lüttge, U. (1977). Nectar composition and membrane transport of sugars and amino acids: a review of the present state of nectar research. *Apidologie*, **8**, 305–19.

Lüttge, U. & Schnepf, E. (1976). Elimination processes by glands. Organic substances. In *Encyclopedia of Plant Physiology*, vol. II B, ed. U. Lüttge & M. G. Pitman, pp. 244–77. Berlin: Springer-Verlag.

Mengel, K. & Judel, G.-K. (1981). Effect of light intensity on the activity of starch synthesizing enzymes and starch synthesis in developing wheat grains. *Physiologia Plantarum*, **51**, 13–18.

Outlaw Jr., W. H. & Fisher, D. B. (1975). Compartmentation in *Vicia faba* leaves. I. Kinetics of ^{14}C in the tissues following pulse labeling. *Plant Physiology*, **55**, 699–703.

Outlaw Jr., W. H., Fisher, D. B. & Christy, A. L. (1975). Compartmentation in *Vicia fava* leaves. II. Kinetics of ^{14}C-sucrose redistribution among individual tissues following pulse labeling. *Plant Physiology*, **55**, 704–11.

Pate, J. S. & Gunning, B. E. S. (1972). Transfer cells. *Annual Review of Plant Physiology*, **23**, 173–96.

Pate, J. S., Gunning, B. E. S. & Briarty, L. G. (1969). Ultrastructure and functioning of the transport system of the leguminous root nodule. *Planta*, **85**, 11–34.

Pate, J. S., Kuo, J. & Hocking, P. J. (1978). Functioning of conducting elements of phloem and xylem in the stalk of the developing fruit of *Lupinus albus* L. *Australian Journal of Plant Physiology*, **5**, 321–36.

Patrick, J. W. (1976). Hormone-directed transport of metabolites. In *Transport and Transfer Processes in Plants*, ed. I. F. Wardlaw & J. B. Passioura, pp. 433–46. New York: Academic Press.

Patrick, J. W. & McDonald, R. (1980). Pathway of carbon transport within developing ovules of *Phaseolus vulgaris* L. *Australian Journal of Plant Physiology*, **7**, 671–84.

Pickard, W. F., Minchin, P. E. H. & Troughton, J. H. (1978). Real time studies of carbon-11 translocation in moonflower. *Journal of Experimental Botany*, **29**, 993–1001.

Preiss, J. & Levi, C. (1980). Starch biosynthesis and degradation. In *Biochemistry of Plants*, vol. 3, ed. J. Preiss, pp. 371–423. New York: Academic Press.

Prokofyev, A. A., Zhadanova, L. P. & Sobolev, A. H. (1957). Certain regularities in the flow of substances from leaves to reproductive organs. *Fiziologiya Rastenii*, **4**, 425–31.

Rost, T. L. & Lersten, N. R. (1970). Transfer aleurone cells in *Setaria lutescens* (*Gramineae*). *Protoplasma*, **71**, 403–8.

Saftner, R. A. & Wyse, R. E. (1980). Alkali cation/sucrose co-transport in the root sink of sugarbeet. *Plant Physiology*, **66**, 884–9.

Sakri, F. A. K. & Shannon, J. C. (1975). Movement of ^{14}C-labeled sugars into kernels of wheat *Triticum aestivum* L. *Plant Physiology*, **55**, 881–9.

Scott, R. K. (1976). The scientific basis for high cereal yields. In *Crop Production – The Way Ahead.* Sixteenth Annual Conference of the East of Scotland College of Agriculture. Edinburgh: East of Scotland College of Agriculture.

Shannon, J. C. (1972). Movement of ^{14}C-labeled assimilates into kernels of *Zea mays* L. I. Pattern and rate of sugar movement. *Plant Physiology*, **49**, 198–202.

Shannon, J. C. & Dougherty, C. T. (1972). Movement of ^{14}C-labeled assimilates into kernels of *Zea mays* L. II. Invertase activity of the pedicel and placento-chalazal tissues. *Plant Physiology*, **49**, 203–6.

Silvius, J. E., Chatterton, J. N. & Kremer, D. F. (1979). Photosynthate partitioning in soybean leaves at two irradiance levels. Comparative responses of acclimated and unacclimated leaves. *Plant Physiology*, **64**, 872–5.

Silvius, J. E., Kremer, D. F. & Lee, D. R. (1978). Carbon assimilation and translocation in soybeans at different stages of leaf development. *Plant Physiology*, **62**, 54–8.

Swanson, C. A., Hoddinott, J. & Sij, J. W. (1976). The effect of selected sink leaf parameters on translocation rates. In *Transport and Transport Processes in Plants*, ed. I. F. Wardlaw & J. B. Passioura, pp. 347–56. New York: Academic Press.

Tanner, W. (1980). On the possible role of ABA on phloem unloading. *Berichte der Deutschen botanischen Gesellschaft*, **93**, 349–51.

Thorne, J. H. (1980). Kinetics of ^{14}C-photosynthate uptake by developing soybean fruit. *Plant Physiology*, **65**, 975–9.

Thorne, J. H. (1981). Morphology and ultrastructure of maternal seed tissues of soybean in relation to the import of photosynthate. *Plant Physiology*, **67**, 1016–25.

Thorne, J. H. (1982*a*). Temperature and oxygen effects on ^{14}C-photosynthate unloading and accumulation in developing soybean seeds. *Plant Physiology*, **69**, 48–53.

Thorne, J. H. (1982*b*). Characterization of the active sucrose transport system of immature soybean embryos. *Plant Physiology*, **70**, 953–8.

Turgeon, R. & Webb, J. A. (1976). Leaf development and phloem transport in *Cucurbita pepo*: maturation of the minor veins. *Planta*, **129**, 265–9.

Walker, A. J. & Ho, L. C. (1977*a*). Carbon translocation in the tomato: carbon import and fruit growth. *Annals of Botany*, **41**, 813–23.

Walker, A. J. & Ho, L. C. (1977*b*). Carbon translocation in the tomato: effects of fruit temperature on carbon metabolism and the rate of translocation. *Annals of Botany*, **41**, 825–32.

Wardlaw, I. F. (1965). The velocity and pattern of assimilate translocation in wheat plants during grain development. *Australian Journal of Biological Science*, **18**, 269–81.

Wardlaw, I. F. & Moncur, L. (1976). Source, sink, and hormonal control of translocation in wheat. *Planta*, **128**, 93–100.

Whipps, J. M. & Lewis, D. H. (1981). Patterns of translocation, storage and interconversion of carbohydrates. In *Effects of Disease on the Physiology of the Growing Plant*, ed. P. G. Ayres, pp. 47–83. Cambridge: Cambridge University Press.

Whittingham, C. P., Keys, A. J. & Bird, I. F. (1979). The enzymology of sucrose synthesis in leaves. In *Photosynthesis II*, vol. 6, ed. M. Gibbs & E. Latzko, pp. 313–26. Berlin: Springer-Verlag.

Wolswinkel, P. (1974*a*). Complete inhibition of setting and growth of fruits of *Vicia faba* L., resulting from the draining of the phloem system by *Cuscuta* species. *Acta Botanica Neerlandica*, **23**, 48–60.

Wolswinkel, P. (1974*b*). Enhanced rate of ^{14}C-solute release to the free space by the phloem of *Vicia fava* stems parasitized by *Cuscuta. Acta Botanica Neerlandica*, **23**, 177–88.

Wolswinkel, P. (1975). The active role of the host (*Vicia faba* L.) in the transfer of nutrient elements from the phloem to the parasite (*Cuscuta* species): metabolically controlled K^+ and Mg^{++} release to the free space. *Acta Botanica Neerlandica*, **24**, 211–24.

Wyse, R. E. (1979). Sucrose uptake by sugar beet tap root tissue. *Plant Physiology*, **64**, 837–41.

Wyse, R. E., Daie, J. & Saftner, R. (1980). Hormonal control of sink activity in sugar beet. *Plant Physiology*, **65**, S–662.

Wyse, R. E. & Saftner, R. A. (1982). Reduction in sink-mobilizing ability following periods of high carbon flux. *Plant Physiology*, **69**, 226–8.

Zee, S. Y. & O'Brien, T. P. (1970). A special type of tracheary element associated with 'xylem discontinuity' in the floral axis of wheat. *Australian Journal of Biological Science*, **23**, 783–91.

C. J. POLLOCK

4 Physiology and metabolism of sucrosyl-fructans

Introduction

Although the major route for the synthesis of reserve polysaccharides in vascular plants is by the action of enzymes utilizing nucleotide sugars as glycosyl donors, linkages between glycosyl groups can be formed without the direct participation of sugar nucleotide intermediates. For example α-1,4-linked glucose polymers can be formed by the action of phosphorylase (Reaction 4.1) (Hanes, 1940).

$$\text{glucose-1-P} + (\text{glucosyl})_n \rightleftharpoons \text{Pi} + (\text{glucosyl})_{n+1} \quad (4.1)$$

In addition, a variety of hydrolytic enzymes can catalyse the synthesis of higher oligosaccharides from both mono- and disaccharide substrates at high concentration (Peat, Whelan & Hinson, 1952; Edelman, 1956). Reaction 4.2, catalysed by yeast invertase, is of this type:

$$\text{sucrose} + \text{sucrose} \rightleftharpoons \text{glucose} + \text{glucose-fructose-fructose} \quad (4.2)$$

Although such transferase reactions have been demonstrated utilizing a variety of substrates, oligosaccharide formation from sucrose is particularly common. This is probably due to the high free energy of hydrolysis of the glucosidic linkage in sucrose which is close to that of UDP glucose (see Table 1.2 of Chapter 1).

Most of the enzymes catalysing such transferase reactions function almost entirely as hydrolases *in vivo*. However, high-relative-molecular-weight polymers can be produced by similar reactions in which there are successive transfers of saccharide residues on to the same acceptor molecule. The products of these reactions can be polymers of glucose, such as the bacterial dextrans (Reaction 4.3), or of fructose as in the bacterial levans or the fructans of higher plants (Reaction 4.4) (Hirst, 1957; Deonder, 1972).

$$m \text{ sucrose} + (\text{glucosyl})_n \rightleftharpoons (\text{glucosyl})_{n+m} + m \text{ fructose} \quad (4.3)$$

$$m \text{ sucrose} + (\text{fructosyl})_n \rightleftharpoons (\text{fructosyl})_{n+m} + m \text{ glucose} \quad (4.4)$$

Sucrose itself is a product of reactions requiring sugar nucleotides as intermediates (see Chapter 2), so ultimately the energy driving the synthesis of polymers from sucrose derives from these intermediates. However, since sucrose is the major translocated sugar in most higher plants and may also be stored within the cell, biosynthesis of polysaccharides derived from it in the manner of equations (4.3) and (4.4) can be separated both spatially and temporally from the production of sugar nucleotides; in contrast to the synthesis of other reserve polysaccharides where sugar nucleotides form the actual substrates of the synthetic reaction.

In this account, I propose to discuss whether or not the experimental evidence supports the operation of this type of mechanism in those species of vascular plants which accumulate fructans (see Chapter 1) and to examine how such a mechanism might be regulated to give the different types of polymer structure and size, and patterns of metabolism. An understanding of the factors controlling fructan metabolism would improve our knowledge of the physiology of a number of important crop plants, including temperate forage grasses and cereals, where fructans form the major reserve carbohydrate in the vegetative organs. In addition an understanding of the selective advantage of accumulation of fructan in species which also synthesize starch might provide more general information on factors limiting starch accumulation and the ways in which such limitations can be overcome.

Biochemical aspects of fructan metabolism

The problems of studying fructans

There is much conflicting evidence on the structure and metabolism of fructans in vascular plants. Frequently this is due to problems associated with their extraction, purification and analysis. These problems are similar to those associated with the study of any polydisperse polymer but are compounded by the high water solubility and the extreme lability of fructans. Solubility in water can be as high as 85% (Archbold, 1940) and there are many reports of the rapid hydrolysis of fructans at low pH (McDonald, 1946) or elevated temperature (Hirst, 1957). These properties lead to difficulties in obtaining complete separation from other neutral components of plant extracts and avoiding partial degradation of the polymer during extraction.

Early purifications were based upon the limited solubility of fructans in ethanol (Archbold, 1940) but the proportion soluble in different concentrations of ethanol can vary markedly with the source of the preparation (Smith & Groteleuschen, 1966). The use of repeated precipitation by ethanol to obtain preparations of constant optical rotation has been used as a criterion of purity (Schlubach & Peitzner, 1937) but it is probable that such preparations represent only a proportion of the range of fructan molecules present within the original sample.

Paper chromatography has been used with considerable success to separate individual oligofructosides from a simple aqueous extract (Bacon & Edelman, 1951) but complex and time-consuming variations on this technique are necessary to fractionate the high molecular-weight components. The advent of gel permeation chromatography has simplified the procedure for obtaining representative fructans. By the appropriate choice of resin, fructans can be separated from mono- and disaccharides (Pollock & Ruggles, 1976) and the individual oligosaccharide components resolved (Darbyshire & Henry, 1978). Thus, high yields of product can be obtained readily by separation of the neutral components of an aqueous extract by ion exchange chromatography and subsequent fractionation of this material by gel filtration (Pollock, Hall & Roberts, 1979). For most purposes purity can be established by the use of fructose-specific colorimetric reagents or by separation and identification of the products of hydrolysis (Bell, 1955). The absence of reducing power in the final preparation is a good indication that partial hydrolysis has not occurred during extraction, since this would liberate free reducing groups. Molecular size can be estimated from the elution pattern during gel filtration relative to markers of known relative molecular weight, or by determining the ratio of terminal glycosyl residues to those within the chain. This ratio can be calculated after separation of the products of methylation analysis by gas–liquid chromatography (Pollock *et al.* 1979).

Substantial advances in understanding the metabolism and structure of fructans were made prior to the advent of the more modern techniques but many of the discrepancies which exist in the literature with reference to variation in molecular size, the presence or absence of specific linkages and the physical properties of the final preparation may be due to the selective methods used in the preparation or to the occurrence of hydrolysis during extraction or purification.

Metabolism of fructans in the Compositae

Enzymes capable of synthesizing fructans were first isolated from bacterial cultures (Hestrin, Avineri-Shapiro & Aschner, 1943) and it was some time before similar preparations were obtained from vascular plants. Trisaccharides containing 2,1- or 2,6-linked fructosyl residues were formed by the action of fungal invertases on strong sucrose solutions (Albon, Bell, Blanchard, Gross & Rundell, 1953; Bacon & Bell, 1953), and transfer of fructosyl residues from inulin to sucrose was demonstrated in extracts from tubers of Jerusalem artichoke (Edelman & Bacon, 1951). Subsequently, a substantial body of evidence has accrued concerning the biochemistry of fructan metabolism in Jerusalem artichoke, mainly owing to the studies of Edelman and co-workers. They have proposed a model to account for their findings (Edelman & Jefford, 1968), which is the most comprehensive description yet produced of fructan metabolism in higher plants. This model will be considered in detail below and its applicability discussed in relation to other species which accumulate fructans.

Two enzymes from tubers of Jerusalem artichoke are together capable of synthesizing fructan. The overall reaction is analogous to the bacterial levansucrases but the two activities can be separated using DEAE cellulose chromatography (Scott, 1968). The first, sucrose-sucrose 1-fructosyl transferase (SST) catalyses the irreversible synthesis of the trisaccharide 1^F-fructosyl sucrose (also known as isokestose or 1-kestose) from two molecules of sucrose (Reaction 4.5)

$$\text{G-1, 2-F} + \text{G-1, 2-F} \rightarrow \text{G-1, 2-F-1, 2-F} + \text{G} \qquad (4.5)$$

This reaction differs from Reaction 4.2 in its apparent irreversibility. The enzyme exhibits high specificity for sucrose and has little or no activity with the trisaccharide as a fructosyl donor or with glucose or water as an acceptor. Free fructose is not a product of the reaction since the enzyme shows no fructofuranosidase (invertase) activity (Scott, Jefford & Edelman, 1966; Edelman & Jefford, 1968). This inability to catalyse fructosyl transfer with water as an acceptor appears to be characteristic of the synthetic system from vascular plants, since bacterial levansucrase also catalyses the hydrolysis of fructan (Deonder, 1972).

This enzyme alone cannot catalyse polymerization above the trisaccharide level since oligofructosides of high DP are inactive as acceptors of fructosyl residues from sucrose (Scott, 1968). Synthesis of polymers is catalysed by the action of a second enzyme β-(2,1)-fructan:β-(2,1)-fructan 1-fructosyl transferase (FFT) (Edelman & Dickerson, 1966) (Reaction 4.6):

$$\text{G-F-(F)}_n + \text{G-F-(F)}_m \rightleftharpoons \text{G-F-(F)}_{n-1} + \text{G-F-(F)}_{m+1} \qquad (4.6)$$

This enzyme mediates the transfer of terminal β-1,2-linked fructosyl residues between two oligosaccharide molecules, with the size and concentration of the various sugars determining which is the donor and which the acceptor. Sucrose acts only as a fructosyl acceptor with this enzyme but is very efficient in this role. The smallest molecule which can act as a fructosyl donor is 1^F-fructosyl sucrose. Incubation of this compound with fructans of high DP leads to rapid transfer of fructosyl residues to the high DP oligosaccharides, with a concomitant release of sucrose. In contrast, incubation of FFT with 1^F-fructosyl sucrose alone leads to little elaboration of material of higher DP since any sucrose liberated as a result of fructosyl transfer competes very efficiently as an acceptor in further transfer reactions. This results in a dynamic equilibrium where fructose residues are continually transferred with no overall change in polymer size. This can be seen clearly if equal concentrations of radioactively labelled sucrose and unlabelled trisaccharide are incubated in the presence of FFT. Rapid equilibration of radioactivity occurs (Reaction 4.7) with no change in concentration of either reactant (Edelman & Jefford, 1968):

$$\text{G-F-F} + \overset{*}{\text{G}}\text{-}\overset{*}{\text{F}} \rightleftharpoons \text{G-F} + \overset{*}{\text{G}}\text{-}\overset{*}{\text{F}}\text{-F} \qquad (4.7)$$

Purified oligofructosides of DP 4–9 will act as both donors and acceptors in the presence of FFT. The final products of such a reaction are mixtures of oligosaccharides with relative concentrations similar to those observed in extracts of growing tubers (Edelman & Dickerson, 1966). As with SST, this enzyme cannot utilize water as a fructosyl acceptor, so the concerted action of the two enzymes is only in the direction of synthesis.

Jerusalem artichoke tubers are able to convert the free glucose liberated during polymer synthesis to sucrose (Edelman & Jefford, 1968). Thus, the distribution of fructans and other sugars observed during tuber growth can be explained by the combined action of the two enzymes described above, together with the normal enzymes of sucrose biosynthesis (see Chapter 2). Since the available evidence favours the vacuolar accumulation of fructan in this tissue, Edelman & Jefford (1968) suggested that FFT acts vectorially across the tonoplast, transferring fructosyl residues from trisaccharides in the cytoplasm to higher DP acceptors in the vacuole, as summarized in Fig. 4.1*a*. Dickerson & Edelman (1966) demonstrated sequential appearance of radioactivity in sucrose, 1^F-fructosyl sucrose and eventually higher oligofructosides extracted from tubers in the two days following incorporation of $^{14}CO_2$ by the leaves of intact plants. Such a pattern of labelling is consistent with the operation of these enzyme systems *in vivo*.

Specific fructan hydrolases have also been extracted from artichoke tubers (Edelman & Jefford, 1964). Two different activities, separable by chromatography on DEAE cellulose, both catalyse the stepwise hydrolysis of the terminal fructoside linkage according to Reaction 4.8:

$$\text{G-F-(F)}_n + H_2O \rightarrow \text{G-F-(F)}_{n-1} + \text{F} \qquad (4.8)$$

These preparations show no transfructosylase activity, nor will they hydrolyse the glycoside bond in sucrose.

By a succession of these reactions inulin is converted irreversibly to a mixture of fructose and sucrose. Hydrolysis is presumed to occur in the vacuole, with the products passing into the cytoplasm for subsequent metabolism (Edelman & Jefford, 1968). The proposed scheme for fructan breakdown is summarized in Fig. 4.1*b*.

Measurement of the activities of the above enzymes in tubers at different stages of growth support the operation of the scheme outlined in Fig. 4.1. SST activity was detectable only in tubers where fructan accumulation was occurring (Scott, 1968). In contrast, FFT was detectable at all stages of growth and fructan hydrolase activity increased during depolymerization induced by storage at low temperatures (Edelman & Jefford, 1968). Free glucose, a product of the SST reaction, could only be detected in actively growing tubers whereas there was an accumulation of free fructose when

Fig. 4.1 Scheme for (*a*) synthesis and (*b*) breakdown of fructan in tubers of Jerusalem artichoke. Modified from Edelman & Jefford (1968). Enzymes catalysing reactions shown are as follows: (1) sucrose–sucrose fructosyl transferase (SST); (2) fructan–fructan fructosyl transferase (FFT); (3) hexokinase; (4) UDPG pyrophosphorylase; (5) phosphoglucomutase; (6) phosphohexose isomerase; (7) sucrose phosphate synthetase; (8) sucrose phosphate phosphatase; (9) fructan hydrolase.

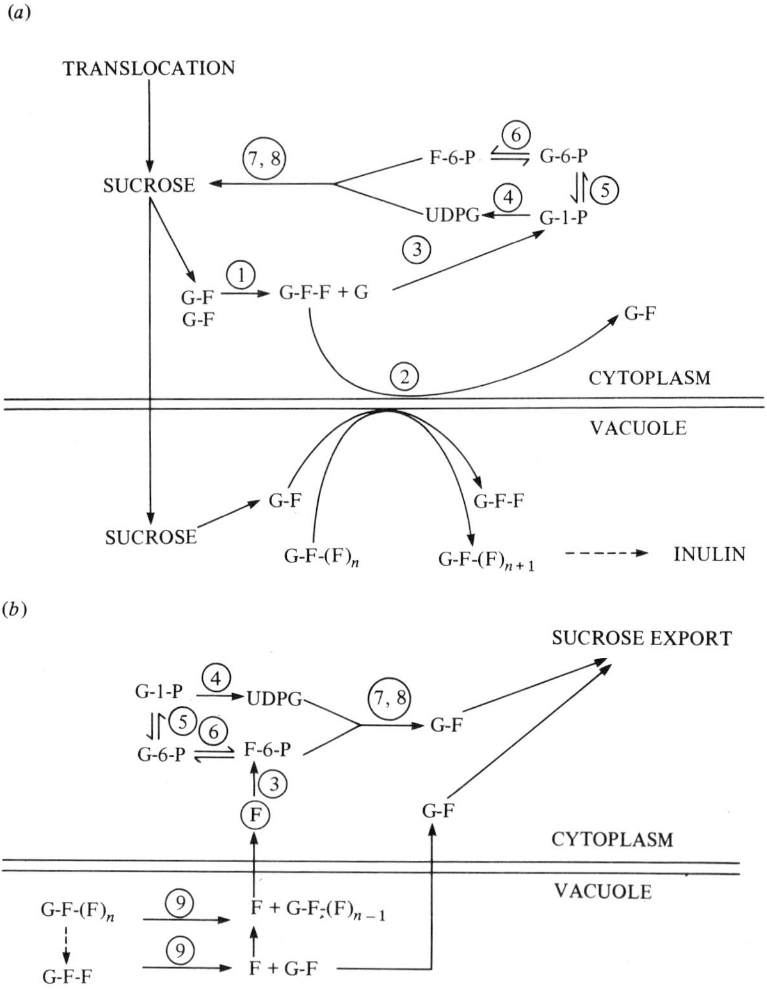

fructan breakdown was initiated by sprouting of the tubers (Jefford & Edelman, 1961).

In other members of the Compositae, *de novo* fructan synthesis can be induced by floating leaf disks on sugar solutions (Chandorkar & Collins, 1972). In leaves of lettuce, such induction is accompanied by the appearance of detectable SST activity and this activity is similar to that isolated from petiolar tissue of normally grown lettuce (Chandorkar & Collins, 1974*a*). Incubation of leaf disks in solutions of radioactive sugars leads to the incorporation of radioactivity into the newly synthesized fructan. The distribution of radioactivity within the individual oligofructosides conforms to the pattern predicted on the basis of a mechanism similar to that of Jerusalem artichoke tubers (Chandorkar & Collins, 1974*b*). These observations support the hypothesis of Edelman & Jefford and also indicate that the mechanism may occur in other members of the same family.

Other studies on fructan metabolism

Since no such comprehensive study exists for other species which accumulate fructans, the model discussed above has formed the basis for much of the subsequent work. Although not all the observations support the model, no satisfactory theory exists at present to explain the discrepancies. Some of these observations are discussed in more detail below.

Metabolism of fructans in Allium *spp.* The structure and metabolism of fructans in the genus *Allium* have been investigated in some detail, with particular reference to onion. The fructans of onion were characterized by Bacon (1957, 1959) and the distribution within different tissues studied by Darbyshire & Henry (1978). They consist of a series of oligofructosides similar to that found in Jerusalem artichoke, except that the mean molecular size is smaller and the trisaccharide fraction consists of two components, 1^F-fructosyl sucrose and 6^G-fructosyl sucrose (neokestose).

The synthetic enzymes from onion, isolated and partially separated by Scott (1968), consist of two activities catalysing similar reactions to those of artichoke SST and FFT. Complete separation of the two activities was obtained by Henry & Darbyshire (1980). Preparations of SST from both sources show similar properties but FFT from onion differs from the artichoke enzyme in that it is unable to utilize inulin as a fructosyl acceptor. In addition, the onion enzyme will catalyse the formation of 6^G-fructosyl sucrose from a mixture of sucrose and 1^F-fructosyl sucrose (Henry & Darbyshire, 1980). The bulbs also contain fructan hydrolase activity (Henry & Darbyshire, 1979).

The relative amount of the individual oligofructosides found *in vivo* differs between onion and artichoke tissue (Edelman & Jefford, 1968; Darbyshire & Henry, 1978) but it has not been demonstrated whether or not such differences can be accounted for by the different properties of the two FFT enzymes. However, FFT from onion and garlic, species with widely differing distributions of molecular size among constituent oligofructosides, show similar insensitivity to the presence of high DP acceptors *in vitro* (Darbyshire & Henry, 1981). The implication of this study is that the kinetic properties of the FFT enzyme are not the only determinant of the distribution of size of polymer *in vivo*.

Activity of SST in other species. Activity of SST has been detected and at least partially characterized in a number of other species which accumulate fructan. These include daffodil (Scott, 1968), chicory (Singh & Bhatia, 1971), *Agave vera-cruz* (Bhatia, Satyanarayana & Srinivasan, 1955), lettuce (Chandorkar & Collins, 1974*a*) and asparagus (Shiomi, Yamada & Izawa, 1979). In all cases, the products of the reaction appeared to be 1^F-fructosyl sucrose and glucose. If this is so, then the wide variation in size of fructan observed in the different accumulating species is not due to differences in the properties of this enzyme. At present, the way in which the various synthetic and degradative enzymes interact to produce this range of degrees of polymerization is unknown.

Metabolism of fructans in the Gramineae. Little direct evidence exists concerning the mechanisms of metabolism of fructan in the ecologically and agriculturally important family, the Gramineae, despite the relative abundance of structural and agronomic studies. As discussed in Chapter 1, a wide range of size of polymer occurs in different species and there are suggestions of differences in structure, not only between species but also in different organs of the same species (Neumuller, 1958). There are also indications that the synthetic mechanisms of at least some temperate grasses may differ markedly from that proposed for other groups. In particular, the homologous series of oligofructosides of increasing DP found in Jerusalem artichoke appears to be absent in some grasses, generally those such as cocksfoot which accumulate fructans of high relative molecular weight (Pollock, 1979). In this and many other species of grass, accumulation of fructan in vegetative plants is concentrated in the leaf bases, below the ligule. When this tissue was analysed 4 h subsequent to exposure of intact plants to $^{14}CO_2$, the majority of the radioactivity was found in sucrose. Over the next 48 h there was a progressive decline in the radioactivity of sucrose and a concomitant rise in that associated with fructan of high DP. There was little or no accumulation of radioactivity in material of intermediate DP, implying direct transfer of

Fig. 4.2 Separation of fructo-oligosaccharides from leaf base and stem material of S.24 perennial ryegrass. Plants were harvested in (*a*) May (flowering tillers) and (*b*) September (vegetative tillers). In each case the neutral components of the water-soluble material were fractionated on a 150×1.5 cm column of Bio-Gel P2. Peak 1 is unresolved high molecular-weight fructan, peak 2 sucrose and peak 3 fructose. Partially resolved oligofructosides of DP 3–10 elute between peaks 1 and 2.

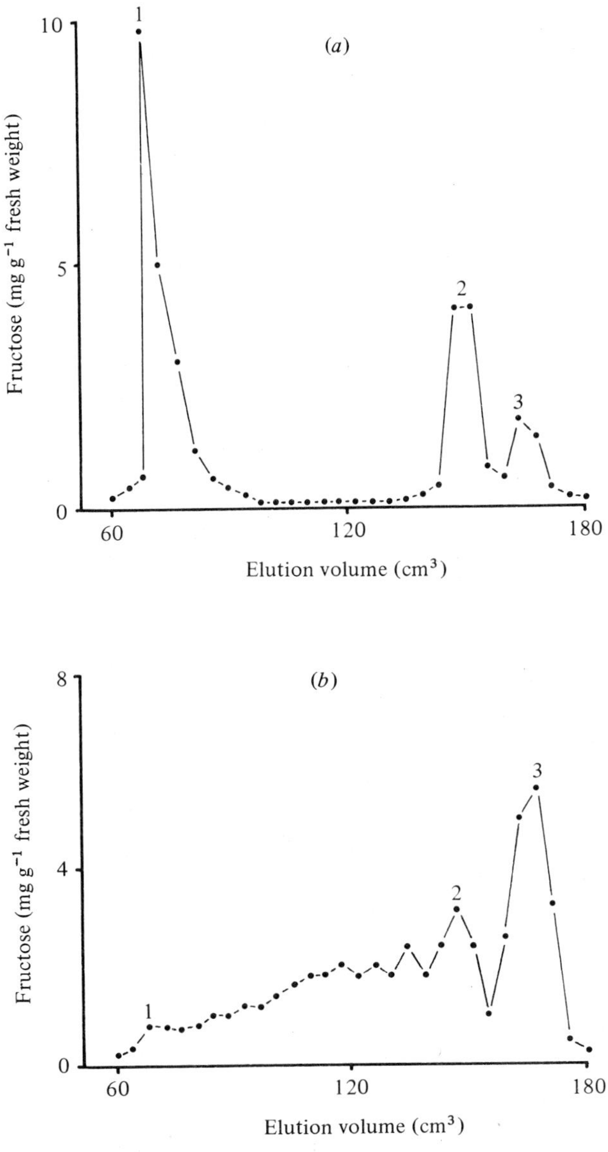

fructosyl residues from sucrose to an acceptor of high relative molecular weight (Pollock, 1979).

Trisaccharides have, however, been isolated from some grasses. Extracts of Italian ryegrass contained three trisaccharides, 1^F-, 6^F- and 6^G-fructosyl sucrose (Bacon, 1959). Incubation of leaf sections of this species with radioactive sugars leads to the formation of radioactive trisaccharides and fructans but not of radioactive oligofructosides (DP 9–15) (Rocher, 1967).

It does appear, however, that the patterns of distribution of fructan observed in grasses may show a pronounced seasonal variation. Temperate forage grasses show two peaks of accumulation – during flowering in the spring and early summer, and during vegetative growth in the autumn and winter (Pollock & Jones, 1979). Extracts of stems and leaf bases of perennial ryegrass and meadow fescue, harvested at anthesis, showed almost no oligosaccharides (Fig. 4.2*a*). In contrast, similar extracts from material harvested in the autumn contained considerable amounts (Fig. 4.2*b*). It is not known whether these represent a homologous series of intermediates equivalent to those of Jerusalem artichoke tubers.

At present, the enzyme systems which yield these patterns have not been characterized*. Fructans of high DP have been produced *in vitro* by the action of purified invertase on concentrated solutions of sucrose-containing grass fructans of lower DP (Schlubach, 1957) but it seems unlikely that this represents the mechanism of synthesis *in vivo*. Attempts to detect activity of fructosyl transferase in extracts of cocksfoot leaves were not successful (Pollock, 1979). Fructan hydrolase activity has been detected in extracts of tall fescue (Smith, 1976) and timothy (Mino & Maeda, 1974) but has not been separated from accompanying invertase activity.

Thus, until active enzyme preparations from grasses are obtained and characterized, we cannot determine whether or not the model proposed by Edelman & Jefford (1968) represents the only mechanism regulating the metabolism of fructan in the tissues of vascular plants.

Sugar nucleotides in the biosynthesis of fructan. Where the activities have been estimated, sufficient SST and FFT exist in extracts to catalyse the observed *in vivo* rates of accumulation of fructan (Scott, 1968; Darbyshire & Henry, 1981). However, by analogy with other polysaccharide synthesizing systems in vascular plants, there have been suggestions that sugar nucleotides may be involved in fructan biosynthesis. UDP-fructose has been detected in extracts of vascular plant tissues (Gonzalez & Pontis, 1963; Umemura, Nakamura & Funahashi, 1967). Efforts to implicate these intermediates in

* *Note added in proof*: See C. J. Pollock (1984, *New Phytologist*, **96**, 527–34) for a discussion of important recent work by N. Shiomi and colleagues on fructosyltransferases and the synthesis of distinct homologous series of fructans in asparagus.

synthesis of inulin have failed (Pontis, 1966) and no evidence was found of nucleotide fructose intermediates during biosynthesis of fructan in cocksfoot (Pollock, 1979). At present, therefore, the evidence is against the direct participation of these compounds.

Physiological aspects of metabolism of fructans

Accumulation of fructan can be affected both by the developmental state of the plant and by the prevailing environmental conditions. Accumulation often occurs in specific organs or at specific stages in the annual growth cycle but, whenever it occurs, it is limited by the capacity of the plant to produce photosynthate in excess of that required for growth. Indeed, treatments which reduce growth often lead to an increase in the proportion of current photosynthate diverted into fructan. The interaction of environmental and developmental factors which affect metabolism of fructans forms the subject of this section.

Developmental modulation of accumulation of fructan

The effects of developmental processes such as flowering or seed development on the patterns of accumulation of reserves have been well documented for many species. Amongst those which accumulate fructans, grasses and cereals have been studied most intensively. In these species, the polymers are concentrated in the vegetative organs, although small amounts have been isolated from seeds (Archbold, 1940). Accumulation occurs during reproductive development in both perennial grasses and annual cereals, with the stem internodes being the major site of storage (Archbold, 1940; Smith, 1973). The lower internodes contain more fructan than the upper ones (Smith, 1967; Pollock & Jones, 1979). There has been doubt over the fate of this material. Early studies on cereals suggested that extensive relocation into the developing grains occurred (Archbold, 1940) but current view is that only around 10% of final grain weight is contributed by mobilization of stem reserves (Stoy, 1979). In perennial grasses, however, such material may be of considerable significance in the establishment of the new vegetative tillers which grow from buds on the lower nodes of the old flowering stems (Davies, 1977).

Material translocated to the developing grain, either from the flag leaf or as mobilized stem reserves may accumulate to a small extent as fructan, but the concentration declines sharply as the grain matures (Archbold, 1940). This decline, together with the distinctive structures of these small, highly branched fructans (Chapter 1, this volume; Medcalf & Cheung, 1971), has led to proposals that these fructans represent transient products of transferase activity by predominantly hydrolytic enzymes acting on the high concentrations of sucrose in the developing grain (Meredith & Jenkins, 1973). If this is so, then metabolism of fructan in grains can be regarded as being distinct and separate from that in the vegetative organs.

The vegetative tillers produced after flowering by perennial grasses accumulate high concentrations of fructan and this process continues during the winter (Pollock & Jones, 1979). Perennial temperate grasses do not exhibit true dormancy (Pollock, Riley, Stoddart & Thomas, 1980) and photosynthetic activity persists throughout this period. Breakdown of fructan occurs prior to the onset of rapid spring growth, together with a transient accumulation of low-relative-molecular-weight sugars (Pollock & Jones, 1979). This mobilization of fructan does not appear to be mediated by current environmental conditions such as temperature (C. J. Pollock, unpublished observation) but rather represents a developmentally induced change. In perennial ryegrass grown under natural conditions, the rate of leaf expansion after transfer to controlled environments increased rapidly in plants transferred from January onwards (Parsons & Robson, 1980). These separate observations may indicate a major developmental change occurring in temperate grasses prior to the onset of spring growth.

Accumulation of fructan in the Compositae shows a different pattern. Accumulation in tubers of Jerusalem artichoke is most marked during late summer, with sucrose from the leaves as the major substrate (Dickerson & Edelman, 1966). During storage of the tubers, depolymerization of the fructan occurs without significant loss of total carbohydrate and this process is accelerated by low temperatures (Jefford & Edelman, 1963). Sprouting of the tubers, which occurs naturally during the subsequent spring, is accompanied by a decline in the carbohydrate content of the tuber as reserves are used to support growth of the new plant (Jefford & Edelman, 1961). Thus, in contrast to temperate grasses, no accumulation occurs during the winter and low temperatures induce depolymerization. The extent to which this is analogous to the accumulation of sugar induced by low temperature in many starch-storing plants is uncertain, as is the overall significance of the accumulation of low-relative-molecular-weight sugars in response to low temperatures (ap Rees, Dixon, Pollock & Franks, 1981) (cf. Chapter 2).

Environmental effects on fructan metabolism

There have been many reports on the effects of different environmental conditions on accumulation of fructan, particularly among grasses and cereals. These studies have been thoroughly reviewed by Archbold (1940) and Smith (1973). Generally, accumulation is enhanced by treatments which reduce growth and hence lower the requirements for assimilates of the rest of the plant. Such treatments include nutrient depletion, exposure to low temperatures or excision of competing organs. Treatments which enhance photosynthetic carbon fixation, such as high irradiance or long photoperiods also lead to increased accumulation. Such increases may also lead to the formation of larger polymers (Smith, 1968).

Equally, treatments leading to enhanced competition for assimilate, such as defoliation by cutting or grazing, or more vigorous tillering induced by heavy application of fertilizer, cause a reduction in accumulation. Thus, accumulation of fructan in cocksfoot was reduced by 60% by defoliation just before ear emergence which restricted photosynthesis to that occurring in the stems and leaf sheaths without affecting demand for assimilate by the developing ear (C. J. Pollock & T. Jones, unpublished observations).

Conclusion – why fructans?

The species which accumulate fructans are taxonomically varied and occupy a wide range of habitats. Significant differences exist in structure, metabolism and physiology between accumulating species. Are there, therefore, any conclusions to be drawn on the selective advantages which might accrue from possession of the ability to synthesize and degrade these compounds? There have been proposals that accumulation could be concerned in frost resistance (Eagles, 1967) or in the regulation of osmotic potential during cell expansion (Darbyshire & Henry, 1978) but one presumes such constraints exist in many species which do not accumulate fructans.

The significant features which distinguish fructans from other storage polysaccharides are their high solubility in water, the ease of interconversion of the different molecular species, their apparently distinctive synthetic mechanism and the lack of specialized storage organelles. It must be noted, however, that many of these features have not been established unequivocally for all accumulating species. It is not clear whether or not they can be correlated with any common characteristics of the various habitats occupied by fructan-accumulators. The range of species examined is relatively small and, in some cases, the methods used to identify or quantify fructans are open to objection (Archbold, 1940).

Accumulation of fructan does appear, however, to be more prevalent among temperate species and ecotypes of grass species found in northerly latitudes often show more pronounced accumulation of fructans than those from Mediterranean areas (Eagles, 1967). To what extent this pattern of distribution is concerned directly with tolerance of low temperature, rather than the ability to fix and store carbon in a readily accessible state under conditions of low temperature and short daylength, is not known (Pollock, Riley, Stoddart & Thomas, 1980).

The proposed synthetic mechanism for fructans emphasizes the fact that their metabolism must be considered in conjunction with that of sucrose, and hence with the interconversion of sucrose and starch within photosynthetic organs. Hopefully, the recent improvements in our understanding of the regulation of sucrose and starch metabolism will provide impetus for the further study of fructans, and this, in turn will lead to a greater understanding of the factors controlling accumulation of reserves in vascular plants.

References

Albon, N., Bell, D. J., Blanchard, P. H., Gross, D. & Rundell, J. T. (1953). Kestose, a trisaccharide formed from sucrose by yeast invertase. *Journal of the Chemical Society*, 24–7.

ap Rees, T., Dixon, W. L., Pollock, C. J. & Franks, F. (1981). Low temperatures sweetening of higher plants. In *The Biochemistry of Fruit and Vegetables*, ed. J. Friend, pp. 41–61. New York & London: Academic Press.

Archbold, H. K. (1940). Fructosans in the monocotyledons, a review. *New Phytologist*, **39**, 185–219.

Bacon, J. S. D. (1957). The water soluble carbohydrates of the onion, *Allium cepa* L. *Biochemical Journal*, **67**, 5p.

Bacon, J. S. D. (1959). The trisaccharide fraction of some monocotyledons. *Biochemical Journal*, **73**, 507–14.

Bacon, J. S. D. & Bell, D. J. (1953). A new trisaccharide produced from sucrose by mold invertase. *Journal of the Chemical Society*, 2528–30.

Bacon, J. S. D. & Edelman, J. (1951). The carbohydrates of Jerusalem artichoke and other Compositae. *Biochemical Journal*, **48**, 114–26.

Bell, D. J. (1955). Mono- and oligosaccharides and acidic monosaccharide derivatives. In *Modern Methods of Plant Analysis*, vol. 2, ed. K. Paech & M. V. Tracey, pp. 1–53. Berlin: Springer-Verlag.

Bhatia, I. S., Satyanarayana, M. N. & Srinivasan, M. (1955). Transfructosidase from *Agave vera-cruz* Mill. *Biochemical Journal*, **61**, 171–4.

Chandorkar, K. R. & Collins, F. W. (1972). *De novo* synthesis of fructo-oligosaccharides in leaf disks of certain Asteraceae. *Canadian Journal of Botany*, **50**, 295–303.

Chandorkar, K. R. & Collins, F. W. (1974*a*). Enzymological aspects of *de novo* synthesis of fructo-oligosaccharides in leaf disks of certain Asteraceae. IV. The activity of sucrose-sucrose 1-fructosyl transferase. *Canadian Journal of Botany*, **52**, 1369–77.

Chandorkar, K. R. & Collins, F. W. (1974*b*). The mechanism of *de novo* synthesis of fructo-oligosaccharides in leaf disks of certain Asteraceae. III. *Canadian Journal of Botany*, **52**, 181–8.

Darbyshire, B. & Henry, R. J. (1978). The distribution of fructans in onions. *New Phytologist*, **81**, 29–34.

Darbyshire, B. & Henry, R. J. (1981). Differences in fructan content and synthesis in some *Allium* species. *New Phytologist*, **87**, 249–56.

Davies, A. (1977). The structure of the grass sward. In *Proceedings of the International Meeting on Animal Production from Temperate Grassland*, ed. B. Gilsenan, pp. 36–44. Dublin: An Foras Talúntais.

Deonder, R. (1972). Role and mechanisms of transglycosylation reactions. In *Biochemistry of the Glycosidic Linkage*, ed. R. Piras & H. G. Pontis, pp. 21–78. New York: Academic Press.

Dickerson, A. G. & Edelman, J. (1966). The metabolism of fructose polymers in plants. 6. Transfructosylation in living tissue of *Helianthus tuberosus* L. *Journal of Experimental Botany*, **17**, 612–19.

Eagles, C. F. (1967). Variation in the soluble carbohydrate content of climatic races of *Dactylis glomerata* (cocksfoot) at different temperatures. *Annals of Botany*, **31**, 645–51.

Edelman, J. (1956). The formation of oligosaccharides by enzymic transglycosylation. *Advances in Enzymology*, **17**, 189–232.

Edelman, J. & Bacon, J. S. D. (1951). Transfructosidation in extracts of the tubers of *Helianthus tuberosus* L. *Biochemical Journal*, **49**, 529–40.

Edelman, J. & Dickerson, A. G. (1966). The metabolism of fructose polymers in plants. Transfructosylation in tubers of *Helianthus tuberosus* L. *Biochemical Journal*, **98**, 787–94.

Edelman, J. & Jefford, T. G. (1964). The metabolism of fructose polymers in plants. 4. β-Fructofuranosidases of tubers of *Helianthus tuberosus* L. *Biochemical Journal*, **93**, 148–61.

Edelman, J. & Jefford, T. G. (1968). The mechanism of fructosan metabolism in higher plants as exemplified in *Helianthus tuberosus*. *New Phytologist*, **67**, 517–31.

Gonzalez, N. S. & Pontis, H. G. (1963). Uridine diphosphate fructose and uridine diphosphate acetylgalactosamine from dahlia tubers. *Biochimica et Biophysica Acta*, **69**, 179–81.

Hanes, C. S. (1940). The reversible formation of starch from glucose-1-phosphate catalysed by potato phosphorylase. *Proceedings of the Royal Society of London, Series B*, **129**, 174–208.

Henry, R. J. & Darbyshire, B. (1979). The distribution of fructan metabolising enzymes in the onion plant. *Plant Science Letters*, **14**, 155–8.

Henry, R. J. & Darbyshire, B. (1980). Sucrose: sucrose fructosyl transferase and fructan: fructan fructosyl transferase from *Allium cepa*. *Phytochemistry*, **19**, 1017–20.

Hestrin, S., Avineri-Shapiro, S. & Aschner, M. (1943). The enzymic production of levan. *Biochemical Journal*, **37**, 450–6.

Hirst, E. L. (1957). Some aspects of the chemistry of the fructosans. *Proceedings of the Chemical Society*, 193–204.

Jefford, T. G. & Edelman, J. (1961). Changes in content and composition of the frucrose polymers of *Helianthus tuberosus* L. during growth of daughter plants. *Journal of Experimental Botany*, **12**, 177–87.

Jefford, T. G. & Edelman, J. (1963). The metabolism of fructose polymers in plants. 2. Effect of temperature on the carbohydrate changes and morphology of stored tubers of *Helianthus tuberosus* L. *Journal of Experimental Botany*, **14**, 56–62.

McDonald, E. J. (1946). The polyfructosans and difructose anhydrides. *Advances in Carbohydrate Chemistry*, **2**, 253–77.

Medcalf, D. G. & Cheung, P. W. (1971). Composition and structure of glucofructans from durum wheat flour. *Cereal Chemistry*, **48**, 1–8.

Meredith, P. & Jenkins, L. D. (1973). Saccharides and saccharidases of developing wheat grains. *New Zealand Journal of Science*, **16**, 599–616.
Mino, Y. & Maeda, K. (1974). Metabolism of sucrose and phlein in the haplocorm of timothy. *Journal of Japanese Grassland Science*, **20**, 6–10.
Neumuller, G. (1958). Pentosans and hexosans (excluding cellulose). In *Encyclopedia of Plant Physiology*, Vol. VI, ed. W. Ruhland, pp. 252–67. Berlin: Springer-Verlag.
Parsons, A. J. & Robson, M. J. (1980). Seasonal changes in the physiology of S.24 perennial ryegrass (*Lolium perenne* L.). 1. Response of leaf extension to temperature during the transition from vegetative to reproductive growth. *Annals of Botany*, **46**, 435–44.
Peat, S., Whelan, W. J. & Hinson, K. A. (1952). Synthetic action of almond emulsin. *Nature*, **170**, 1056–7.
Pollock, C. J. (1979). Pathway of fructosan synthesis in leaf bases of *Dactylis glomerata*. *Phytochemistry*, **18**, 777–9.
Pollock, C. J., Hall, M. A. & Roberts, D. P. (1979). Structural analysis of fructose polymers by gas–liquid chromatography and gel filtration. *Journal of Chromatography*, **171**, 411–15.
Pollock, C. J. & Jones, T. (1979). Seasonal patterns of fructan and metabolism in forage grasses. *New Phytologist*, **83**, 8–15.
Pollock, C. J., Riley, G. J. P., Stoddart, J. L. & Thomas, H. (1980). The biochemical basis of plant response to temperature limitations. *Annual Report of the Welsh Plant Breeding Station*, 1979, pp. 227–46.
Pollock, C. J. & Ruggles, P. A. (1976). Cold-induced fructosan synthesis in leaves of *Dactylis glomerata*. *Phytochemistry*, **16**, 1643–6.
Pontis, H. G. (1966). Observations on the *de novo* synthesis of fructosans *in vivo*. *Archives of Biochemistry and Biophysics*, **116**, 416–24.
Rocher, J. P. (1967). Les lévanes de *Lolium italicum*. Synthèse dans les organes végétatifs. *Physiologie Végétale*, **5**, 71–80.
Schlubach, H. H. (1957). Uber den Kohlenhydratstoffweschsel der Grasarten. *Biochimia*, **22**, 96–101.
Schlubach, H. H. & Peitzner, H. (1937). Untersuchungen über Fructose-Anhydride. XVIII. Die Konstitution des Triticins. *Justus Liebigs Annalen der Chemie*, **530**, 120–30.
Scott, R. W. (1968). Transfructosylation in higher plants containing fructose polymers. 244 pp. University of London: Ph.D. Thesis.
Scott, R. W., Jefford, T. G. & Edelman, J. (1966). Sucrose fructosyltransferase from higher plant tissues. *Biochemical Journal*, **100**, 23 p.
Shiomi, N., Yamada, J. & Izawa, M. (1979). Synthesis of several fructo-oligosaccharides by asparagus fructosyltransferases. *Agricultural and Biological Chemistry*, **43**, 2233–44.
Singh, R. & Bhatia, I. S. (1971). Substrate specificity of fructosyl transferase from chicory roots. *Phytochemistry*, **10**, 2037–9.
Smith, A. E. (1976). β-Fructofuranosidase and invertase activity in tall frescue culm bases. *Journal of Agricultural and Food Chemistry*, **24**, 476–8.
Smith, D. (1967). Carbohydrates in grasses. II. Sugar and fructosan composition of the stem bases of bromegrass and timothy at several growth stages and in different plant parts at anthesis. *Crop Science*, **7**, 62–7.

Smith, D. (1968). Carbohydrates in grasses. IV. Influence of temperature on the sugar and fructosan composition of timothy plant parts at anthesis. *Crop Science*, **8**, 331–4.

Smith, D. (1973). The nonstructural carbohydrates. In *The Biochemistry of Herbage*, vol. 2, ed. G. W. Butler & R. W. Bailey, pp. 105–55. New York: Academic Press.

Smith, D. & Groteleuschen, R. D. (1966). Carbohydrates in grasses. 1. Sugar and fructosan composition of the stem bases of several northern-adapted grasses at seed maturity. *Crop Science*, **6**, 263–6.

Stoy, V. (1979). The storage and remobilisation of carbohydrates in cereals. In *Crop Physiology and Cereal Breeding*, ed. J. H. J. Spiertz & Th. Kramer, pp. 55–9. Wageningen: Center for Agricultural Publishing and Documentation.

Umemura, Y., Nakamura, M. & Funahashi, S. (1967). Isolation and characterisation of uridine diphosphate fructose from tubers of Jerusalem artichoke (*Helianthus tuberosus* L.). *Archives of Biochemistry and Biophysics*, **199**, 240–52.

O. KANDLER AND H. HOPF

5 Biosynthesis of oligosaccharides in vascular plants

Introduction

This chapter deals with the biosynthesis of the so-called 'primary oligosaccharides', with the exception of the sucrose and the kestose series (fructan) which are discussed in Chapters 2 and 4 respectively. The primary oligosaccharides comprise those oligosaccharides which are synthesized *de novo* by the transfer of a glycosyl residue to a mono- or oligosaccharide. They occur freely in many plants in significant amounts and are of metabolic relevance. The so-called 'secondary oligosaccharides', which arise by the hydrolysis of higher oligosaccharides, polysaccharides or heterosides, are not discussed here.

As a first hypothesis, it was reasonable to regard the synthesis of oligosaccharides as the reverse of their hydrolysis by glycosidases on the basis of the potential reversibility of all enzyme reactions. This assumption was supported by the successful synthesis of various oligosaccharides by glycosidases in the presence of high concentrations of the monomers (Hassid, 1958; Nikaido & Hassid, 1971; Courtois, Petek & Dong, 1961). Although such mechanisms may be responsible for the occurrence of a diversity of oligosaccharides in decaying or fermenting plant material and in honey (Siddiqui, 1970), they are probably of no significance *in vivo*, since the concentrations of the free monosaccharides are usually much too low to allow reactions to proceed in the direction of synthesis.

Today, it is well established that oligo- and polysaccharides are formed by the transfer of a glycosyl residue from a donor molecule to an acceptor molecule by specific glycosyl transferases. In these reactions, the values of the negative free energy of hydrolysis of the glycosydic linkages in the donor molecule are significantly higher than those of the newly formed glycosidic bonds. Examples of glycosidic linkages exhibiting different 'energy levels' have been given in Table 1.2 (Chapter 1).

After the discovery of UDPG (Cardini, Caputto, Paladini & Leloir, 1950) and the establishment of its role as glucosyl donor in the synthesis of sucrose (see Chapter 2), it was widely assumed that nucleotide-activated monosaccharides would also be donors in the biosynthesis of other oligosaccharides. Although this is true in many cases, disaccharides or oligosaccharides of a low degree of polymerization also serve as glycosyl-donors in the synthesis of the various series of sucrosyl oligosaccharides. This occurs in the case of fructans of the kestose series (see Chapter 4) and in the raffinose series, where sucrose and galactinol, respectively, serve as glycosyl donors.

The nature of the acceptor molecule varies considerably. In the case of the non-reducing oligosaccharides, either monosaccharides phosphorylated at C-6 or disaccharides and oligosaccharides of low degree of polymerization serve as glycosyl acceptors but the free monosaccharides do not. The sucrose synthase reaction, the only reaction involving free fructose as acceptor, physiologically operates in the reverse direction, supplying ADP(UDP)-glucose for the synthesis of starch and other polymers (see Chapter 2 and Preiss & Levi, 1980).

Although reducing di- and oligosaccharides of plant origin are commonly derived by the hydrolysis of polysaccharides or heterosides, it has recently been shown that the reducing disaccharides, maltose (Linden, Schilling, Brackenhofer & Kandler, 1975; Schilling & Kandler, 1975) and gentiobiose (H. Hopf & O. Kandler, unpublished) are synthesized *de novo in vivo.* α-D-Glucose 1-phosphate or UDPG, respectively, act not only as donors but also as acceptors (or at least as the precursor of the acceptor). The mechanisms of these reactions remain to be elucidated, however.

I. Biosynthesis of non-reducing oligosaccharides with the exception of sucrose and the kestose series of fructans

(1) *α,α-Trehalose* [*O*-α-D-glucopyranosyl-(1-1)-α-D-glucopyranoside] *and selaginose* [*O*-α-D-glucopyranosyl-(1-1)-α-D-glucopyranoside-(2-1)-α-D-glucopyranoside]. The biosynthesis of trehalose resembles that of sucrose. As shown by Leloir & Cabib (1953) and Cabib & Leloir (1958) for yeast, trehalose is synthesized by the combined action of a UDP-α-D-glucose: glucose 6-phosphate 1-glucosyltransferase (trehalose 6-phosphate synthase) and a trehalose 6-phosphate phosphatase in accordance with Equations 5.1 and 5.2.

$$\text{UDPG} + \text{G-6-P} \xrightarrow{\text{transferase}} \alpha,\alpha\text{-trehalose-6-P} + \text{UDP} \tag{5.1}$$

$$\alpha,\alpha\text{-trehalose 6-P} \xrightarrow{\text{phosphatase}} \alpha,\alpha\text{-trehalose} + \text{Pi} \tag{5.2}$$

The transferase is specific for UDPG as donor and the reaction is not reversible. Both reactions have been demonstrated in enzyme preparations from pollen of *Lilium longiflorum* (Gussin & McCormack, 1970). Also, labelled trehalose 6-phosphate has been identified in extracts of *Selaginella kraussiana* after photosynthesis in $^{14}CO_2$ (Fischer, 1973). Both findings indicate that trehalose is synthesized by the same reactions in yeast as in plants.

Selaginose, a trisaccharide found only in a few species of *Selaginella* (Fischer & Kandler, 1975), may be synthesized by the glucosylation of trehalose at C-2. This is indicated by the kinetics of labelling of the two compounds in *S. kraussiana* during a period of photosynthesis in $^{14}CO_2$, followed by a chase period in $^{12}CO_2$. Incorporation of ^{14}C into trehalose clearly precedes that into selaginose during the application of $^{14}CO_2$ but trehalose loses and selaginose continues to gain label during the subsequent chase period (Fischer, 1973). No information concerning the *in vitro* synthesis of selaginose is yet available.

(2) *Raffinose series* [*O*-α-D-galactopyranosyl-(1-6)-*O*-α-D-glucopyranosyl-(1-2)-β-D-fructofuranoside]. Early reports on the biosynthesis of raffinose, employing crude enzyme preparations from ripe seeds of *Vicia faba* (Bourne, Walter & Pridham, 1965; Pridham & Hassid, 1965) indicated that UDP-galactose may act as the galactosyl donor and sucrose as acceptor in accordance with Equation 5.3.

$$\text{UDP-D-galactose} + \text{sucrose} \rightarrow \text{raffinose} + \text{UDP} \tag{5.3}$$

The same reaction was catalysed by a crude enzyme preparation from pea chloroplasts. This preparation also synthesized galactinol when *myo*-inositol was added in addition to UDP-D-galactose (Imhoff, 1973). However, none of the enzymes involved have been purified and their kinetics have not been studied in greater detail. Thus, the reproducibility of these enzyme preparations and the relevance of their activities *in vivo* remain uncertain.

Sufficient *in vivo* and *in vitro* evidence has accumulated, however, to describe the biosynthesis of the raffinose series as a sequential action of a set of α-D-galactosyltransferases, catalysing the transfer of a galactosyl residue from UDP-α-D-galactose first to *myo*-inositol to yield galactinol and, then, from galactinol to either sucrose to yield raffinose, or to raffinose and its higher homologues resulting in the formation of the various members of the raffinose series, as described in Equations 5.4 to 5.8.

$$\text{UDP-}\alpha\text{-D-galactose} + myo\text{-inositol} \rightleftharpoons \text{galactinol} + \text{UDP} \tag{5.4}$$

$$\text{galactinol} + \text{sucrose} \overset{\text{GST}}{\rightleftharpoons} \text{raffinose} + myo\text{-inositol} \tag{5.5}$$

$$\text{galactinol} + \text{raffinose} \overset{\text{GRT}_1}{\rightleftharpoons} \text{stachyose} + myo\text{-inositol} \tag{5.6}$$

$$\text{galactinol} + \text{stachyose} \overset{\text{GRT}_2}{\rightleftharpoons} \text{verbascose} + myo\text{-inositol} \tag{5.7}$$

$$\text{galactinol} + \text{verbascose} \overset{\text{GRT}_3}{\rightleftharpoons} \text{ajugose} + myo\text{-inositol} \tag{5.8}$$

(The abbreviations for the enzymes involved are explained later in the text.)

The first evidence that galactinol is the galactosyl donor for the raffinose series was derived from two major *in vivo* observations: first, galactinol is found only in plants which contain members of the raffinose series (Senser & Kandler, 1967*a*) and, second, during photosynthesis in $^{14}CO_2$, the kinetics of its labelling and the intramolecular distribution of ^{14}C in galactinol exhibit all the characteristics of a precursor of the galactosyl moiety of raffinose, as depicted in Figs. 5.1 and 5.2 (Senser & Kandler, 1967*b*; Tanner, 1969). The galactosyl moiety of galactinol was labelled much faster than was the *myo*-inositol moiety. The latter exhibited a lower specific activity than the former, even after 4 h photosynthesis in $^{14}CO_2$. This indicates that the galactosyl moiety derived from the photosynthetic carbon cycle is rapidly transferred to raffinose, via galactinol, whereas the *myo*-inositol moiety is newly synthesized at a much lower rate and serves only as the galactosyl carrier.

The synthesis of galactinol from UDP-α-D-galactose and L-*myo*-inositol by the UDP-α-D-galactose:L-*myo*-inositol-1-galactosyltransferase in accordance with Equation (5.4) was demonstrated in enzyme preparations from peas by Frydman & Neufeld (1963). It is very specific with respect to the galactosyl donor but is of low specificity with respect to the various isomers of inositol. All the isomers tested *in vitro* acted as acceptors, whereas as yet only α-D-galactopyranosyl-L-*myo*-inositol has been found to act *in vivo*.

Fig. 5.1 Kinetics of labelling of various oligosaccharides in a detached leaf of *Catalpa bignonioides* during photosynthesis in $^{14}CO_2$ + air, followed by a chase period in $^{12}CO_2$ + air (after Senser & Kandler, 1967*b*).

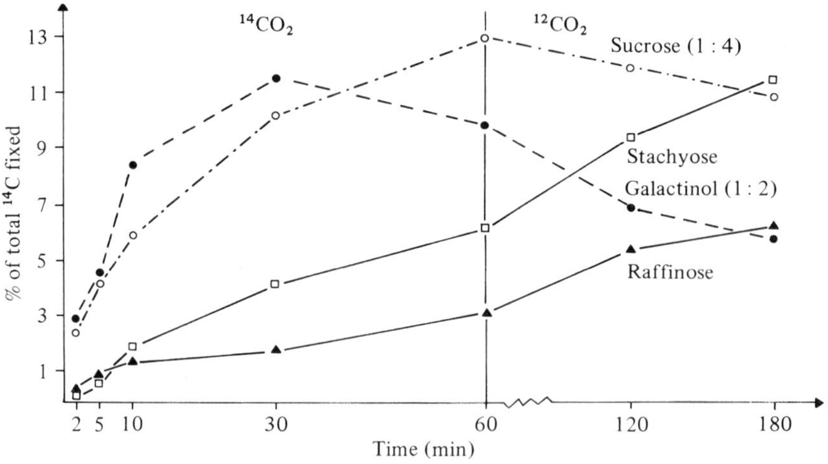

The enzyme catalysing the reaction of Equation (5.5) [galactinol: sucrose-6-galactosyltransferase (GST)] has been isolated from seeds of broad bean and from wheat germ (Lehle, Tanner & Kandler, 1970; Lehle & Tanner, 1973) and been characterized. It was also found in crude extracts of leaves of *Cerastium arvense* (Schwarzmaier, 1973). It shows a pH optimum of 6.8 and is different from α-D-galactosidase. The rate of transfer is five times greater than that of the hydrolysis of galactinol. The K_m for sucrose is 9×10^{-4} mol l^{-1}, that for galactinol is 7×10^{-3} mol l^{-1}. The enzymes display a high specificity for the acceptor. Out of 10 possible acceptors tested, only sucrose gave a positive result. Since no tri- or tetrasaccharide could serve as acceptor, GST is obviously unable to carry on the polymerization beyond the trisaccharide level.

Fig. 5.2 Intramolecular distribution of ^{14}C within galactinol, raffinose and stachyose after different periods of photosynthesis in $^{14}CO_2$ + air by a detached leaf of *Lamium maculatum* (after Senser & Kandler 1967*b*).

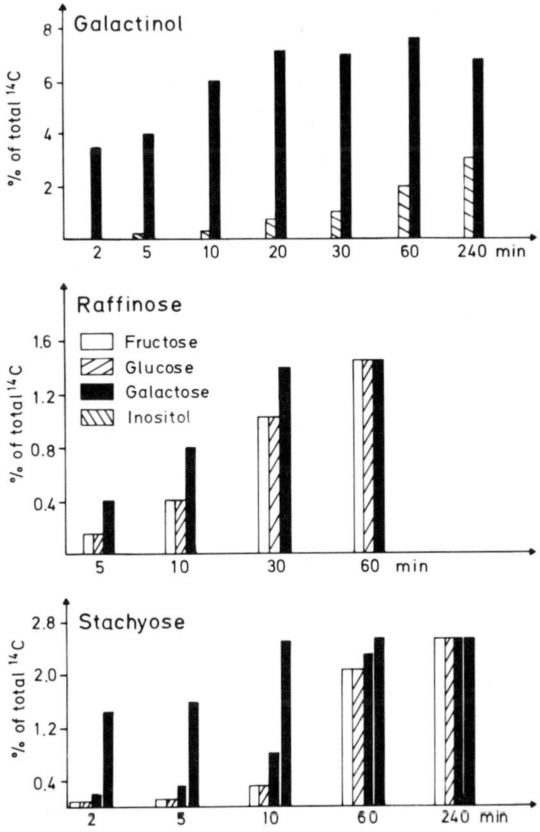

Galactinol could not be replaced as a donor by UDP-α-D-galactose but *p*-nitrophenyl-α-D-galactopyranoside and raffinose also serve as donors to some extent. The reaction with raffinose as donor corresponds to the exchange reaction between raffinose and sucrose (Equation 5.9) described by Moreno & Cardini (1966) in crude extracts of wheat germ.

$$\text{raffinose} + [^{14}\text{C}]\text{sucrose} \rightleftharpoons [^{14}\text{C}]\text{raffinose} + \text{sucrose} \qquad (5.9)$$

This exchange reaction may result in a rapid labelling of raffinose, either during photosynthesis of leaves in $^{14}CO_2$ or during metabolism of exogenously applied, easily assimilated labelled monosaccharides by plant tissue, even under conditions where only little or no net synthesis of raffinose occurs. In this case, however, the distribution of ^{14}C within the raffinose molecule is reversed compared with that described by Senser & Kandler (1967*b*), i.e., in leaves of *Cerastium arvense*, the sucrose moiety of raffinose is more strongly labelled than the galactosyl residue (Schwarzmaier, 1973).

The synthesis of higher homologues of the raffinose series is catalysed by a series of α-D-galactosyltransferases which use only raffinose, or raffinose and one or more of its higher homologues, as acceptors but do not react with sucrose. The first member of this series, the galactinol:raffinose-6-galactosyltransferase (GRT_1), was first isolated from the seeds of French bean (Tanner & Kandler, 1966) and was later also found in leaves of *Cucurbita* sp. (Webb & Pathak, 1970). By analogy with Equation (5.9), the GRT_1 catalyses the exchange reaction of Equation (5.10).

$$\text{stachyose} + [^{14}\text{C}]\text{raffinose} \rightleftharpoons [^{14}\text{C}]\text{stachyose} + \text{raffinose} \qquad (5.10)$$

A detailed description of GRT_1 is given by Tanner & Kandler (1968) and Lehle & Tanner (1973). They found an equilibrium constant [stachyose] × [*myo*-inositol]/[raffinose] × [galactinol] of about 4, indicating that the reaction of Equation (5.6) favours the synthesis, and that the negative free energy of hydrolysis of galactinol is about 4 kJ larger than that of the glycosidic bond of the terminal galactosyl residue of stachyose. The transgalactosylation reaction was 20 times greater than the hydrolysis of the D-galactosyl donor, galactinol. The transferase isolated from French bean exhibited a high substrate specificity: galactinol but neither UDP-α-D-galactose nor ADP-α-D-galactose are utilized as donor and raffinose but neither sucrose nor stachyose serve as acceptors. The latter specificity is in agreement with the absence of verbascose from the seeds of French bean. The transferase (GRT_2) from seeds of broad bean which do contain verbascose in addition to stachyose, exhibited, however, a significant affinity not only for raffinose but also for stachyose as an acceptor and synthesized verbascose *in vitro* (Tanner, Lehle & Kandler, 1967).

Only a slow synthesis of ajugose was observed when verbascose was supplied as acceptor. Synthesis of ajugose was distinctly more rapid with enzyme preparations from the seeds of pea and *Vicia sativa*, which contain more ajugose than do the seeds of *V. faba* (O. Preuss, W. Tanner & O. Kandler, unpublished), thus indicating the presence of a third variety of transferase (GRT_3). In the course of the enrichment of the transferase from these three species, the relative amounts of stachyose, verbascose, or ajugose formed from the corresponding lower acceptors remained unchanged. This indicates that one and the same enzyme catalyses the two or three different transfer reactions. The formation of the higher homologues of the raffinose series may therefore be expressed by Equation (5.11).

$$\text{galactinol} + \text{sucrose-Gal}_n \overset{GRT_n}{\rightleftharpoons} \text{sucrose-Gal}_{n+1} + \textit{myo}\text{-inositol} \quad (5.11)$$

where *n* may be any number from one to five.

The GRT_{1-5} catalysing these reactions differ with respect to their specificity towards acceptors. This depends on the species from which they are isolated. Even *in vitro*, they do not transfer galactosyl residues from galactinol to any homologue of raffinose of an order higher than that actually found in the tissues from which the enzymes were derived. Hence, the observed qualitative difference in the composition of sucrosyl oligosaccharides in the various leguminous seeds is not caused by metabolic regulation but by differences in the acceptor sites of the respective GRT. Therefore, while GRT_{1-5} may be considered to be the products of a set of alleles of the same gene, GST, which differs from GRT not only by its different specificity towards acceptors, but also in its distinct sensitivity to oxygen, is likely to be coded by a different gene.

(3) *Lychnose series* [*O*-α-D-galactopyranosyl-(1-6)-*O*-α-D-glucopyranosyl-(1-2)-β-D-fructofuranosyl-(1-1)-α-D-galactopyranoside]. The kinetics of labelling of the oligosaccharides in leaves of various Carophyllaceae during photosynthesis in $^{14}CO_2$ showed that sucrose, galactinol and raffinose become labelled much more rapidly than lychnose and isolychnose (Schwarzmaier, 1973). The latter contain significant amounts of ^{14}C only after several hours and their higher homologues only after 18 h. This finding indicates that, at first, raffinose is synthesized from sucrose and galactinol and that lychnose or isolychnose are formed later by the transfer of a galactosyl residue to the fructose moiety of raffinose.

The intramolecular distribution of ^{14}C in lychnose and isolychnose of leaves of *Cerastium arvense* showed that, in summer when there is only little or no net synthesis of the lychnoses, the raffinose moiety but not the galactosyl residues at the F^1 or F^3 hydroxyls of raffinose became significantly labelled during the first hour of photosynthesis in $^{14}CO_2$. This indicates an extensive exchange in accordance with Equation (5.12).

$$\text{galactosyl raffinose} + [^{14}\text{C}]\text{raffinose} \rightleftharpoons \text{galactosyl-}[^{14}\text{C}]\text{raffinose} + \text{raffinose} \quad (5.12)$$

The galactosyl residue in lychnose or isolychnose, however, also became labelled when the experiment was carried out in autumn and winter, when a high net synthesis of these oligosaccharides takes place. Even then, the ^{14}C content of the galactosyl residues at F^1 or F^3 did not exceed but, at most, equalled that of the galactosyl residue of the raffinose moiety.

Hence, it is unlikely that the galactosyl residues of lychnose and isolychnose at the F^1 or F^3 positions are derived from a donor such as UDP-galactose or galactinol whose specific activity should be distinctly higher than that of the galactosyl residue of raffinose, at least after short periods of photosynthesis in $^{14}CO_2$ (Schwarzmaier, 1973).

These findings are compatible with the assumption that lychnose and isolychnose are formed by a galactosyl transfer from one raffinose molecule to another in accordance with Equation (5.13). In this way, the synthesis of the higher homologues resembles that of the kestose series from fructosyl-sucrose (see Chapter 4).

$$\text{raffinose} + \text{raffinose} \rightleftharpoons \text{lychnose} + \text{sucrose} \quad (5.13)$$

Direct evidence for this mechanism *in vitro* has recently been obtained (H. Hopf and O. Kandler, unpublished). Activity of a raffinose:raffinose 1^F-α-galactosyltransferase and a raffinose:raffinose 3^F-α-galactosyltransferase could be detected in enzyme preparations from whole plants of *Cerastium arvense* during the winter season. Both enzymes show low activity so that incubation times of several hours are necessary in order to obtain measurable synthesis of lychnose or isolychnose. The enzymes are highly specific for raffinose as both galactosyl donor and acceptor; D-galactose, D-galactose 1-P, UDP-D-galactose or galactinol do not serve as galactosyl donors. Formation of the higher homologues of lychnose or isolychnose, using raffinose as the galactosyl donor and lychnose or isolychnose as the acceptor, has yet to be demonstrated.

(4) *Umbelliferose* [*O*-α-D-galactopyranosyl-(1-2)-*O*-α-D-glucopyranosyl-(1-2)-β-D-fructofuranoside]. The synthesis of umbelliferose *in vitro* from sucrose as acceptor and UDP-D-galactose as donor occurs with enzyme preparations from leaves of *Aegopodium podagraria* (Hopf & Kandler, 1974).

$$\text{UDP-D-galactose} + \text{sucrose} \rightleftharpoons \text{umbelliferose} + \text{UDP} \quad (5.14)$$

To date, it has not proved possible to demonstrate a reversal of the reaction by applying either labelled sucrose or labelled UDP-D-galactose. The enzyme, an UDP-D-galactose:sucrose 2-D-galactosyltransferase, exhibits a pH optimum of 7.5. It is sensitive to oxygen and requires protection by dithioerythritol. Galactinol or D-galactose 1-phosphate cannot replace UDP-D-galactose as donor. The synthesis of umbelliferose as described in Equation (5.14) is compatible with the following *in vivo* observations (Hopf & Kandler, 1976). First, no labelled galactinol was found after photosynthesis in $^{14}CO_2$ in mature leaves of *Ae. podagraria*, which do not form raffinose but only umbelliferose. In young leaves, however, which contain small amounts of raffinose in addition to umbelliferose, galactinol became labelled under the same experimental conditions. Second, after short periods of photosynthesis of mature leaves of *Ae. podagraria* in $^{14}CO_2$, the galactosyl residue of umbelliferose showed a specific activity much higher than that of the sucrose moiety, as is expected if the galactosyl residue is derived rather directly from intermediates of the photosynthetic cycle without being diluted by passage through a large pool of glycosyl residues, for example, sucrose.

(5) *Planteose* [*O*-α-D-glucopyranosyl-(1-2)-*O*-β-D-fructofuranosyl-(6-1)-α-D-galactopyranoside]. Synthesis of planteose from sucrose and free galactose by an α-D-galactosidase from *Plantago ovata* was reported by Courtois *et al.* (1961). However, it seems unlikely that a process requiring high concentrations of galactose can function in cells which contain only traces of free galactose. Evidence for the pathway of planteose biosynthesis was obtained with partially purified enzyme preparations from seeds of *Sesamum indicum* (Hopf, Spanfelner & Kandler, 1984). Labelled planteose was formed with sucrose as acceptor and ^{14}C-UDP-D-galactose as donor in accordance with Equation (5.15).

$$\text{UDP-D-[}^{14}\text{C]galactose} + \text{sucrose} \rightleftharpoons [^{14}\text{C]planteose} + \text{UDP} \qquad (5.15)$$

No synthesis of planteose was observed, however, with galactinol, D-galactose 1-phosphate, raffinose or stachyose as galactosyl donors and [^{14}C]sucrose as acceptor.

The enzyme preparation also contained a very high activity of UDP-D-glucose 4-epimerase, an enzyme epimerizing 0.1 μmol of UDP-D-galactose to UDP-D-glucose per mg protein in 10 min, so stopping the synthesis of planteose. The activity of the epimerase could be approximately halved by adding 10^{-2} mol l^{-1} of Mn^{2+}. The UDP-D-galactose:sucrose 6^F-α-galactosyl transferase shows a pH optimum of pH 6.2 and does not need protection by SH-reagents; it is not inhibited by UDP-D-glucose, D-glucose 1-P or D-galactose 1-P. The *in vitro* synthesis of sesamose and of the other higher homologues of planteose has not yet been achieved.

(6) *Gentianose* [*O*-β-D-glucopyranosyl-(1-6)-*O*-α-D-glucopyranosyl-(1-2)-β-D-fructofuranoside]. The intramolecular distribution of ^{14}C in gentianose after short intervals of photosynthesis by leaves of *Gentiana lutea* in $^{14}CO_2$ showed that the glucosyl residue bound to the glucose moiety of sucrose exhibited a higher specific activity than sucrose (Heiniger & Franz, 1977). This indicated that the glucosyl donor rapidly gains a high specific activity upon the onset of photosynthesis as do the nucleotide sugars. However, all attempts to demonstrate synthesis of gentianose *in vitro* with nucleotide-activated glucose or D-glucose 1-phosphate as donor have so far failed (G. Franz, U. Heiniger, H. Hopf, N. Schilling, personal communication). Also the suggestion that, by analogy with galactinol, glucinol [β-D-glucopyranosyl-(1-1)-*myo*-inositol] may serve as glucosyl donor was disproved by the finding that glucinol did not become significantly labelled during photosynthesis in $^{14}CO_2$ in leaves of various species of *Gentiana*, although labelled gentianose was formed in these leaves (Hopf & Kandler, 1980).

II. Reducing disaccharides

(1) *Maltose* [*O*-α-D-glucopyranosyl-(1-4)-D-glucopyranose]. The formation of maltose and its higher homologues by the action of β-amylase is well known (see Preiss & Levi, 1980, and Chapter 10). These compounds may also arise by the action of a glucan-glucosyltransferase present in wheat germ (Edelman & Keys, 1961) and in spinach leaves (Linden, Tanner & Kandler, 1974), which catalyses the reactions in Equations 5.16–5.20.

$$\emptyset^* + 0{-}\emptyset \rightleftharpoons 0{-}\emptyset^* + \emptyset \qquad (5.16)$$

$$\emptyset^* + 0{-}0{-}0{-}0{-}0{-}0\text{ - - -} \rightleftharpoons 0{-}\emptyset^* + 0{-}0{-}0{-}0{-}0\text{ - - -} \qquad (5.17)$$

$$\emptyset^* + 0{-}0{-}0{-}0{-}0{-}0\text{ - - -} \rightleftharpoons 0{-}0{-}\emptyset^* + 0{-}0{-}0{-}0\text{ - - -} \qquad (5.18)$$

$$\emptyset^* + 0{-}0{-}0{-}0{-}0{-}0\text{ - - -} \rightleftharpoons 0{-}0{-}0{-}\emptyset^* + 0{-}0{-}0\text{ - - -} \qquad (5.19)$$

$$\emptyset^* + 0{-}0{-}0{-}0{-}0{-}0\text{ - - -} \rightleftharpoons 0{-}0{-}0{-}0{-}\emptyset^* + 0{-}0\text{ - -} \qquad (5.20)$$

In these equations, 0 represents a glucosyl unit, ∅ a reducing glucosyl unit, 0* a ^{14}C-labelled glucosyl unit, – an α1–4 linkage, and - - - the continuation of such linkages as in amylose.

These reactions yield smaller oligomers of a glucan by inserting free glucose. Reaction (5.16) merely represents an exchange reaction leading to labelling of glucose at the reducing end of the chain when free [^{14}C]glucose is present. The *in vivo* function of this enzyme may be the formation of primers for starch synthesis from preformed glucan.

Evidence for *de novo* synthesis of maltose from early intermediates of the photosynthetic cycle, for example sugar phosphates, has been derived from the kinetics of its labelling during photosynthesis of leaves in $^{14}CO_2$. Maltose behaves like an early product of photosynthesis, rather than a storage product such as starch and its degradation products. The maximum percentage of fixed ^{14}C in maltose is reached after only 2–3 min as observed with sugar phosphates (Kandler, 1967). In addition, the intramolecular distribution of ^{14}C in maltose occurring freely in the cell is different from that liberated by the action of β-amylase from starch derived from the same leaf. After short periods of photosynthesis in $^{14}CO_2$, the non-reducing end of free maltose is more heavily labelled than the reducing end (Table 5.1). The label tends to become uniformly distributed within 10 min as photosynthesis in $^{14}CO_2$ proceeds. The maltose obtained from starch of the same leaf, however, is uniformly labelled even after the shortest interval investigated (Linden *et al.* 1975). Direct evidence for a *de novo* synthesis of maltose from α-D-glucose 1-phosphate was obtained with enzyme preparations from spinach leaves (Schilling & Kandler, 1975). It was suggested that the reaction might involve the transfer of a glucosyl residue from glucose 1-phosphate to free glucose, similar to the maltose phosphorylase reaction described by Fitting & Doudoroff (1952), which, however, utilizes β-D-glucose 1-phosphate instead of α-D-glucose 1-phosphate employed in the experiments with the spinach enzyme. Further studies (Schilling, 1982) revealed that the addition of free

Table 5.1 *Changes in the distribution of label in free maltose and in maltose, derived from starch by the action of β-amylase, which were labelled during photosynthesis of leaves in $^{14}CO_2$. The quotients ^{14}C in the non-reducing end/^{14}C in the reducing end of maltose are given* (*Linden* et al. *1975*)

	Sycamore Maltose		Gentian Maltose		Spinach Maltose		Common catalpa Maltose
Time (min)	free	from starch	free	from starch	free	from starch	free
1	5.2	0.9	n.d.	n.d.	1.5	1.1	n.d.
2	4.7	1.0	10.0	1.15	1.6	1.2	1.9
3	2.9	1.05	8.0	1.15	n.d.	n.d.	n.d.
4	2.7	0.95	3.8	0.85	n.d.	n.d.	n.d.
5	1.7	n.d.	2.4	n.d.	1.2	1.1	1.5
7	1.2	n.d.	1.1	1.00	n.d.	n.d.	n.d.
10	1.0	n.d.	n.d.	n.d.	1.2	1.2	1.05
20	1.1	n.d.	1.0	1.1	n.d.	n.d.	1.05
60	n.d.	n.d.	1.05	1.1	n.d.	n.d.	n.d.

n.d. = not determined

glucose is not necessary and maltose is formed exclusively from α-D-glucose 1-phosphate in accordance with Equation (5.23). In addition, some maltotriose and maltotetraose were formed.

$$2\alpha\text{-D-glucose 1-phosphate} \rightarrow \text{maltose 1-phosphate} + \text{Pi} \quad (5.21)$$

$$\text{maltose 1-phosphate} \rightarrow \text{maltose} + \text{Pi} \quad (5.22)$$

$$2\alpha\text{-D-glucose 1-phosphate} \rightarrow \text{maltose} + 2\text{Pi} \quad (5.23)$$

$$\text{G-1-P} + \text{E} \rightarrow \text{G-E} + \text{Pi} \quad (5.24)$$

$$\text{G-1-P} + \text{X} \rightarrow \text{G-X} + \text{Pi} \quad (5.25)$$

$$\text{G-E} + \text{G-X} \rightarrow \text{G-G-X} + \text{E} \quad (5.26)$$

$$\text{G-G-X} \rightarrow \text{G-G} + \text{X} \quad (5.27)$$

$$\text{G-G-X} + \text{G-E} \rightarrow \text{G-G-G-X} + \text{E} \quad (5.28)$$

$$\text{G-G-G-X} \rightarrow \text{G-G-G} + \text{X} \quad (5.29)$$

X may be an unknown glucosyl carrier, a second but different site of the enzyme (E), or a second enzyme different from E.

As discussed by Schilling (1981), the most simple mechanism of maltose synthesis would be the formation of maltose 1-phosphate (Equation 5.21) and its subsequent splitting by the action of a phosphatase (Equation 5.22). However, it was not possible to detect maltose 1-phosphate in the reaction mixture nor to trap a glucosyl-enzyme intermediate, although one would expect to find such an intermediate because the configuration of the anomeric carbon is retained, and an exchange is observed between ^{32}P and D-glucose 1-phosphate. Both observations suggest a double displacement mechanism (Equations 5.24–5.26) as has been demonstrated in the case of sucrose 6-phosphate synthase (Voet & Abeles, 1970).

The unique feature of maltose synthesis from 2 molecules of D-glucose 1-phosphate is the use of a relatively energy-rich compound (see Table 1.1 in Chapter 1) as acceptor. This may indicate that it is not D-glucose 1-phosphate itself that is the final acceptor but rather a glucosyl residue attached to a carrier (Equation 5.25) at the expense of 'phosphate energy'. The glucosyl residue of the donor molecule is then transferred to the C-4 position of the bound acceptor (Equation 5.26). The carrier-bound maltose thus formed may then either be liberated by hydrolysis (Equation 5.27) or serve as acceptor for another transglucosylation step yielding maltotriose (Equations 5.28 and 5.29). Such a process would allow the synthesis of a series of higher homologues of maltose which could serve as primers for starch synthesis. The relative rates of hydrolysis and transglucosylation respectively would control the ratio between maltose and the various homologues formed. The enzyme preparations used in the study of the so-called 'unprimed' starch synthesis (Lavintman & Cardini, 1973; Lavintman, Tandecarz, Carceller, Mendiara & Cardini, 1974) and glycogen synthesis (Krisman & Barengo, 1975) may also have contained the reported maltose synthase activity (Schilling, 1982) but in a mixture with other enzymes which carry on chain elongation beyond the limitations of the chain length that is typical of maltose synthase.

The enzyme was obtained only from young and not from mature leaves, and, to be stabilized, requires the presence of D-glucose 1-phosphate during the extraction procedure. A K_m of 1–2 mM for D-glucose 1-phosphate was observed at pH 6.8; Pi, ATP and D-gluconolactone act as competitive inhibitors.

When only D-[^{14}C]glucose 1-phosphate is applied, both glucose moieties of the synthesized maltose are equally labelled, independent of the reaction time, whereas the reducing end of maltose tends to become diluted as the reaction proceeds when unlabelled free glucose is added to the reaction mixture together with D-[^{14}C]glucose 1-phosphate (Table 5.2). The dilution of the label is probably caused by the exchange reaction (Equation 5.16) between uniformly labelled maltose, formed from D-[^{14}C]glucose 1-phosphate, and unlabelled free glucose. In fact, the enzyme preparation still exhibited strong glucan-glucosyltransferase activity which catalyses this exchange, although the maltose-synthesizing activity was more than 200-fold enriched. The enzyme preparation also showed significant phosphatase and α-glucosidase activity and catalysed an exchange between ^{32}P and D-glucose 1-phosphate. It is, however, not known whether the various activities stem from one and the same enzyme or from different ones. When we view the present data on *de novo* maltose synthesis, two distinct discrepancies between the *in vivo* observations and the *in vitro* evidence become obvious: firstly, the unequal

Table 5.2 *Distribution of ^{14}C within maltose synthesized by an enzyme preparation from spinach leaves. The enzyme assay contained as precursor either 1.0 mmol l^{-1}* [D-^{14}*C*]*glucose 1-phosphate alone* (*A*), *or in combination with 20 mmol l^{-1} unlabelled glucose* (*B*) (*after Schilling, 1982*)

Incubation time (min)	Per cent in the reducing end of maltose	
	A	B
3	53.2	51.1
15	47.3	47.9
30	48.7	41.6
45	52.3	36.5
60	50.9	29.8

labelling of maltose after short periods of photosynthesis in $^{14}CO_2$ is incompatible with the synthesis of maltose from 2 molecules of D-glucose 1-phosphate; secondly maltose synthase activity is found only in very young leaves, whereas the labelling kinetics of maltose during photosynthesis in $^{14}CO_2$ do not show such a dependence on the developmental stage. Therefore, the details of the *in vivo* synthesis of maltose and its role in starch synthesis and carbohydrate metabolism in plants in general are still open questions.

(2) *Gentiobiose.* Until recently, gentiobiose was known only as a breakdown product of gentianose. It has, however, been found in considerable quantities in the seed coat of the ripening seeds of ivy and in the bleeding sap of birch, although no gentianose could be detected in these plants (H. Hopf, O. Kandler, unpublished). When ^{14}C-monosaccharides or [^{14}C]sucrose were fed to ripening seeds of ivy, labelled gentiobiose was formed but no gentianose. Gentiobiose synthesis *in vivo* could be demonstrated with a dialysed enzyme preparation from ripening seeds of ivy, when UDP-D-[^{14}C]glucose was employed as precursor (H. Hopf & O. Kandler, unpublished). Residues of the gentiobiose thus formed were equally labelled even in the presence of unlabelled glucose or sucrose, indicating the following reaction (Equation 5.30).

$$2\ \text{UDP-D-glucose} \rightarrow \text{gentiobiose} + 2\ \text{UDP} \quad (5.30)$$

The enzyme preparation has optimal activity at pH 6.0. It is sensitive to oxygen and, for protection, requires 0.002 mol l^{-1} $Na_2S_2O_4$, which is not replaceable by mercaptoethanol or dithioerythritol. Only UDP-D-glucose,

Fig. 5.3 Pathways of biosynthesis of primary oligosaccharides. Gal-? or Glc-? = unknown glycosyl donors; (—P) = phosphorylated intermediates; --?-- = suggested reaction.

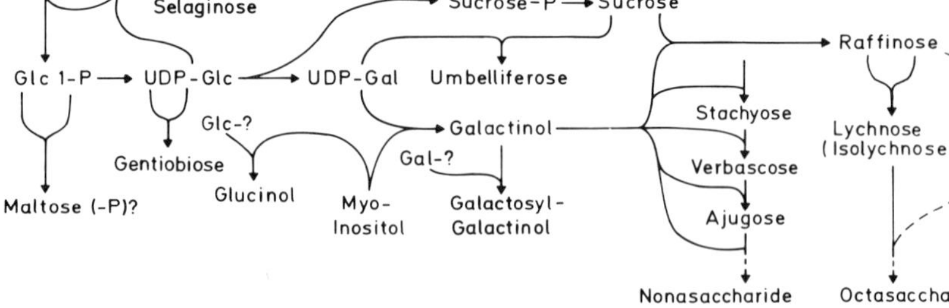

but not glucose or D-glucose 1-phosphate, can serve as donor and acceptor. No exchange between gentianose and UDP-D-glucose, [^{14}C]sucrose or [^{14}C]glucose was observed.

The mechanism of synthesis of gentiobiose thus resembles that of maltose, in that both the glucosyl donor and acceptor contain a relatively energy-rich phosphate bond, although no phosphorylated intermediary disaccharide phosphate can be detected. This suggests the formation of a bound acceptor (Equations 5.25–5.27). In contrast to maltose synthesis, however, no exchange between the glucosyl donor and the end product is observed and the configuration of the anomeric carbon becomes inverted. Both findings suggest a single displacement mechanism.

Concluding remarks

The present view of the biosynthesis of oligosaccharides in respect of the donor and acceptor requirements is depicted in Fig. 5.3. The important role of the phosphate- and nucleotide-activated monosaccharides as glycosyl donors for the synthesis of disaccharides, the unique function of galactinol as galactosyl donor for the raffinose series, and the central position of sucrose, both as donor and acceptor in the biosynthesis of the higher oligosaccharides, are obvious. A number of question marks in the above scheme remain to be resolved and the details of the reaction mechanisms are, with the exception of sucrose synthesis, unknown. Such knowledge will, however, be necessary before we can fully understand the regulation of carbohydrate metabolism and the storage processes on a molecular basis. Hence, the biochemistry of oligosaccharides deserves more attention than it has received during the last decade.

References

Akazawa, T. & Okamoto, K. (1980). Biosynthesis and metabolism of sucrose. In *The Biochemistry of Plants*, vol. 3, ed. J. Preiss, pp. 199–200. New York: Academic Press.

Bourne, E. J., Walter, M. W. & Pridham, J. B. (1965). The biosynthesis of raffinose. *Biochemical Journal*, **87**, 802–6.

Cabib, E. & Leloir, L. F. (1958). The biosynthesis of trehalose phosphate. *Journal of Biological Chemistry*, **231**, 259.

Cardini, C. E., Caputto, R., Paladini, A. C. & Leloir, L. F. (1950). Uridine diphosphate glucose: the coenzyme of the galactose-glucose phosphate isomerization. *Nature*, **165**, 191–2.

Courtois, J. E., Petek, F. & Dong, T. (1961). Synthèse de plantéose par action transférante de l'α-galactosidase des graines de *Plantago*. *Bulletin de la Société Chimique et Biologique*, **43**, 1189–96.

Edelman, J. & Keys, A. J. (1961). A maltose–glucose transglucosylase from wheat germ. *Biochemical Journal*, **79**, 12*P*.

Feingold, D. S. & Avigad, G. (1980). Sugar nucleotide transformation in plants. In *The Biochemistry of Plants*, vol. 3, ed. J. Preiss, pp. 102–70. New York: Academic Press.

Fischer, M. (1973). Identifizierung, Vorkommen und Umsatz der Oligosaccharide bei Seleginellaceen. Thesis, University of Munich.

Fischer, M. & Kandler, O. (1975). Identifizierung von Selaginose und deren Verbreitung in der Gattung Selaginella. *Phytochemistry*, **14**, 2629–33.

Fitting, C. & Doudoroff, M. (1952). Maltose phosphorylase in *Neisseria meningitidis*. *Journal of Biological Chemistry*, **199**, 153–7.

Frydman, R. B. & Neufeld, E. F. (1963). Synthesis of galactosyl inositol by extracts from peas. *Biochemical and Biophysical Research Communications*, **12**, 121–5.

Gussin, A. E. S. & McCormack, J. H. (1970). Trehalase and the enzymes of trehalose biosynthesis in *Lilium longiflorum* pollen. *Phytochemistry*, **9**, 1915–20.

Hassid, W. Z. (1958). The synthesis and transformation of the oligosaccharides in plants, including their hydrolysis. In *Encyclopaedia of Plant Physiology*, vol. VI, ed. W. Ruhland, pp. 125–36. Berlin, Goettingen, Heidelberg: Springer-Verlag.

Heiniger, U. & Franz, G. (1977). Biosynthesis of the trisaccharide gentianose. *Plant Physiology*, **59**, Supplement p. 100.

Hopf, H. & Kandler, O. (1974). Biosynthesis of umbelliferose in *Aegopodium podagraria*. *Plant Physiology*, **54**, 13–14.

Hopf, H. & Kandler, O. (1976). Physiologie der Umbelliferose. *Biochemie und Physiologie der Pflanzen*, **169**, 5–36.

Hopf, H. & Kandler, O. (1980). *O*-β-D-Glucopyranosyl-(1-1)-myo-inositol (glucinol) in higher plants. *Zeitschrift für Pflanzenphysiologie*, **100**, 189–95.

Hopf, H., Spanfelner, M. & Kandler, O. (1984). Planteose synthesis in seeds of *Sesamum indicum* L. *Zeitschrift für Pflanzenphysiologie*, **114**, 485–92.

Imhoff, V. (1973). Synthesis of galactosides by chloroplasts isolated from pea leaves. *Hoppe-Seylers Zeitschrift für Physiologische Chemie*, **354**, 1550–4.

Kandler, O. (1967). Biosynthesis of poly- and oligosaccharides during photosynthesis in green plants. In *Harvesting the Sun*, ed. A. San Pietro, F. A. Grear & T. J. Army, pp. 131–52. New York: Academic Press.

Krisman, C. R. & Barengo, R. (1975). A precursor of glycogen biosynthesis: 1,4-glucan-protein. *European Journal of Biochemistry*, **52**, 117–23.

Lavintman, N. & Cardini, C. R. (1973). Particulate UDP-glucose:protein transglucosylase from potato tuber. *FEBS Letters*, **29**, 43–6.

Lavintman, N., Tandecarz, J., Carceller, M., Mendiara, S. & Cardini, C. E. (1974). Role of uridine diphosphate glucose in the biosynthesis of starch. *European Journal of Biochemistry*, **50**, 145–55.

Lehle, L. & Tanner, W. (1973). The function of myo-inositol in the biosynthesis of raffinose. *European Journal of Biochemistry*, **38**, 103–10.

Lehle, L., Tanner, W. & Kandler, O. (1970). Myo-inositol, a cofactor in the biosynthesis of raffinose. *Hoppe-Seylers Zeitschrift für Physiologische Chemie*, **351**, 1494–8.

Leloir, L. F. & Cabib, E. (1953). The enzymatic synthesis of trehalose phosphate. *Journal of the American Chemical Society*, **75**, 5445–6.

Linden, J. C., Schilling, N., Brackenhofer, H. & Kandler, O. (1975). Asymmetric labelling of maltose during photosynthesis in $^{14}CO_2$. *Zeitschrift für Pflanzenphysiologie*, **76**, 176–81.

Linden, J. C., Tanner, W. & Kandler, O. (1974). Properties of glucosyltransferase and glucan transferase from spinach. *Plant Physiology*, **54**, 752–7.

Moreno, A. & Cardini, C. E. (1966). A raffinose–sucrose transgalactosidase from wheat germ. *Plant Physiology*, **41**, 909–10.

Nikaido, H. & Hassid, Z. (1971). Biosynthesis of saccharides from glucopyranosyl esters of nucleoside pyrophosphates ('sugar nucleotides'). In *Advances of Carbohydrate Chemistry and Biochemistry*, ed. R. S. Tipson & D. Horton, vol. 26, pp. 351–76. New York: Academic Press.

Preiss, J. & Levi, C. (1980). Starch biosynthesis and degradation. In *The Biochemistry of Plants*, vol. 3, ed. J. Preiss, pp. 371–423. New York: Academic Press.

Pridham, J. B. & Hassid, W. Z. (1965). Biosynthesis of raffinose. *Plant Physiology*, **40**, 984–6.

Schilling, N. (1982). Characterization of maltose biosynthesis from α-D-glucose-1-phosphate in *Spinacia oleracea* L. *Planta*, **154**, 87–93.

Schilling, N. & Kandler, O. (1975). α-Glucose-1-phosphate, a precursor in the biosynthesis of maltose in higher plants. *Biochemical Society Transactions*, **3**, 985–7.

Schwarzmaier, G. (1973). Untersuchungen über den Stoffwechsel der Saccharosegalactoside bei Caryophyllaceen. Thesis, University of Munich.

Senser, M. & Kandler, O. (1967*a*). Vorkommen und Verbreitung von Galactinol in Blättern Höherer Pflanzen. *Phytochemistry*, **6**, 1533–40.

Senser, M. & Kandler, O. (1967*b*). Galactinol, ein Galactosyl-donor für die Biosynthese der Zucker der Raffinosefamilie in Blättern. *Zeitschrift für Pflanzenphysiologie*, **57**, 376–88.

Siddiqui, L. R. (1970). The sugars of honey, In *Advances of Carbohydrate Chemistry and Biochemistry*, ed. R. S. Tipson & D. Horton, vol. 25, pp. 285–309, New York: Academic Press.

Tanner, W. (1969). The function of myo-inositol glycosides in yeasts and higher plants. *Annals of the New York Academy of Science*, **165**, 726–42.

Tanner, W. & Kandler, O. (1966). Biosynthesis of stachyose in *Phaseolus vulgaris*. *Plant Physiology*, **41**, 1540–2.

Tanner, W. & Kandler, O. (1968). Myo-inositol, a cofactor in the biosynthesis of stachyose. *European Journal of Biochemistry*, **4**, 233–9.

Tanner, W., Lehle, L. & Kandler, O. (1967). Myo-inositol, a cofactor in the biosynthesis of verbascose. *Biochemical and Biophysical Research Communications*, **29**, 166–71.

Voet, J. & Abeles, R. H. (1970). The mechanism of action of sucrose phosphorylase. *Journal of Biological Chemistry*, **245**, 1020–31.

Webb, J. A. & Pathak, S. (1970). Biosynthesis of stachyose in *Curcurbita*. *Plant Physiology*, **46**, Supplement p. 27.

EDWARD A. DREW

6 Physiology and metabolism of cyclitols

Introduction

Despite their $[CH_2O]_n$ formulation, the cyclitols lack the reactivity of the sugars, having no aldehyde or ketone groups. They are chemically more similar to, but even less reactive than, the straight-chain polyols which also possess only hydroxyl groups. Consideration of the cyclitols as carbohydrates may indeed be questionable but, as their metabolism in plants appears to be directly associated with that of mono-, oligo- and polysaccharides, they are included in this book.

Studies of cyclitol metabolism in vascular plants centre around the synthesis of *myo*-inositol, the most widespread of these compounds, and its subsequent epimerization to other inositols, methylation to *O*-methyl ethers, incorporation into cell-wall components, and the accumulation and utilization of *myo*-inositol hexaphosphate or phytic acid. (A stricter nomenclature is *myo*-inositol hexakisphosphate – Cosgrove, 1980) The incorporation of *myo*-inositol into the phosphoinositides of membranes will not be considered here.

The chemical structure and taxonomic distribution of the naturally occurring cyclitols have been summarized in Chapter 1 and many specific occurrences will be noted later. In addition to the topics listed above, recent studies on the diverse cyclitols accumulated in seagrasses will be described and their possible phylogenetic significance will be discussed.

Quantitative significance

It is important to establish at the outset what concentrations of the numerous cyclitols occur in vascular plants. As an initial guide, quantitative data gleaned from Plouvier's wide ranging surveys, published between 1947 and 1974, revealed the maximum values set out in Table 6.1. Although it is unwise to attempt definition of a storage compound solely in terms of the quantity accumulated, significant accumulation is intrinsic to function in storage and it seems reasonable to set an arbitrary lower level of 1% dry weight for such a definition. Only two pentols, two mono-*O*-methyl hexols and one dimethyl ether qualify as storage products on that basis in this listing.

The additional, more recent analyses listed in Table 6.2 show that, at least in specific instances including certain non-vascular plants, several other cyclitols achieve this 1% criterion.

Unfortunately, much of the recent detailed work on cyclitol metabolism in plants has been purely biochemical and no indications of quantities involved have been given. Nevertheless, it appears from Tables 6.1 and 6.2 that at least one representative from each group of cyclitols considered – tetrols, pentols, hexols and mono- and di-*O*-methyl hexols – can accumulate to levels in excess of 1% dry weight and therefore represent a considerable store of carbohydrate-type material.

Biosynthesis

Hexols

It is generally considered that most inositols, and consequently their methylated derivatives, are derived in both plants and animals by epimerization of *myo*-inositol. Since this compound is apparently ubiquitous in living

Table 6.1 *Maximum cyclitol content of plants according to Plouvier*

Cyclitol	% Dry weight	Plant species	Organ	Reference
Tetrols				
l-leucanthemitol	0.15	*Chrysanthemum leucanthemum*	flower	Plouvier, 1962*a*
Pentols				
d-quercitol	2.00	*Quercus* spp.	leaves?	Plouvier, 1955*b*
l-viburnitol	1.75	*Menispermum canadense*	leaves	Plouvier, 1956*a*
Hexols				
myo-inositol	0.45	*Fagus sylvatica*	leaves	Plouvier, 1973
l-*chiro*-inositol	0.20	*Mallotus phillipensis*	bark	Plouvier, 1964
scyllo-inositol	0.71	*Hernandia sonora*	leaves	Plouvier, 1971
Mono-*O*-methyl ethers				
d-bornesitol	0.18	*Pleiocarpa mutica*	leaves	Plouvier, 1973
l-bornesitol	0.33	*Lathyrus vernus*	leaves	Plouvier, 1955*d*
sequoyitol	0.17	*Metasequoia glyptostroboides*	leafy green stems	Plouvier, 1962*b*
d-ononitol	0.68	*Lathyrus vernus*	leaves	Plouvier, 1973
d-pinitol	3.30	*Amorpha microphylla*[a]	leaves	Plouvier, 1955*d*
l-pinitol	0.86	*Artemisia dracunculus*	leaves	Plouvier, 1956*b*
l-quebrachitol	4.00	*Shepherdia argentea*[a]	leaves	Plouvier, 1955*a*
Di-*O*-methyl ethers				
dambonitol	0.54	*Nerium oleander*	leaves	Plouvier, 1960
liriodendritol	1.45	*Liriodendron tulipifera*	leaves?	Plouvier, 1955*c*

[a] several other species contained over 1% dry weight.

Table 6.2 *Additional data on maximum cyclitol content in plants*

Cyclitol	% Dry weight	Plant species	Organ	Reference
Tetrols				
d-1,4/2,5 cyclohexanetetrol	7.00	*Monochrysis lutheri*	(alga)	Craigie, 1974
Hexols				
myo-inositol	3.60	*Zostera noltii*	leaves	Drew, 1980
l-chiro-inositol	12.30	*Cymodocea rotundata*	leaves	Drew, 1983
muco-inositol	8.50	*Thalassodendron ciliatum*	leaves	Drew, 1983
Mono-*O*-methyl ethers				
d-pinitol	1.49	*Lotus corniculatus*	leaves?	Riggs & Strong, 1967
sequoyitol	0.42	*Cycas circinalis*	leaves?	Riggs & Strong, 1967
l-quebrachitol	5.70	Maple syrup	syrup	Stinson *et al.* 1967
d-1-*O*-methyl-*muco*-inositol	7.40	*Araucaria excelsa*[a]	leaves	Dittrich, Gietl & Kandler, 1972
methyl-*O*-*muco*-inositol	6.30	*Amphibolis antarctica*	leaves	Drew, 1983
Phosphates				
myo-inositol hexaphosphate	0.37[b]	*Lemna* spp.	leaves	Weiner, 1979

[a] 3.10 to 7.40% in 9 other gymnosperms.
[b] equivalent to 0.10% free *myo*-inositol.

organisms, its biosynthesis has been widely studied. Fungi, algae, vascular plants and animals all use a basically similar mechanism involving cyclization of D-glucose. This mechanism, predicted as early as 1900 by Maquenne, is readily acceptable simply because D-glucose has the same configuration as *myo*-inositol on C2 to C5 (see Fig. 6.1).

Formation of [^{14}C]*myo*-inositol photosynthetically from $^{14}CO_2$ was first demonstrated by Richardson & Axelrod (1957) whilst Kindl, Scholda & Hoffmann-Ostenhof (1966) reported as much as 3.5% of fixed ^{14}C in *myo*-inositol in *Chrysanthemum leucanthemum*. Detailed studies of the incorporation of ^{14}C from specifically labelled [^{14}C]glucose substrates, such as those of Loewus & Kelly (1962) using *Petroselinum*, clearly indicated that direct cyclization was involved. As shown in Fig. 6.1*a*, C1 of glucose became C6 of *myo*-inositol, C2 became C5 and C6 became C1, with no indication of chain fragmentation.

Accumulation of *myo*-inositol 1-phosphate during enzyme-mediated cyclization of glucose in preparations of *Candida utilis* (Chen & Charalampous, 1964) suggested that phosphorylated intermediates are involved and the enzyme *myo*-inositol 1-phosphate synthase has recently been purified to homogeneity from *Lemna gibba* by Ogunyemi, Pittner & Hoffmann-Ostenhof

Fig. 6.1 Biosynthesis of cyclitols from D-glucose. (*a*) Stereochemistry of D-glucose cyclization to *myo*-inositol. (*b*) Proposed aldol condensation mechanism for *myo*-inositol biosynthesis from D-glucose. (*c*) Proposed stereochemistry of D-glucose cyclization to *l*-quercitol in *Quercus robur*. (*d*) Proposed biosynthetic route from D-glucose to tetrols in plants.

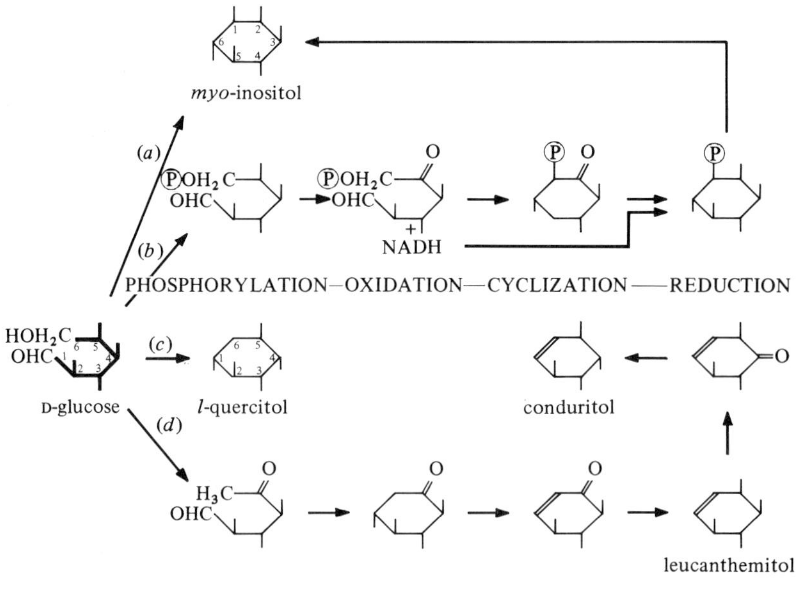

(1978). This enzyme has been tentatively classified as EC 5.5.1.4 by Loewus, Loewus, Maiti & Rosenfield (1978). Such enzymes have elsewhere been called cycloaldolase and appear to be totally NAD^+ dependent, suggesting involvement of a redox step in their action. The aldol condensation route (Fig. 6.1*b*), proposed by Kindl *et al.* (1966), seems to fit the facts reasonably and has been substantiated further by the work of Pittner & Hoffmann-Ostenhof (1978).

Animal and plant *myo*-inositol 1-phosphate synthases appear to be fundamentally different (Ogunyemi *et al.* 1978), the latter being smaller, having 500 times the specific activity of the enzyme from rat testes, and being a metalloprotein probably involving Mn^{2+}.

Apart from a suggestion by Drew (1978) (see later) that the seagrass, *Cymodocea nodosa*, may synthesize other hexols directly during cyclization reactions from D-glucose since the final configuration resulting from C1 and C6 is not predetermined, no direct biosynthesis of other inositols has been suggested.

Pentols

Kindl (1969) reported that neither *myo*-inositol nor any other known cyclitol could act as a precursor for pentols. Only D-glucose was suitable. Despite experimental difficulties including considerable randomization of ^{14}C from specifically labelled [1-^{14}C]glucose or [6-^{14}C]glucose, he suggested that, in *Quercus robur*, *l*-quercitol was synthetized by cyclization of D-glucose in a manner analogous to that of synthesis of *myo*-inositol (Fig. 6.1*c*).

Tetrols

Biosynthesis of leucanthemitol in *Chrysanthemum leucanthemum* probably follows a route similar to that for *l*-quercitol although rearrangements at the hexose level, before cyclization, must occur (Kindl, 1969) as suggested in Fig. 6.1*d*. Again *myo*-inositol is not a suitable precursor. Epimerization of leucanthemitol in *Marsdenia erecta* would then lead to formation of conduritol.

Interconversions

Epimerization

Hexols and their methyl ethers are synthesized by epimerization either of *myo*-inositol to the parent inositol before methylation or of an already methylated inositol. The mechanism probably involves formation of an inosose at the appropriate carbon atom. Such epimerization was first suggested by Hellcu (1957) in humans where feeding of large quantities of *myo*-inositol results in accumulation and excretion of *scyllo*-inositol (scyllitoluria).

The mechanism illustrated in Fig. 6.2*h* has been demonstrated in plants (*Calycanthus occidentalis*) by Scholda, Billek & Hoffmann-Ostenhof (1964) using [^{14}C]*myo*-inositol and [^{14}C]glucose as substrates. *myo*-Inositol was converted to *scyllo*-inositol via *myo*-inosose-2 in which the subsequent orientation at C2 is obviously flexible. Several such epimerizations require both NAD and NADP (Kindl, 1969) which indicates involvement of different dehydrogenases in formation and subsequent reduction of the inososes.

Methylation

Hoffmann-Ostenhof (1969) reported that *myo*-inositol is always a good precursor for *O*-methyl-inositols in plants and that (except for *d*-pinitol) a specific methylation of the parent inositol, itself formed by epimerization if other than *myo*-inositol, occurs. His team isolated and partially purified a pair of enzymes, from different plants, capable of methylating *myo*-inositol to either *l*-bornesitol or *d*-bornesitol (Fig. 6.2*c*, *d*). *S*-adenosyl-methionine, the

Fig. 6.2 Epimerizations and methylations of plant cyclitols derived from *myo*-inositol. Plants studied: (*a*) *Artemisia vulgaris*; (*b*) *Artemisia dracunculus*; (*c*) *Vinca minor*; (*d*) *Pisum sativum*; (*e*) *Trifolium incarnatum*; (*f*) *Juniperus communis*; (*g*) *Chlorella fusca*; and (*h*) *Calycanthus occidentalis*.

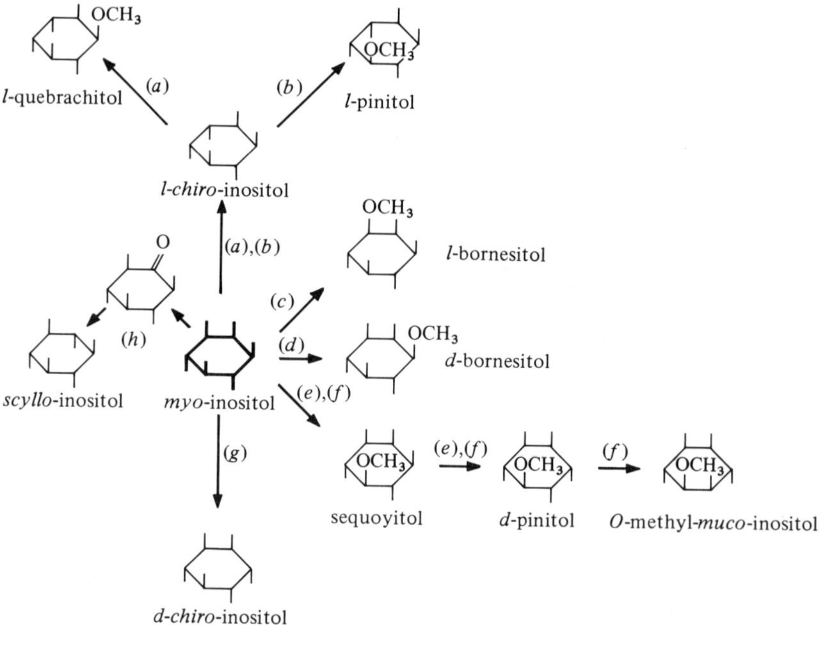

widespread methyl donor, was also functional here and the two enzymes had identical properties except for their stereo-specificity. They are therefore an enantiogenetic pair of enzymes. At least seven methyl transferases of this type, differing only in positional specificity, are required to account for all known mono-methyl inositols.

The mechanism of formation of *C*-methyl-inositols, which are known only from algae, appears to be different. The work of Kindl (1969) and Wober & Hoffmann-Ostenhof (1970) has shown that L-*gluco*-heptulose acts as an intermediate between D-glucose and laminitol in *Chlorella fusca*, with no involvement of *myo*-inositol. L-arabinose may also act as a direct precursor of the heptulose. Epimerization of laminitol then leads to mytilitol formation.

Interconversions in plants

The range of epimerization and methylation reactions involving cyclitols so far identified and studied in plants is shown in Fig. 6.2. Most of the instances noted are from the work of Hoffmann-Ostenhof and his team whilst the formation of methyl-*muco*-inositol in *Juniperus communis* was reported by Dittrich, Gietl & Kandler (1972).

Utilization

Cell-wall biosynthesis

Loewus, Kelly & Neufield (1962) first demonstrated that exogenous [^{14}C]*myo*-inositol was readily incorporated into pectin, hemicellulose and uronic acids. This process has since been shown to be widespread in plants, especially in actively growing tissues such as root tips of *Zea mays* (Roberts, Deshusses & Loewus, 1968), germinating lily pollen (Chen & Loewus, 1977) and germinating wheat (Maiti & Loewus, 1978). It has become clear that oxidative cleavage of *myo*-inositol to glucuronic acid is the primary route of such biosyntheses. However, Loewus *et al.* (1978) suggested that sources other than *de novo* synthesis of *myo*-inositol by cyclization of D-glucose may be involved. Thus, phytic acid hydrolysis, epimerization of other inositols, and release from other esters, ethers, glycosides and phosphoinositides could provide the substrate for the *myo*-inositol oxidation pathway they have traced into plant cell-wall components.

The stereochemistry of oxidative *myo*-inositol cleavage is shown in Fig. 6.3. The quantitative significance of this route for carbon into cell-wall components is not clear. Chen, Loewus & Loewus (1977) showed that inhibition with a *myo*-inositol antagonist, MMO (= 2-*O*,*C*-methylene-*myo*-inositol) seriously impaired growth of pollen tubes of lily, suggesting such a route may be very important therein, as also does the relatively high concentration of *myo*-inositol in pollen – 0.6% dry weight according to Maiti, Rosenfield & Loewus (1978).

However, Verma & Dougall (1979) suggested that the *myo*-inositol oxidation pathway was relatively unimportant in synthesis of pentosyl and galacturonosyl cell-wall components in carrot cell cultures, with the more usual sugar-nucleotide oxidation pathway predominant in these cells.

Thus, whilst various studies, reviewed by Loewus (1974), have indicated that the *myo*-inositol oxidation pathway is functional in plants, no conclusion can yet be drawn concerning its overall contribution to cell-wall synthesis, although Roberts & Cetorelli (1973) concluded that *myo*-inositol oxidation was concerned with primary cell-wall synthesis and that sugar nucleotide oxidation contributed to secondary wall synthesis. The two pathways are illustrated in Fig. 6.3.

Roberts, Shah & Loewus (1967) have suggested another interesting role for *myo*-inositol in certain plants such as *Lemna* and *Petroselinum* which contain high levels of the branched sugar apiose in their insoluble carbohydrates. *myo*-Inositol was a much better precursor for apiose than either D-glucose or, according to Beck & Kandler (1966), glucuronic acid. However, opinion still seems to be that decarboxylation of glucuronic acid occurs prior to internal rearrangements of the resulting pentose sugar. This branched-chain sugar also occurs in significant amounts in a range of seagrasses (van

Fig. 6.3 Alternative pathways for biosynthesis of cell-wall components in plants involving oxidation of *myo*-inositol or sugar nucleotides.

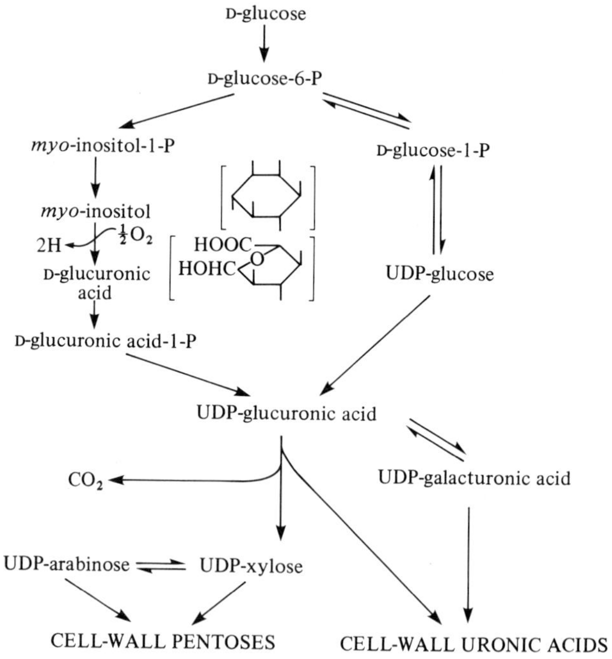

Beusekom, 1967), most of which accumulate significant amounts of *myo*-inositol and other inositols (Drew, 1980).

Phytic acid

myo-Inositol hexaphosphate, generally found as its calcium or magnesium salts, occurs in most plant parts but especially seeds where, in rice grains for instance, it can amount to 80% of total plant phosphate (Asada, Tanaka & Kasai, 1969). The aquatic plants, *Lemna* and *Wolffiella*, contain particularly high concentrations of phytic acid in vegetative tissues (see Table 6.2 and Roberts & Loewus, 1968). Apart from some *myo*-inositol monophosphate, other *myo*-inositol phosphates are generally absent and both Roberts & Loewus (1968) and Asada *et al.* (1969) report incorporation of [^{14}C]*myo*-inositol, [^{3}H]*myo*-inositol and [^{32}P]phosphate into only the mono- and hexaphosphates, even during the very rapid accumulation of phytic acid which occurs during grain maturation.

Kindl & Hoffmann-Ostenhof (1966) reported incorporation of [^{14}C]glucose into phytic acid in *Sinapis*, *Chrysanthemum* and *Phaseolus* in a manner suggesting direct cyclization and subsequent phosphorylation. They proposed the scheme shown in Fig. 6.4. *X* may represent a complex containing *myo*-inositol 1-phosphate, with subsequent, *in situ*, phosphorylations mediated by the phospho-inositol kinase and *myo*-inositol hexaphosphate phosphotransferase systems described by Chakrabarti & Majumber (1978) and Biswas, Biswas, Chakrabarti & De (1978) from mung bean. The activity of the former enzyme in seeds peaked about 20 days after flowering, suggesting direct involvement in accumulation of phytic acid *in vivo* in ripening seeds. Biswas *et al.* (1978) proposed the scheme set out in Fig. 6.5 for biosynthesis and degradation of phytic acid in ripening and germinating mung bean seeds.

Fig. 6.4 Proposed scheme for biosynthesis of phytic acid from D-glucose.

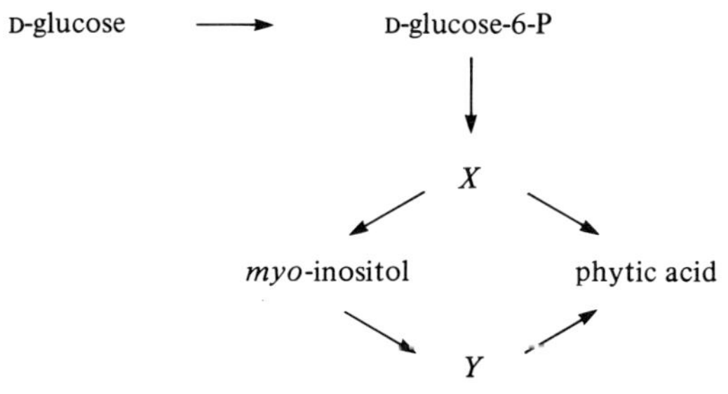

myo-Inositol-IAA esters

An active growth-controlling role has recently been indicated for *myo*-inositol as it is readily enzymatically esterified with indole-3-acetic acid (IAA) in *Zea mays* kernels (Kopcewicz, Ehmann & Bandurski, 1974). Most of the IAA present in plant tissues generally appears to be covalently linked to other compounds (Bandurski & Schulze, 1977), either as amides as in the legumes or as esters, often of *myo*-inositol or glucose, in the monocotyledons. Bandurski & Schulze (1974) found 24 μg kg^{-1} fresh weight free IAA in *Zea mays* plus 328 μg kg^{-1} esterified IAA, much presumably in *myo*-inositol-IAA and the arabinoside and galactoside of that complex found in *Zea mays* by Ueda, Ehmann & Bandurski (1970). Such IAA esters may act as a reserve of IAA, as a sink for excess IAA and as an aid in IAA transport since they are moved faster. In addition, Bandurski & Schulze (1978) have shown that *myo*-inositol-IAA is resistant to the peroxidase from *Zea mays* which rapidly degrades free IAA, thus adding protection to the list of possible roles.

Fig. 6.5 Proposed scheme for biosynthesis and degradation of phytic acid in *Phaseolus* seeds. MI = *myo*-inositol. Enzymes involved: (1) *myo*-inositol 1-P-synthase; (2) phosphoinositol-kinase; (3) phytase; (4) *myo*-inositol hexaphosphate-ADP-phosphotransferase; and (5) proposed new enzyme system.

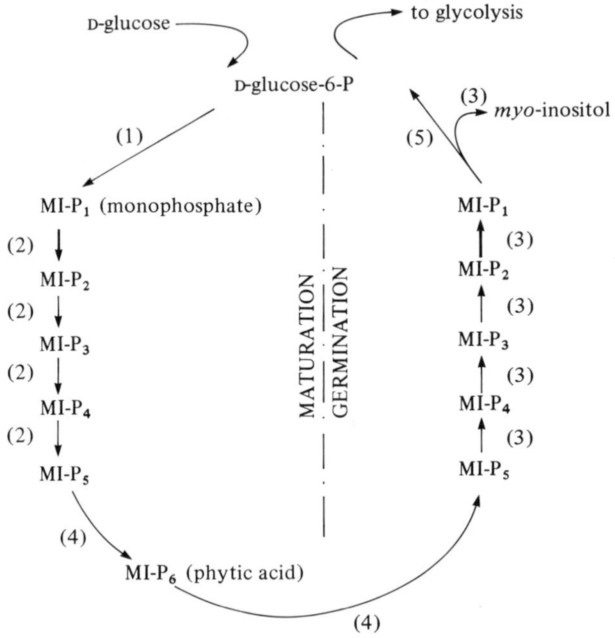

myo-Inositol is therefore implicated in the maintenance of low levels of free IAA together with considerable reserves of protected, but readily activated, complexes; these are features essential for the regulatory role of IAA.

myo-*Inositol glycosides*

Tanner & Kandler (1968) and Tanner (1969) showed clearly that in vascular plants such as *Catalpa* and *Phaseolus*, a *myo*-inositol galactoside, galactinol, could donate a galactosyl unit to raffinose to form stachyose and more complex oligosaccharides. This subject is reviewed in Chapter 5.

Kemp, Loughman & Ephritikhine (1978) have presented evidence for the synthesis in mung bean of *myo*-inositol glucosides from UDP-glucose and a protein containing bound *myo*-inositol. The glucoside apparently then received further glucosyl units for the synthesis of callose glucans (see scheme in Fig. 6.6) whilst still attached to the protein. The polyglucosides were readily hydrolysed off the protein intact and appeared to become insoluble in water when they contained more than eight glucosyl units.

Such examples of free or protein-bound *myo*-inositol glucosides involved in oligo- and polysaccharide syntheses are the only known occurrences of such glycosides not concerned with phosphoinositides. Complex phosphoinositides isolated from peanut contain arabinose and galactose in addition to *myo*-inositol, glycerol, ethanolamine, fatty acids and phosphoric acid, whilst maize seeds and soybeans contain phytosphingosine phosphatides with galactose, arabinose, mannose, a hexuronic acid and glucosamine, some of which were possibly combined in *myo*-inositol glycoside form.

Other cyclitols

Apart from the widespread involvement of *myo*-inositol in cell-wall biosynthesis, phytic acid accumulation and in specific examples of oligosaccharide biosynthesis, few instances of cyclitol involvement in plant biochemistry, other than their biosynthesis, have been reported.

Fig. 6.6 Biosynthesis of callose glucans. MI, *myo*-inositol; G, glucose.

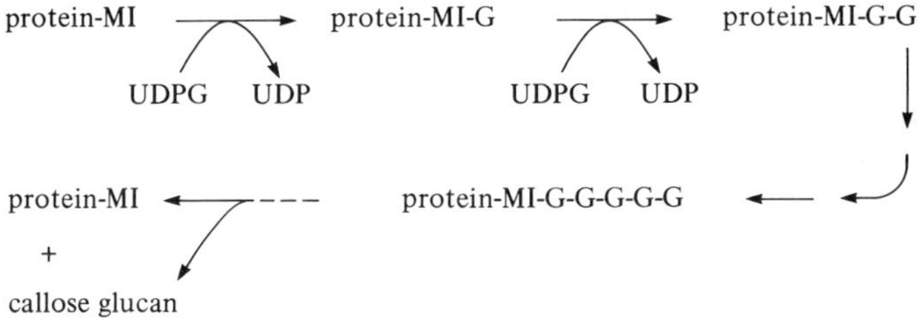

Table 6.3(*a*) *Cyclitol content and distribution within zannichelliacean seagrasses*

		% Extracted dry weight			
	Organ	*myo*-Inositol	*l-chiro*-Inositol	*muco*-Inositol	*O*-methyl-*muco*-inositol
Amphibolis antarctica	leaves	0.1	0.5	0.4	1.7
Cymodocea nodosa	leaves	0.8	5.7	1.4	–
	rhizomes	t	1.2	2.2	–
	roots	t	2.8	2.4	–
C. rotundata	leaves	0.5	10.6	–	–
	rhizomes	t	2.6	–	–
	roots	0.1	2.3	–	–
C. serrulata	leaves	t	2.0	–	–
	rhizomes	t	2.8	–	–
Halodule wrightii	leaves	1.0	–	–	–
H. uninervis	leaves	2.5	–	–	–
	rhizomes and roots	0.3	–	–	–
Syringodium isoetifolium	leaves	0.8	6.3	P	–
	rhizomes and roots	t	0.7	–	–
Thalassodendron ciliatum	leaves	0.6	5.3	6.1	–
	stems	t	1.8	1.2	–
	rhizomes	t	1.0	1.5	–
	roots	t	0.4	0.8	–

t = < 0.1% extracted dry weight.
P = present, > 0.1% but not quantified accurately.

Craigie (1969) reported a tetrol (*d*-1,4/2,5 cyclohexanetetrol) involved as a major osmoregulatory compound in the unicellular alga, *Monochrysis lutheri*. Photosynthetic production of the tetrol in response to increased salinity doubled the cells' content of this cyclitol to 7% dry weight within 4 h, whilst decreased salinity caused appropriate loss to the medium within a few minutes. However, such a dramatic but rare occurrence has little relevance to carbohydrate metabolism in vascular plants.

Drew (1978, 1980, 1983) has shown that probably the highest concentrations of cyclitol so far reported in vascular plants occur in a particular group of seagrasses.* Available information about biosynthesis and utilization of cyclitols in these plants is summarized below.

Seagrass cyclitols

Accumulation

The twelve genera of seagrasses recognized by den Hartog (1970) are widely distributed in tropical, temperate and arctic seas. They are monocotyledonous plants able to survive in the totally marine environments of the intertidal and sublittoral zones of the oceans. Drew (1980) analysed the

Table 6.3(*b*) myo-*Inositol content in other seagrasses*

	% Extracted dry weight		
Genus	leaves	rhizomes	roots
Hydrocharitaceae			
Enhalus acroides	t	t	t
Halophila stipulacea	−t	−t	NA
Thalassia hemprichii	0.8	0.5	0.2
Posidoniaceae			
Posidonia australis	0.0	NA	NA
P. oceanica	t	NA	NA
Zosteraceae			
Heterozostera tasmanica	0.2	NA	NA
Phyllospadix torreyi	2.5	0.6[a]	[a]
Zostera marina	2.9	3.0	3.8
Z. noltii	1.8	3.6	1.8

NA = not analysed; t = < 0.1%; −t = minute trace.
[a] rhizomes and roots analysed together.

* *Footnote added in proof:* Popp (1984) reported the occurrence of relatively large amounts of *chiro*-inositol (up to 1.2% dry wt), pinitol (1.3%) and quebrachitol (3.5%) in leaves of a range of mangrove trees. She speculated that these compounds, plus the acyclic polyol mannitol (up to 8.0% dry wt), may have an osmoregulatory role in these plants which are able to cope with the osmotic stress of living in seawater, a situation they share with the seagrasses.

soluble carbohydrates of ten seagrass genera from various parts of the world. He found significant amounts of fructose, glucose, sucrose and *myo*-inositol in them all whilst, in one group, he found other cyclitols. These seagrasses, classed in the Zannichelliaceae by Hutchinson (1959), contained considerable amounts of *l-chiro*-inositol, *muco*-inositol and, in one genus, an *O*-methyl-*muco*-inositol (Drew, 1978). None of these three unusual cyclitols had been

Fig. 6.7 GLC analyses of TMS-ethers of soluble carbohydrates in zannichelliacean seagrasses, using methods described by Drew (1978). Species: (*a*) *Amphibolis antarctica*; (*b*) as (*a*) but extract boiled under reflux with hydriodic acid, destroying sugars and demethylating 1a to 3 with little change in 4 and 5; (*c*) *Cymodocea nodosa*; (*d*) *Syringodium isoetifolium*; (*e*) *Thalassodendron ciliatum* (initial peak of 2 masked by large peak of 3); and (*f*) *Halodule uninervis*. Compounds: 1, fructose; 1a, *O*-methyl-*muco*-inositol; 2, α- and β-glucose; 3, *muco*-inositol; 4, *l*-chiro-inositol; 5, *myo*-inositol; and 6, sucrose.

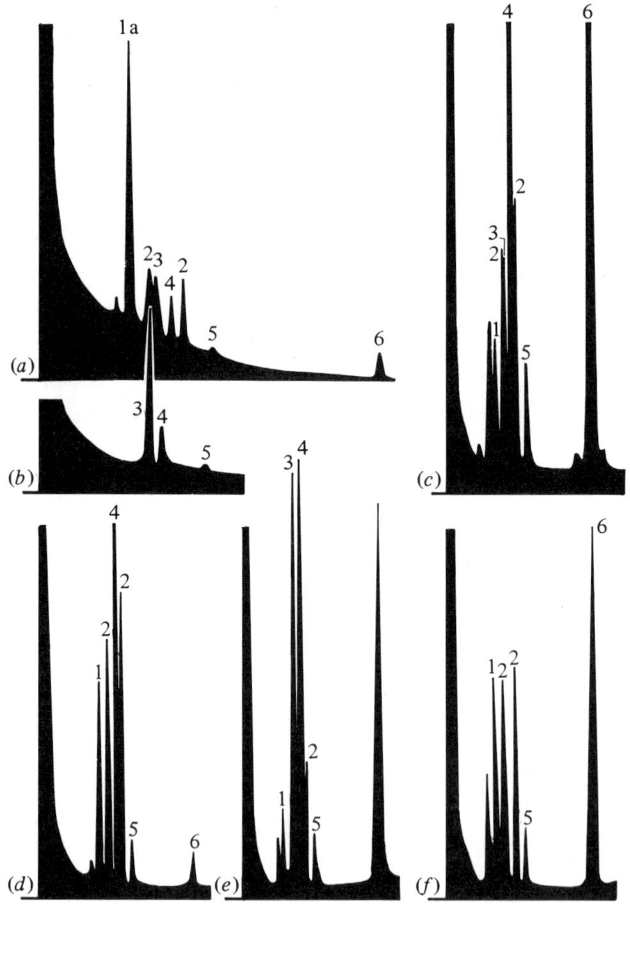

reported from monocotyledonous plants before and they could not be detected even as traces in the other seagrass groups. Quantitative data are set out in Table 6.3 and representative GLC scans in Fig. 6.7.

Biosynthesis

The much reduced concentrations of cyclitols found in rhizomes and roots of these seagrasses as compared with leaves suggest that they may be direct products of photosynthesis but that they are neither readily translocated nor synthesized from appropriate substrates elsewhere in the plant. Drew (1978) demonstrated the incorporation of ^{14}C from $NaH^{14}CO_3$ into *myo*-inositol and *l-chiro*-inositol in *Cymodocea nodosa* leaves (Table 6.5) although in small amounts compared with incorporation into sucrose, the major initial photosynthetic product in that plant.

In *C. nodosa* leaves, both *myo*-inositol and *l-chiro*-inositol also accumulated ^{14}C from exogenous [^{14}C]glucose at similar rates, but exogenous supply of [^{14}C]*myo*-inositol led only to [^{14}C]*myo*-inositol in the soluble fraction and massive incorporation into insoluble materials (Drew, 1978). Data are included in Table 6.5. During radiochromatography, *muco*-inositol was not separated from sucrose so that any accumulation of ^{14}C from [^{14}C]glucose into the former compound was masked by the large amounts in sucrose. However, these observations do suggest that *myo*-inositol was not the direct precursor of other cyclitols in *C. nodosa* but rather that cyclization of glucose produced similar amounts of both *myo*- and *l-chiro*-inositol and possibly some *muco*-inositol since this was also present in the leaves. As the products of cyclization depend only on the orientation of the newly specified hydroxyl groups from C1 and C6, there are 4 possible products, as shown in Fig. 6.8, although one possibility, *epi*-inositol, is unknown in living organisms and its formation may be prevented by the steric hindrance of having all but one hydroxyl in the *cis* orientation.

Thus, cyclization of glucose by a less-specific cyclo-aldolase than that present in other plants, which produces only *myo*-inositol, may occur in

Fig. 6.8 Possible cyclitol products of glucose cyclization. Compounds: (*a*) D-glucose; (*b*) *epi*-inositol; (*c*) *l-chiro*-inositol; (*d*) *muco*-inositol; and (*e*) *myo*-inositol.

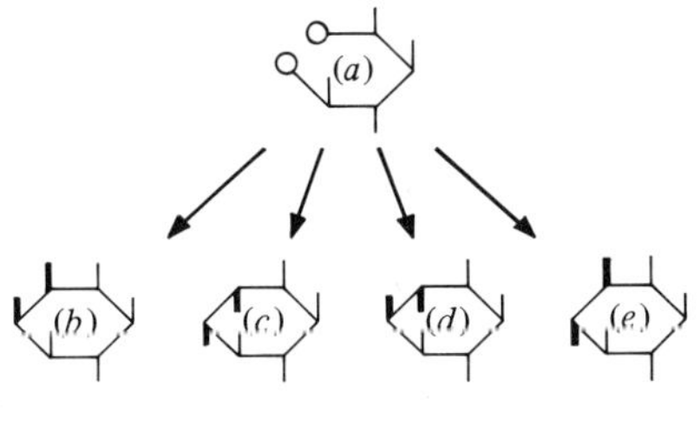

C. nodosa and other related seagrasses. The occurrence of approximately equal amounts of *muco*- and *l-chiro*-inositol in *Thalassodendron ciliatum* (Table 6.3*a*, Fig. 6.7*e*) lends support to this hypothesis, and it may be that, in both these seagrasses, concentrations of *myo*-inositol are kept low by that compound's involvement in cell-wall biosynthesis despite its production at a rate similar to the other cyclitols. In *C. nodosa* leaves, *l-chiro*-inositol biosynthesis could only be detected in summer. In spring, before the plant had begun rapid growth, the only cyclitol produced was *myo*-inositol. This suggests that the less specific cyclo-aldolase may function only when large amounts of *myo*-inositol are required for cell-wall biosynthesis.

Utilization

Amounts of cyclitol remained unchanged in leaves of *C. nodosa* and *Th. ciliatum* during several days of dark starvation, even after sucrose and other soluble carbohydrates were exhausted (Fig. 6.9). The fact that the cyclitols were not readily utilized even under metabolic stress may explain how *l-chiro*-inositol can accumulate in considerable amounts in *C. nodosa* despite incorporation of only about 1% of photosynthetically fixed carbon into that compound. The unusual cyclitols accumulated in these seagrass leaves may

Fig. 6.9 Utilization of soluble carbohydrates in leaves of seagrasses during dark starvation. (*a*) *Cymodocea nodosa*; (*b*) *Thalassodendron ciliatum*. Compound identifications as in Fig. 6.7. Values expressed as the percentage of the dry material remaining after extraction with 80% ethanol.

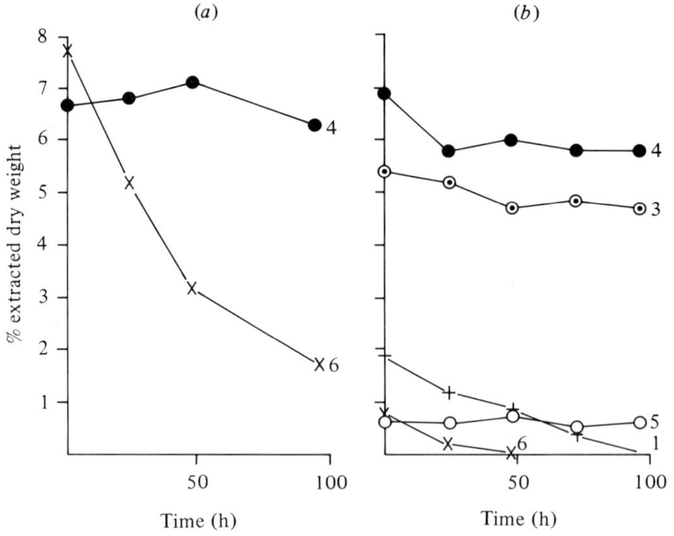

thus be merely metabolic byproducts of a non-specific cyclization mechanism, whilst the leaves may lack effective epimerization mechanisms to convert them to *myo*-inositol for which catabolic pathways appear widespread in plants.

Phylogeny

Within the zannichelliacean seagrasses, a series can be identified according to the complexity of cyclitol content. This ranges from *Amphibolis* with four cyclitols to *Halodule* with only *myo*-inositol. The latter condition is also found in all the other seagrasses studied and only traces of *myo*-inositol occur in some genera. This series is illustrated in Table 6.4.

Whether the condition in *Amphibolis* or *Halodule* is more primitive within the Zannichelliaceae, and hence in seagrasses in general, is difficult to assess. On the basis of morphological, anatomical, geographical and fossil criteria, den Hartog (1970) postulated that *Amphibolis* and *Thalassodendron*, with their stiff erect stems and sympodial rhizomes, were more primitive and closest to the salt-tolerant shrub he envisaged as ancestral to the seagrasses. This viewpoint is diametrically opposed to that of Arber (1920) who

Table 6.4 *Cyclitol distribution in seagrass taxa*

	Cyclitol content[a]			
Genus	*O*-Methyl-*muco*-inositol	*muco*-Inositol	*l*-*chiro*-Inositol	*myo*-Inositol
Zannichelliaceae				
Amphibolis	++++	+	+	+
Thalassodendron	–	++++	++++	+
Cymodocea	–	+	++++	+
Syringodium	–	+	++++	+
Halodule	–	–	–	++++
Zosteraceae				
Phyllospadix	–	–	–	++++
Zostera	–	–	–	++++
Heterozostera	–	–	–	+
Hydrocharitaceae				
Thalassia	–	–	–	t
Enhalus	–	–	–	t
Halophila	–	–	–	t
Posidoniaceae				
Posidonia	–	–	–	t

[a] ++++, > 1% extracted dry weight (major component); +, 0.1–0.99% extracted dry weight (minor component); t, < 0.1% extracted dry weight (trace component).

postulated a transition from terrestrial to marine via fresh and brackish waters.

The novel mechanism for biosynthesis of cyclitols found in most zannichelliacean seagrasses, the absence of most genera possessing this mechanism from the Caribbean which was not isolated from other tropical oceans until the Eocene, the restricted distribution of *Thalassodendron* and the very restricted distribution of *Amphibolis* (southern Australia only), as illustrated by den Hartog (1970), all argue for a recent evolution of the multiple cyclitol condition by acquisition of a second but less specific cyclase. This would supplement the specific enzyme forming *myo*-inositol in all other plants and animals.

Conclusions

Despite their $[CH_2O]_n$ formulation, cyclitols and their methylated derivatives are seldom considered carbohydrates in the plant physiological sense. Nevertheless, they occur in considerable amounts in a number of plants and at least *myo*-inositol (the most studied) is actively involved in a number of important biochemical and physiological processes. There is no evidence to suggest similar involvement of other cyclitols and, even in the zannichelliacean seagrasses, wherein exotic cyclitols appear to accumulate as byproducts from a novel non-specific glucose cyclization enzyme system, there seems to be little further metabolic involvement and the plants apparently require only the additional *myo*-inositol formed at the same time.

myo-Inositol itself does not accumulate to even 1% dry weight in most plants, and, when storage of phytic acid occurs, this implies high levels of phosphate relative to normal but still low *myo*-inositol content (see Table 6.2). Furthermore, large amounts of *myo*-inositol are not involved in oligo- and polysaccharide synthesis even if the products do, as with cell-wall pectins etc., accumulate in considerable amounts.

Table 6.5 *Distribution of* ^{14}C *within the soluble carbohydrates of* Cymodocea nodosa *leaves after assimilation of various* ^{14}C*-labelled substrates*

		% soluble carbohydrate ^{14}C in				
Substrate	Conditions	*myo*-Inositol	*l-chiro*-Inositol	Sucrose	Glucose	Fructose
$NaH^{14}CO_3$	Light	1.2	1.1	89.3	6.3	2.1
$[^{14}C]$glucose	Dark	13.4	23.2	63.4	–	–
$[^{14}C]$*myo*-inositol	Dark	100.0	–	–	–	–

Thus, cyclitols do not seem to act as usable stores of carbon within vascular plants but mainly as intermediates in important pathways or as useless accumulations of byproducts from other important processes. The reason for complex interconversions of cyclitols and their methyl ethers in some plants remains obscure although the speculations of Loewus (1965) concerning their subsequent oxidation to specifically methylated glucuronic acids found in cell-wall polysaccharides require further investigation. Modern analytical methods allow easy separation, identification and quantitation of the inositols and their derivatives. They will surely, therefore, be found in increasing numbers of plant biochemical and physiological studies and researchers should always bear in mind their possible involvement in carbohydrate interconversions.

References

Arber, A. (1920). *Water Plants, a Study of Aquatic Angiosperms*, pp. 123–5. Cambridge: Cambridge University Press.

Asada, K., Tanaka, K. & Kasai, Z. (1969). Formation of phytic acid in cereal grains. *Annals of the New York Academy of Sciences*, **165**, 801–14.

Bandurski, R. S. & Schulze, A. (1974). Concentrations of indole-3-acetic acid and its esters in *Avena* and *Zea*. *Plant Physiology*, (*Bethesda*), **54**, 257–62.

Bandurski, R. S. & Schulze, A. (1977). The concentrations of indole-3-acetic acid and its derivatives in plants. *Plant Physiology* (*Bethesda*), **60**, 211–13.

Bandurski, R. S. & Schulze, A. (1978). The bound auxins: protection of indole-3-acetic from peroxidase-catalysed oxidation. *Planta*, **139**, 203–8.

Beck, E. & Kandler, O. (1966). Isotopenstudien zur Biosynthese von Apiose in *Lemna*. *Zeitschrift für Pflanzenphysiologie*, **55**, 71–84.

Beusekom, C. F. van (1967). Ueber einige Apiose-vorkommnisse den *Helobiae*. *Phytochemistry*, **6**, 573–6.

Biswas, B. B., Biswas, S., Chakrabarti, S. & De, B. P. (1978). A novel metabolic cycle involving *myo*-inositol phosphates during formation and germination of seeds. In *Cyclitols and Phosphoinositides*, ed. W. W. Wells & F. Eisenberg, pp. 57–68. New York: Academic Press.

Chakrabarti, S. & Majumber, A. L. (1978). Phosphoinositol kinase from plant and avian sources. In *Cyclitols and Phosphoinositides*, ed. W. W. Wells & F. Eisenberg, pp. 69–81. New York: Academic Press.

Chen, I. W. & Charalampous, F. C. (1964). Mode of conversion of glucose-6-P to inositol, and the role of DPN (diphosphopyridine nucleotide) and NH_4^+ (ammonia) ions. *Biochemical and Biophysical Research Communications*, **17**, 521–6.

Chen, M. & Loewus, F. A. (1977). *Myo*-inositol metabolism in *Lilium longiflorum* pollen. Uptake of *myo*-inositol-2-^{3}H. *Plant Physiology*, **59**, 653–7.

Chen, M., Loewus, M. W. & Loewus, F. A. (1977). Effect of a *myo*-inositol antagonist, 2-*O*-*C*-methylene-*myo*-inositol, on the metabolism of *myo*-inositol-2-^{3}H and D-glucose-1-^{14}C in *Lilium longiflorum* pollen. *Plant Physiology*, **59**, 658–63.

Cosgrove, D. J. (1980). *Inositol Phosphates*. Amsterdam: Elsevier.

Craigie, J. S. (1969). Some salinity-induced changes in growth, pigments and cyclohexanetetrol in *Monochrysis lutheri*. *Journal of the Fisheries Research Board of Canada*, **26**, 2959–67.

Craigie, J. S. (1974). Cyclitols. In *Algal Physiology and Biochemistry*, ed. W. P. D. Stewart, p. 223. London: Blackwell Scientific Publishers.

Dittrich, P., Gietl, M. & Kandler, O. (1972). D-1-*O*-methyl-*muco*-inositol in higher plants. *Phytochemistry*, **11**, 245–50.

Drew, E. A. (1978). Carbohydrate and inositol metabolism in the seagrass, *Cymodocea nodosa*. *New Phytologist*, **81**, 249–64.

Drew, E. A. (1980). Soluble carbohydrate composition of seagrass. In *Handbook of Seagrass Biology – an Ecosystem Perspective*, ed. R. Phillips & C. P. McRoy, pp. 247–59. New York: Garland STPM Press.

Drew, E. A. (1983). Sugars, cyclitols and seagrass phylogeny. *Aquatic Botany*, **15**, 387–408.

Hartog, C. den (1970). *The Seagrasses of the World*, pp. 14–18, Amsterdam: North-Holland Publishing Co.

Helleu, C. (1957). Étude de l'élimination urinaire du méso-inositol en fonction du régime alimentaire. *Bulletin de la Société de Chimie Biologique*, **39**, 633–40.

Hoffmann-Ostenhof, O. (1969). Enzymes involved in the *O*-methylation of inositols in higher plants: enzymic degradation of cyclitols in higher plants, yeasts and bacteria. *Annals of the New York Academy of Sciences*, **165**, 624–9.

Hutchinson, J. (1959). *The Families of Flowering Plants*, vol. II. *Monocotyledons*, 2nd edn. Oxford: Clarendon Press.

Kemp, J., Loughman, B. C. & Ephritikhine, G. (1978). The role of inositol glucosides in the synthesis of β1,3-glucans. In *Cyclitols and Phosphoinositides*, ed. W. W. Wells & F. Eisenberg, pp. 439–49. New York: Academic Press.

Kindl, H. (1969). Biosynthesis of epimers of *myo*-inositol, cyclohexanepentols, cyclohexenetetrols, and *C*-methyl inositols. *Annals of the New York Academy of Sciences*, **165**, 615–23.

Kindl, H. & Hoffmann-Ostenhof, O. (1966). Untersuchungen über die Biosynthese der Cyclite, XI. Der Ursprung des meso-Inosit-Anteils der Phythinsaure in höhren Pflanzen. *Biochemische Zeitschrift*, **345**, 454–60.

Kindl, H., Scholda, R. & Hoffmann-Ostenhof, O. (1966). The biosynthesis of cyclitols. *Angewandte Chemie, International Edition in English*, **5**, 165–73.

Kopcewicz, J., Ehmann, A. & Bandurski, R. S. (1974). Enzymatic esterification of indole-3-acetic acid to *myo*-inositol and glucose. *Plant Physiology*, **54**, 846–57.

Loewus, F. A. (1965). Inositol metabolism and cell-wall formation in plants. *Federation Proceedings*, **24**, 855–62.

Loewus, F. A. (1974). The biochemistry of *myo*-inositol in plants. *Recent Advances in Phytochemistry*, **8**, 179–207.

Loewus, F. A. & Kelly, S. (1962). The metabolism of *myo*-inositol in plants. *Federation Proceedings*, **21**, 88.
Loewus, F. A., Kelly, S. & Neufield, E. F. (1962). Metabolism of *myo*-inositol in plants: conversion to pectin, hemicellulose, *d*-xylose, and sugar acids. *Proceedings of the National Academy of Sciences of the USA*, **48**, 421–5.
Loewus, F. A., Loewus, M. W., Maiti, I. B. & Rosenfield, C-L. (1978). Aspects of *myo*-inositol metabolism and biosynthesis in higher plants. In *Cyclitols and Phosphoinositides*, ed. W. W. Wells & F. Eisenberg, pp. 249–67. New York: Academic Press.
Maiti, I. B. & Loewus, F. A. (1978). *Myo*-inositol metabolism in germinating wheat. *Planta*, **142**, 55–60.
Maiti, I. B., Rosenfield, C-L. & Loewus, F. A. (1978). *Myo*-inositol content of lily pollen. *Phytochemistry*, **17**, 1185–6.
Maquenne, L. (1900). *Les sucres et leurs principaux dérivés*, pp. 189. Paris: Ganthier-Villars.
Ogunyemi, E. O., Pittner, F. & Hoffmann-Ostenhof, O. (1978). Studies on the biosynthesis of cyclitols, XXXVI. Purification of *myo*-inositol-1-phosphate synthase of the duckweed *Lemna gibba* to homogeneity by affinity chromatography on NAD-sepharose. Molecular and catalytic properties of the enzyme. *Hoppe-Seyler's Zeitschrift für physiologische Chemie*, **359,** 613–16.
Pittner, F. & Hoffmann-Ostenhof, O. (1978). Studies on the biosynthesis of cyclitols, XXXVII. On the mechanism and function of Schiff's base formation as an intermediary reaction step of *myo*-inositol-1-phosphate synthase from rat testicles. *Hoppe-Seyler's Zeitschrift für physiologische Chemie*, **359**, 1395–1400.
Plouvier, V. (1955*a*). Sur la présence de québrachitol chez les Eleagnacées. Sa recherche dans quelques autres Myrtiflorae. *Comptes Rendus Hebdomadaires des Séances de l'Académie des Sciences, Paris*, **232**, 1239–41.
Plouvier, V. (1955*b*). Sur la recherche du *d*-quercitol chez quelques Fagacées et autres Plantes. *Comptes Rendus Hebdomadaires des Séances de l'Académie des Sciences, Paris*, **240**, 113–15.
Plouvier, V. (1955*c*). Sur un cyclitol nouveau, le liriodendritol, isolé des *Liriodendron* (Magnoliacées). *Comptes Rendus Hebdomadaires des Séances de l'Académie des Sciences, Paris*, **241**, 765–7.
Plouvier, V. (1955*d*). Sur deux cyclitols nouveaux – le *l*-bornesitol isolé de *Lathyrus vernus* Borh., le *d*-ononitol isolé d'*Ononis natrix* L. (Légumineuses). *Comptes Rendus Hebdomadaires des Séances de l'Académie des Sciences, Paris*, **241**, 983–5.
Plouvier, V. (1956*a*). Sur la recherche de quelque cyclitols: viburnitol, scyllitol, pinitol. *Comptes Rendus Hebdomadaires des Séances de l'Académie des Sciences, Paris*, **242**, 2389–92.
Plouvier, V. (1956*b*). Sur un cyclitol nouveau, le *l*-pinitol – isolé d'*Artémisia dracunculus* L. (Composées). *Comptes Rendus Hebdomadaires des Séances de l'Académie des Sciences, Paris*, **243,** 1913–15.
Plouvier, V. (1960). Nouvelles recherches de cyclitols dans quelques groupes botaniques: signification phylogénique du séquoyitol. *Comptes Rendus Hebdomadaires des Séances de l'Académie des Sciences, Paris*, **251**, 131–3.
Plouvier, V. (1962*a*). Sur un cyclitol nouveau, le leucanthémitol isolé de la

Grande Marguérite, *Chysanthemum leucanthemum* L. Sa recherche dans quelques autres Composées-Anthemidées. *Comptes Rendus Hebdomadaires des Séances de l'Académie des Sciences, Paris*, **255**, 360–2.
Plouvier, V. (1962*b*). Nouvelles recherches de cyclitols dans quelques groupes botaniques: le L-inositol des Composées, le D-pinitol des Légumineuses. *Comptes Rendus Hebdomadaires des Séances de l'Académie des Sciences, Paris*, **255**, 1770–2.
Plouvier, V. (1964). Recherche des L-inositol, *l*-quebrachitol, et D-pinitol dans quelques groupes botaniques. Présence de l'acide shikimique dans le *Mammea americana* L. (Guttifères). *Comptes Rendus Hebdomadaires des Séances de l'Académie des Sciences, Paris*, **258**, 2921–4.
Plouvier, V. (1971). Nouvella recherche du scyllitol dans quelques groupes botaniques. *Comptes Rendus Hebdomadaires des Séances de l'Académie des Sciences, Série D, Sciences Naturelles, Paris*, **273**, 1625–8.
Plouvier, V. (1973). Recherche de cyclitols dans quelques groupes botaniques: Cryptogames, Polygonacées, Apocynacées. *Comptes Rendus Hebdomadaires des Séances de l'Académie des Sciences, Série D, Sciences Naturelles, Paris*, **277**, 1945–8.
Popp, M. (1984). Chemical composition of Australian mangroves. II. Low molecular weight carbohydrates. *Zeitschrift für Pflanzenphysiologie*, **113**, 411–21.
Richardson, K. E. & Axelrod, B. (1957). Changes in the inositol content during germination and growth of some higher plants. *Plant Physiology*, **32**, 334–7.
Riggs, N. V. & Strong, F. M. (1967). Analysis of cyclitols in Cycad plant tissues. *Analytical Biochemistry*, **19**, 351–6.
Roberts, R. M. & Cetorelli, J. J. (1973). UDP-D-glucuronic acid pyrophosphorylase and the formation of UDP-D-glucuronic acid in plants. In *Biogenesis of Plant Cell Wall Polysaccharides*, ed. F. A. Loewus, pp. 49–68. New York: Academic Press.
Roberts, R. M., Deshusses, J. & Loewus, F. A. (1968). Inositol metabolism in plants. V. Conversion of *myo*-inositol to uronic acid and pentose units of acidic polysaccharides in root-tips of *Zea mays*. *Plant Physiology (Bethesda)*, **43**, 979–89.
Roberts, R. M. & Loewus, F. A. (1968). Inositol metabolism in plants. VI. Conversion of *myo*-inositol to phytic acid in *Wolffiella floridana*. *Plant Physiology (Bethesda)*, **43**, 1710–16.
Roberts, R. M., Shah, R. H. & Loewus, F. A. (1967). Inositol metabolism in plants. IV. Biosynthesis of apiose in *Lemna* and *Petroselenium*. *Plant Physiology (Bethesda)*, **42**, 659–66.
Scholda, R., Billek, G. & Hoffmann-Ostenhof, O. (1964). Untersuchungen über die Biosynthese der Cyclite, VI. Die Bildung von Scyllit in *Calycanthus occidentalis*. *Hoppe-Seyler's Zeitschrift für physiologische Chemie*, **339**, 28–32.
Stinson, E. E., Dooley, C. J., Purcell, J. M. & Ard, J. A. (1967). Quebrachitol – a new component of maple sap and sirup. *Journal of Agricultural and Food Chemistry*, **15**, 394–7.

Tanner, W. (1969). The function of *myo*-inositol glycosides in yeasts and higher plants. *Annals of the New York Academy of Sciences*, **165**, 726–42.
Tanner, W. & Kandler, O. (1968). *Myo*-inositol, a cofactor in the biosynthesis of stachyose. *European Journal of Biochemistry*, **4**, 233–9.
Ueda, M., Ehmann, A. & Bandurski, R. S. (1970). Gas chromatographic analysis of indole-3-acetic acid myoinositol esters in maize kernels. *Plant Physiology (Bethesda)*, **46**, 715–19.
Verma, D. C. & Dougall, D. K. (1979). Biosynthesis of *myo*-inositol and its role as a precursor of cell-wall polysaccharides in suspension cultures of wild-carrot cells. *Planta*, **146**, 55–62.
Weimer, W. C. (1979). Naturally occurring organic phosphorus compounds in aquatic plants. *Environmental Science and Technology*, **13**, 826–9.
Wober, G. & Hoffmann-Ostenhof, O. (1970). Untersuchungen über die Biosynthese der Cyclite, 26 Mitt.: Bildung von Mytilit (*C*-methyl-*l*-*scyllo*-inosit) in *Chlorella fusca. Monatschefte feur Chemie*, **101**, 1861–3.

D. H. LEWIS

7 Physiology and metabolism of alditols

Introduction

General textbooks on the biochemistry or chemotaxonomy of vascular plants often only pay lip-service to alditols (sugar alcohols, acyclic polyols) and usually couple what brief discussion there is with that of cyclitols (inositols). However, these two classes of carbohydrate differ both structurally (see Chapter 1) and biosynthetically (compare below with Chapter 6). Nevertheless, as elaborated below, they may sometimes function physiologically in a similar manner. Figure 7.1 shows how D-glucitol (sorbitol) and *myo*-inositol are linked chemically via D-glucose. Galactitol is similarly structurally related to D-*chiro*-inositol and D-mannitol to *neo*-inositol but there is no suggestion that these acyclic and cyclic polyols are metabolically

Fig. 7.1 Chemical relationships between *myo*-inositol, D-glucose and D-glucitol. In solution, the cyclic and acyclic forms of glucose are in equilibrium with the former predominating. Conversion of the former to *myo*-inositol is an internal oxido-reduction (see Chapter 6); conversion of the latter to glucitol is a reduction (see below in this Chapter).

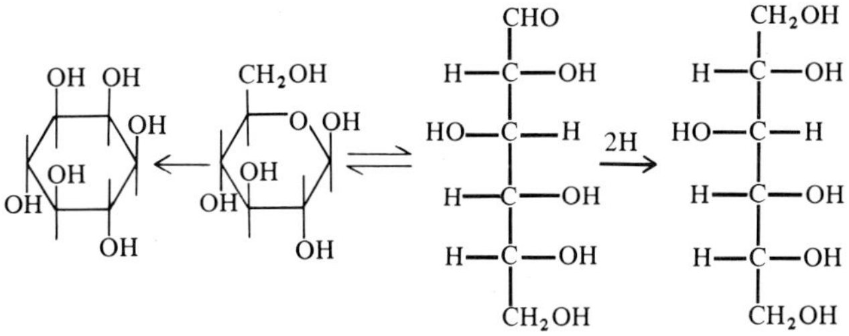

CYCLIC POLYOL (*myo*-INOSITOL)

CYCLIC ALDOSE (D-GLUCOPYRANOSE)

ACYCLIC ALDOSE (D-GLUCOSE)

ACYCLIC POLYOL (D-GLUCITOL) (= SORBITOL)

related via D-galactose and D-mannose respectively. Since no discussion in this chapter concerns L-sugars or alditols, the prefix, D-, can be assumed except, of course, when optically inactive alditols such as galactitol and allitol are being considered.

Table 1.8 of Chapter 1 listed 17 alditols that occur naturally in plants. Of these, only a few, notably mannitol, glucitol, galactitol, allitol and hamamelitol, all of which have six carbon atoms, have been studied in any detail from physiological or biochemical points of view. Virtually nothing is known about the metabolism of the other 6- and the 4-, 5- and 7- or more carbon alditols, which occur only in a few, taxonomically disjunct, genera (see Chapter 1). The discussion here will therefore, of necessity, be restricted to a consideration of the roles played by those 6-C alditols listed above.

Despite the calculation of Bieleski (1982) that about 30% of annual global productivity involves alditols, a statistic which surely indicates their substantial quantitative significance in global carbon budgets, there is only a handful of laboratories which have been concerned with their metabolism in vascular plants. Progress between the time of the appearance of the review of this topic by Lewis and Smith (1967) and of that by Bieleski (1982) has not been spectacular and much remains to be learnt. Part of the problem has been the inability consistently to extract active enzymes of their metabolism from vascular plants. However, this bottle-neck to progress is showing signs of widening and significant advances can be expected in the near future. The topic was also reviewed by Stacey (1974).

In their survey of the roles of alditols in algae, fungi, lichens and vascular plants, Lewis & Smith (1967) discussed their biological functions under six headings: occurrence and quantitative variation; carbohydrate storage; translocation; osmo-regulation; enzymology and control of synthesis; co-enzyme regulation and storage of reducing power. Although these headings could still form a framework within which to assess advances made since 1967, there is substantial overlap between them and the following, hopefully more logical, arrangement has been adopted here: (1) storage of carbon – quantitative variation, synthesis, translocation, degradation; (2) storage of reducing power; (3) osmo-regulation. The first two headings concern essentially biochemical aspects of the physiology of alditols; the third is essentially a biophysical phenomenon but with interesting, unsolved biochemical undertones. Conversely, the understanding of translocation of alditols, included under the first heading, must involve biophysical considerations.

Storage of carbon

Seasonal and diurnal/nocturnal quantitative variation

In Chapter 6, Drew arbitrarily defined a storage carbohydrate as one which accumulates to more than 1% of the dry weight of the tissue in which it resides. On this criterion, the hexitols easily qualify as storage carbohydrates since amounts considerably in excess of 1% are frequently encountered, especially in fruits and leaves (see Wallaart, 1980, for glucitol in Rosaceae). However, as well as this quantitative criterion, implicit in the concept of a storage compound is the property of being available for re-utilization at a time removed from that of deposition.

Seasonal variations of concentration in leaves or bark have been recorded for many alditols e.g. mannitol in olive (Nuccorini, 1930) and *Gardenia* (Asai, 1937); glucitol in apple (Whetter & Taper, 1963; Chong, 1971; Hansen & Grausland, 1973); hamamelitol in *Primula clusiana* (Sellmair & Kandler, 1970); and allitol in *Itea* (Kull & Baitinger-Haardt, 1977). In many cases, highest concentrations occur in winter but precise patterns depend on environmental conditions (Chong & Taper, 1971). Discussion of these winter maxima is deferred to the section on osmo-regulation below. For the moment, it is sufficient to note that there is abundant evidence for a seasonal separation of synthesis (or accumulation of translocated alditol) and re-utilization (localized degradation or export of alditol). A catalogue of the details of the quantitative variations which occur in vegetative tissues is beyond the terms of reference of this review but such details may be obtained from the papers quoted above and other references therein (see also Bieleski, 1982). In fruits and seeds also, periods of deposition are succeeded by ones of utilization, e.g. glucitol in fruits and seeds of various Rosaceae (Donen, 1939; Martin, 1937; Kidd, West, Griffiths & Potter, 1940; Nuccorini & Bartoli, 1932; Whetter & Taper, 1966; Whiting, 1970; Coombe, 1976).

Unlike concentrations of starch and sucrose in leaves which often show marked diurnal/nocturnal fluctuations, most alditols in photosynthetic tissue turn over very slowly e.g. glucitol in plum (Anderson, Andrews & Hough, 1961), hamamelitol and its galactoside, clusianose, in *P. clusiana* (Sellmair, Beck & Kandler, 1969; Sellmair & Kandler, 1970) and dulcitol in *Euonymus* (Bliss, Hamon & Lukaszewski, 1972*a*, *b*). By contrast, allitol and its corresponding ketose, allulose (= psicose, ribohexulose), show reciprocal diurnal/nocturnal fluctuations in concentration (Hough & Stacey, 1966; Kull & Baitinger-Haardt, 1977). [A similar relationship between volemitol and sedoheptulose has been noted in certain leafy liverworts and is related to unusual patterns of dark fixation of carbon dioxide (A. Christie & D. H. Lewis, unpublished).] Glucitol in stag's horn plantain also shows a maximum during the day (Lambers, Blacquière & Stuiver, 1981).

The evidence referred to above clearly demonstrates that alditols can act as stores of carbohydrate. Before proceeding to more specific aspects of this role, it is necessary to mention in passing that alditols are either absent from, or present in low concentrations in, certain tissues or fluids in species where they occur in abundance elsewhere in the plant. When alditols are synthesized in leaves in substantial amounts, they are also translocated in the phloem from these leaves (see below). It might be expected, therefore, that they should also be found in all 'sink' tissues. These include flowers and, in particular, nectaries and developing seeds.

In a survey of ten species of glucitol-containing Rosaceae, representing seven genera, Bieleski & Redgwell (1980) demonstrated that this alditol was almost absent from most of their nectars. In seven species, apricot, cotoneaster, flowering quince, peach, plum, pyracantha and *Prunus campanulata*, glucose and fructose on average comprised 93.5% of total sugar with sucrose a poor third at 4.9%. Glucitol, on average, amounted to merely 1.6% with a flowering quince showing the highest value (5.3%). In other species (apple, hawthorn, pear), sucrose, glucose and fructose were present in about equal concentrations but, again, that of glucitol was very low (average of 0.8%). These figures contrast with those in leaves of apricot where, for example, Bieleski & Redgwell (1977) recorded the following concentrations (nmol mg^{-1} fresh weight); glucitol 139, sucrose 27, glucose 29, fructose 3.

Although glucitol occurs in seeds of apple (Whetter & Taper, 1966), it is absent from those of plantain as is mannitol from those of ash (Jukes & Lewis, 1974), notwithstanding the facts that glucitol and mannitol respectively are products of photosynthesis in these species and that mannitol is translocated in ash (see below). Hamamelitol constitutes about 30% of total sugar in seeds of *P. clusiana* (Sellmair & Kandler, 1970).

Synthesis

In those autotrophic plants which accumulate alditols in leaves, the latter are produced as end products of photosynthesis, usually together with, and often in excess of, sucrose. Table 7.1 provides some representative data for incorporation of ^{14}C from $^{14}CO_2$ in the light into several hexitols relative to that into sucrose. Data from other sources cannot be summarized in this way but ratios equal to or greater than unity for incorporation of ^{14}C into alditol:incorporation into sucrose have been recorded under a variety of conditions for glucitol in apple (4.58, 4.60 – Hansen & Ryugo, 1979; 1.78, 1.37 – Suleiman, A. A., McGee, E. E. M. & Lewis, D. H., unpublished) and plum (1.06 – McGee, E. E. M. & Lewis, D. H., unpublished); and for mannitol in lilac (1.02 from data in Table III of Trip, Krotkov & Nelson, 1963, but higher values are evident from sampling times between June and October from their Fig. 1.)

In contrast to these data, values less than unity (i.e. greater incorporation into sucrose than into alditols) have been found for mannitol in white ash (Trip *et al.* 1963), hamamelitol in *P. clusiana* (Sellmair & Kandler, 1970) and several species of *Plantago* (Suleiman, A. A., McGee, E. E. M. & Lewis, D. H., unpublished). Photosynthetic production of dulcitol has been noted by Bliss *et al.* (1972).

Alditols can also be synthesized by leaf tissue from exogenous hexoses e.g. glucitol from glucose by plum (Anderson, Andrews & Hough, 1962) and pear (Bieleski, 1977), glucitol from fructose by apple (Grant & ap Rees, 1981) and allitol from glucose and fructose by *Itea ilicifolia* (Hough, Shankar Iyer & Stacey, 1973). Other carbohydrates must also be the source of alditols in non-photosynthetic vascular plants e.g. angiospermous parasites of the

Table 7.1 *Relative proportions of ^{14}C incorporated into alditols and sucrose during photosynthesis by several species*

Species	Material exposed to $^{14}CO_2$ in light	Time of exposure (h)	% ^{14}C of soluble fraction in (a) Alditol	(b) Sucrose	References
Plum[a]	Leafy spurs	1.5	28	N.D.	Anderson *et al.* 1961
		15.5	53	N.D.	
Apricot[a]	Laminar strips	0.5	47	14	Bieleski & Redgwell, 1977
		2.0	70	18	
Apple[a]	Whole 12-week-old seedlings	0.5	36	26	Grant & ap Rees, 1981
Itea[b]	Young leafy twigs	0.5	21[c]	12	Kull & Baitinger-Haardt, 1977

Alditol present: [a] = glucitol, [b] = allitol, [c], corresponding value for allulose = 32%. N.D. = not determined but data for glucitol shown to indicate that, after photosynthesis, for 15.5 h, ^{14}C in glucitol must have exceeded that in sucrose.

Table 7.2 *Hexitols in hemiparasitic Scrophulariaceae and holoparasitic Orobanchaceae (data of Hodgson, 1973; cf. Plouvier, 1963)*

Mannitol only:	*Rhinanthus*
Mannitol > Galactitol:	*Euphrasia, Odontites, Pedicularis, Lathraea, Orobanche elatior*
Mannitol = Galactitol:	*Parentucellia, Orobanche minor*
Mannitol < Galactitol:	*Orthocarpus*
Galactitol only:	*Melampyrum*

Orobanchaceae in which mannitol can reach 20% dry weight (Chapter 1). Interestingly, this family and the related hemiparasitic members of the Scrophulariaceae represent species which contain more than one hexitol. As shown in Table 7.2, a series exists from species with mannitol only, via species with mannitol and galactitol to those with galactitol only. Galactitol in *Melampyrum* indicates the origin of its discontinued synonym, melampyrite. In the hemiparasitic species, the hexitols are principal photosynthetic products but show a slow metabolic turnover (Hodgson, 1973).

The leaf is certainly the principal site of synthesis of alditols and, in seedlings of apple, the only part of the plant capable of substantial conversion of exogenous fructose into glucitol. This conversion is not, however, light-dependent (Grant & ap Rees, 1981). These authors showed that, after six hours' exposure to [^{14}C]fructose, leaf disks converted 43% of absorbed fructose to glucitol in the light and 31% in the dark with corresponding values into sucrose of 11 and 8% respectively. By contrast, bark and wood of stems, complete root systems and root apices only incorporated 1–2% into glucitol. The respective corresponding values for incorporation into sucrose for these tissues were 25, 28, 13 and 11%.

As alditols are both products of photosynthesis and are more reduced than their corresponding sugars, and as chloroplasts are the principal site in leaves of the production of the biosynthetically important reductant NADPH, it might be expected that the site of their synthesis is the chloroplast. Indeed, De Villiers (1978) provided experimental evidence, involving isolation of chloroplasts from plum leaves by non-aqueous methods, which was interpreted as supporting a chloroplastic site of synthesis of glucitol. However, critical evidence for lack of contamination of chloroplasts by cytoplasm was not provided and it must be remembered that similar evidence for a chloroplastic site of synthesis of sucrose, formerly accepted as reliable, is now discredited (see Bird, Cornelius, Keys & Whittingham, 1974, and Chapter 2).

Circumstantial evidence that alditols are synthesized outside the chloroplast comes from two sources. Firstly, the polyol-containing Orobanchaceae do not contain fully functional chloroplasts although they may contain pro-plastids still capable of heterotrophic synthesis of alditols. Secondly, in non-polyol-containing plants such as spinach, alditols do not penetrate chloroplastic membranes and glucitol is therefore often used as the stabilizing osmoticum during isolation of chloroplasts (e.g. Stokes & Walker, 1972). As alditols are translocated (see below), their rapid efflux through a chloroplast membrane readily permeable to them would be expected if their site of synthesis was within chloroplasts. This very tenuous evidence in favour of an extra-chloroplastic site is now being supported by two independent kinds of experimental data, the isolation of alditol-synthesizing enzymes from, and the effects of phosphate-sequestering compounds on, polyol-containing leaves.

Evidence concerning the first is discussed in the section on enzymology below. The logic of the second approach is as follows. The principal form by which carbon fixed in photosynthesis is exported from the chloroplast in angiosperms is triose phosphate in a process which involves exchange for inorganic phosphate (see Chapter 8). Experimental lowering of cytoplasmic [Pi] inhibits this exchange of TP and Pi and causes a reduced rate of photosynthesis but an enhanced synthesis of starch (Herold & Lewis, 1977; Herold, McGee & Lewis, 1980). A convenient way of lowering [Pi] *in vivo* is to supply leaf disks with mannose or other analogues of glucose about carbon atom 2 (e.g. glucosamine or 2-deoxyglucose). Like glucose, these three analogues are phosphorylated via hexokinase to their corresponding 6-phosphates which, in many plants, are not further metabolized rapidly. In this way, phosphorus becomes sequestered in organic form (Chen-She *et al.* 1975; Herold, Lewis & Walker, 1976). As far as apportioning of photosynthetic products is concerned, this results in greater retention of fixed carbon in the chloroplast, represented by enhanced synthesis of chloroplastic starch and diminished synthesis of cytoplasmic sucrose. It follows that, if synthesis of alditols is chloroplastic, a larger proportion of fixed carbon should be diverted to them under conditions of cytoplasmic sequestration of phosphate

Fig. 7.2 Alternative sites for synthesis of polyols in photosynthetic tissue in relation to chloroplastic and cytoplasmic triose phosphate (see text for explanation) (TP = triose phosphate; Pi = orthophosphate).

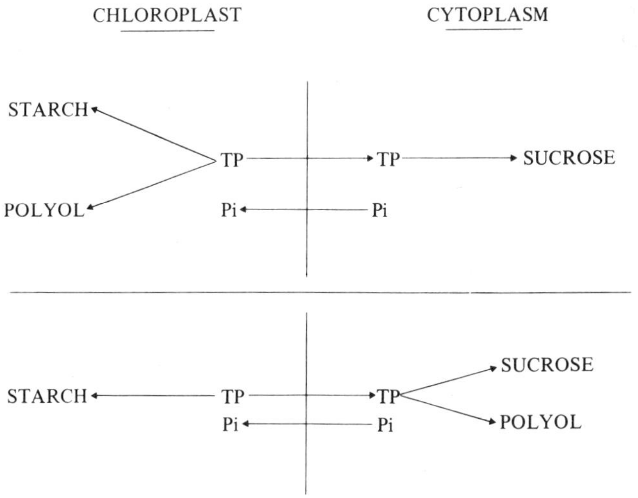

than under control, phosphate-sufficient, conditions. In this way, the incorporation of ^{14}C from $^{14}CO_2$ in the light into alditols should parallel that into starch and sequestration of phosphate should increase the incorporation of ^{14}C into alditol relative to that into sucrose (Fig. 7.2*a*). On the other hand, if synthesis is cytoplasmic, the distribution of ^{14}C between alditol and sucrose should be similar in phosphate-sufficient and phosphate-deficient cytoplasm (Fig. 7.2*b*). When this experimental approach is applied to the glucitol-containing leaves of apple, plum and plantain, the second situation, consistent with a cytoplasmic site of synthesis for this alditol, obtains (Table 7.3). That is, although sequestration of Pi enhances synthesis of starch, the relative syntheses of sucrose and glucitol are not affected. As mannose 6-phosphate appears to be a precursor of mannitol (see below), this approach is not valid for plants containing this alditol. Use of the other analogues should be instructive.

As noted above, there have been few reports of the isolation and characterization of enzymes concerned with the synthesis of alditols. Early

Table 7.3 *Effect of mannose on photosynthesis and partitioning of fixed ^{14}C between soluble and insoluble fractions and between sucrose and glucitol in leaf disks of apple, plantain and plum (Suleiman, A. A., McGee, E. E. M. & Lewis, D. H., unpublished)*

		Photo-synthesis	Insoluble	Soluble	Glucitol: sucrose ratio
Experiment 1					
Apple	Control	100.0	8.1[a]	91.9[a]	1.78
	Mannose (10 mM)	73.9[b]	16.1	83.9	1.47
Plantain	Control	100.0	10.2	89.8	0.39
	Mannose (10 mM)	32.6[b]	20.9	79.1	0.32
Experiment 2					
Apple	Control	100.0	16.4	83.6	1.06[c]
	Mannose (10 mM)	84.3	28.5	71.5	1.08
	Mannose (50 mM)	77.4	45.9	54.1	1.05
Plum	Control	100.0	9.9	90.1	1.37
	Mannose (10 mM)	73.7	22.3	77.7	1.39
	Mannose (50 mM)	38.2	49.3	50.7	1.66

Leaf disks were floated overnight in the dark on water or mannose solution and then exposed to $^{14}CO_2$ in the light for 1 h.

[a] Distribution (%) between insoluble and soluble fractions.

[b] Percentage of control.

[c] Analysis of variance of arc-sine transformed data for proportion in sucrose shows that the treatments in the better replicated experiment 2 ($n = 4$) are not statistically different.

reports suggested that extracts could interconvert alditols and their corresponding ketoses (glucitol ↔ fructose, Edson, 1953; McCorkindale, 1953; Gorrod, 1961; L-arabitol ↔ L-ribulose in the non-alditol-containing plants, tobacco, pea and wheat, Kocourek, Tichá & Kŏstíř, 1964). The frequent co-occurrence of polyol-ketose pairs in plants also encouraged the view that the free compounds were biosynthetically related (Lewis & Smith, 1967) although, by 1967, enzymes from fungi which catalysed interconversion of sugar phosphate and alditol phosphates had been described and Trip, Krotkov & Nelson (1964) had suggested that, in celery, interconversion of fructose and mannitol occurred via phosphorylated intermediates. That phosphate esters were involved gained credence thirteen years later when glucitol phosphate was recorded in leaves of apricot (Bieleski & Redgwell, 1977). A few more years later, they demonstrated that both glucitol 6-phosphate and glucitol 1-phosphate were present in this tissue at approx. 7 μmol kg^{-1} fresh weight and accounted for approx. 4% of the hexose monophosphate fraction (Redgwell & Bieleski, 1978). They studied the kinetics of labelling of components of the hexose-phosphate and free sugar fractions and concluded that their data were consistent with the pathway:

fructose 6-phosphate → glucitol 6-phosphate → glucitol

It should be noted that inclusion of glucose 6-phosphate as the immediate precursor of glucitol 6-phosphate is not inconsistent with their data. They point out that the presence of glucitol 1-phosphate 'is an unexplained anomaly' and suggest that it could arise from G-1-P, F-1-P or via glucitol 1,6-bisphosphate derived from fructose 1,6-bisphosphate. The possibility of a mutase relating it to glucitol 6-phosphate was also raised but they could not suggest a role for its presence. Further evidence that glucitol 6-phosphate is an intermediate in the synthesis of the free alditol came with the discovery of a glucitol 6-phosphate phosphatase in extracts of apple leaves (Grant & ap Rees, 1981). The enzyme in crude extracts was more active with the alditol phosphate than with several hexose phosphates and mannitol 1-phosphate. Ammonium sulphate fractionation improved the ratio of activity with glucitol 6-phosphate to that with glucose 6-phosphate from 1.80 to 8.43. Activity was also detected in pear leaves but not in apical tips of pear roots or complete root systems of apple seedlings.

Although the diurnal/nocturnal reciprocal changes in concentrations of allitol and allulose mentioned above suggest ready interconversion, the idea that aldoses or, more probably, aldose phosphates are precursors of alditols is gaining ground. Firstly, in studies with *Euonymus*, Schradie (1963) (quoted in Bliss *et al.* 1972) could find no tagatose, the ketose equivalent of galactitol. The latter authors suggest that galactose is the immediate precursor but galactose phosphate is also a possibility. More recently, NADPH-dependent aldose 6-phosphate reductases have been reported in loquat fruit (Hirai, 1979), in leaves of several woody Rosaceae (Negm & Loescher, 1981) and in leaves of celery, privet and other mannitol-containing species (Loescher, Redgwell & Bieleski, 1982). In this way, both glucose phosphate- and mannose phosphate-reductases have now been found in plants and possibly a galactose phosphate reductase can be inferred to occur. It must be emphasized that Grant & ap Rees (1981) commented that activities in extracts of fruit recorded by Hirai (1979) were very low. However, in confirmation of a cytoplasmic site of synthesis, Rumpho, Edwards & Loescher (1983) have recently shown that mannose 6-phosphate reductase is a cytoplasmic enzyme in leaves of celery.

The evidence so far in this section shows four important features of alditol metabolism: firstly, alditols are often synthesized in amounts greater than sucrose; secondly, like sucrose, they are synthesized outside the chloroplast; thirdly, again like sucrose, they are synthesized via their phosphates involving specific phosphatases; and, fourthly, the enzymes most likely to be concerned with synthesis are NADPH-dependent aldose phosphate reductases. The cytoplasmic source of sufficient NADPH is an unsolved problem – is the reducing power derived directly from the chloroplast, is an oxidative pentose phosphate pathway in the cytoplasm active in leaves of alditol-containing plants or are other mechanisms involved? Answers to these questions require further research.

Two additional features in common with sucrose link this section with the next two. Sucrose is translocated from its primary site of synthesis in 'source' leaves to distant 'sinks' where it is utilized either via invertase or the mis-named sucrose synthetase (see Chapters 2 and 3). Translocation of alditols is the subject of the next section and degradation in sinks of the one after that. Before embarking on these, however, it is worth recalling that before leaves become 'sources', they are themselves 'sinks' during their juvenile stages. During their period as 'sinks', they have high activities of sucrose-degrading enzymes (e.g. see Greenland & Lewis, 1981*a*, *b* for invertase). It might be expected on comparative grounds that, during their period as 'sinks', leaves of alditol-containing plants would have high activities of alditol-degrading enzymes before high activities of the synthetic ones develop. That this expectation is a valid one has recently been demonstrated by Loescher, Marlow & Kennedy (1982). They showed that, as apple leaves mature, there is a large increase in glucose 6-phosphate reductase and a corresponding decline in glucitol dehydrogenase, an NAD-linked enzyme which converts free glucitol to fructose. The role of such enzymes are discussed further below.

Translocation

Chapters 2 and 3 emphasize that sucrose is the principal form by which carbon is translocated in the phloem of vascular plants. It has been suggested by Arnold (1968) that, as sucrose is a non-reducing disaccharide, it represents a compound by which hexoses are transported in a 'protected', unreactive form. Chemically, alditols may be regarded as non-reducing monosaccharides and, as such, also represent a 'protected', unreactive form of their corresponding aldose or ketose. In this context, therefore, it is not surprising that they are translocated in the phloem. Zeigler (1975) delimited three types of plant on the basis of the composition of sugars in their sieve-tube sap, (a) species with sucrose as the predominant sugar, (b) species which have considerable amounts of oligosaccharides, especially sucrosyl-galactans of the raffinose series, and (c) those which contain considerable quantities of alditols. Most work on translocation has concerned group (a) (see Chapter 3) and, although there have been studies on translocation in species in group (b), more is known of the biosynthesis of the oligosaccharides (see Chapter 5).

Zimmerman & Zeigler (1975) have listed species which have been shown to contain alditols in sieve-tube elements. Almost all belong to the Oleaceae (mannitol), Rosaceae (glucitol) and Celastraceae (galactitol). Concentrations in excess of 20% w/v sucrose equivalents were frequently recorded for all three polyols and this represents as much as 60–90% of carbon exported from the leaves (Bieleski, 1982). It seems probable that all alditols which occur in leaves will be found in sieve-tube exudates i.e. all are translocated. For example, allitol occurs in bark of *Itea* (Kull & Baitinger-Haardt, 1977) and hamamelitol in seeds of *Primula clusiana* (Sellmair & Kandler, 1970) but their presence specifically in sieve tubes has not been demonstrated.

The suggestion of Obaton (1929) that mannitol was translocated in celery has been amply confirmed using tracer techniques. Important papers include, for glucitol, those of Webb & Burley (1962), Bieleski (1969) and the series from the laboratory of Hansen between 1967 and 1979 (see Hansen, 1967, 1970 and 1979 for details of the whole series). Trip, Nelson & Krotkov (1965) describe translocation of mannitol in white ash and lilac.

In view of the detailed discussion of translocation of sucrose in Chapter 3, it is not appropriate to dwell on mechanisms here. Indeed, the limited number of studies of translocation of alditols has not added significantly to the understanding of its mechanisms. Much more information about alditols is needed on all aspects discussed in Chapter 3. Is their transport from photosynthetic cells to minor veins entirely symplastic or is there an apoplastic phase? Does loading of alditols into sieve-tubes involve an ATP-dependent proton co-transport system? Is their long-distance transport by mass-flow? How are they unloaded within sink tissues and transported to the cells which utilize them? Are enzymes which degrade alditols absent from sieve-tubes? It is time for more studies of plants in Zeigler's group (c) Such a sideways glance could well contribute significantly to the formulation of more general principles about translocation.

Degradation

Notwithstanding the precise mechanisms by which alditols get to 'sink' tissues, get there they do and there they are utilized. The capacity to utilize alditols has been studied, for example, in fruits, nectaries and root tips (Hansen, 1970; Bieleski & Redgwell, 1980; Grant & ap Rees, 1981) as well as in young leaves (Loescher *et al.* 1982). A pattern is emerging about which generalizations are possible. The general rule seems to be that 'sinks' have a limited capacity to synthesize alditols from sucrose or hexose but have a greater capacity than mature leaves, the sources of the alditols, for utilizing them. Furthermore, they utilize them by enzymatic mechanisms distinct from those of synthesis. Several observations already mentioned in previous sections are consistent with this pattern of events but a few examples will illustrate the generalities.

Hansen (1970) injected [^{14}C]glucitol into apples and demonstrated that it was converted to sucrose, glucose and fructose. On the other hand, when the last three sugars were injected, they were interconverted but not metabolized to glucitol. As noted above, Bieleski & Redgwell (1980) recorded the very low concentrations of glucitol in nectars of rosaceous trees. They also demonstrated that flowers, including the nectary, could convert glucitol to other sugars. In experiments lasting 16 h, various [^{14}C]sugars and [^{14}C]glucitol were fed to flowers via their pedicels and the distributions of ^{14}C in sugars in the nectar were analysed. Patterns from sucrose, glucose and fructose were similar. In pear, 41.4% of the ^{14}C was present in sucrose, 28.6% in glucose and 29.2% in fructose with a mere 0.2% in glucitol. Very similar values resulted when [^{14}C]glucitol was supplied (sucrose 43.2%, glucose 31.6%, fructose 24.8%, glucitol 0.1%). In plum and apricot, there was much less ^{14}C in sucrose from all sources but, again, although more unchanged glucitol occurred in the nectars in these species, the three sugars and glucitol gave similar patterns, a clear indication of the latter's rapid utilization. The failure of root tips of apple to convert fructose to glucitol has been mentioned above. Grant & ap Rees (1981) did, however, demonstrate metabolism of glucitol in a 'sink' tissue. Whereas, after an incubation time of 6 h, leaf disks in both light and dark retained more than 85% of absorbed [^{14}C]glucitol as unchanged glucitol and complete root systems to the extent of 78%, root tips could metabolize the alditol and only 42% remained unchanged. This metabolism was correlated with an incorporation of ^{14}C into insoluble material (13.7% in root tips compared with 0.6–3.3% in leaf disks and complete root systems). Utilization of glucitol by an artificial 'sink', namely tissue culture of apple, has also been demonstrated (Chong & Taper, 1972; Negm & Loescher, 1979*a*).

It is evident from the above that 'source' leaves can synthesize alditols rapidly both photosynthetically and heterotrophically but, with the exception of allitol in *Itea*, cannot utilize them at significant rates whereas the converse is true for 'sink' tissues. It is also true that tissues which do not normally contain alditols do not metabolize them (Lewis & Smith, 1967). An exception is leaf tissue of a cultivated rose (Bieleski, 1977). The genus *Rosa* belongs to a sub-family of the Rosaceae which does not contain glucitol. Nevertheless, when leaves had been supplied with [^{14}C]glucitol for 4 h, only 6% of the ^{14}C in the alcohol-soluble fraction remained in glucitol. Most was in fructose (34%) with 20.3% and 16.6% in sucrose and glucose respectively. A converse situation occurs in the parasitic angiosperm, witchweed. As noted above, hemiparasitic members of the Scrophulariaceae contain mannitol, galactitol or both. Okonkwo (1964) recorded that witchweed would not grow *in vitro* on mannitol. In one case, the rose has no active enzymes for alditol biosynthesis but possesses those of degradation: in the other, witchweed has those of synthesis but not of degradation. As other members of the Scrophulariaceae do not have the capacity to synthesize alditols either (Plouvier, 1963), it is clear that loss or gain of one or both of the synthetic and degradatory pathways within a single taxonomic family is possible. The evolutionary pressures that bring such changes about are obscure.

As for the enzymes of synthesis, the study of enzymes of degradation of alditols is in its infancy. Most recent success comes from the laboratory of W. H. Loescher. He and his co-workers have now extracted an NAD-dependent glucitol (sorbitol) dehydrogenase from apple and pear mesocarp tissue cultures (Negm & Loescher, 1979*a*, *b*), apple fruits (Marlow & Loescher, 1980) and immature apple leaves (Negm & Loescher, 1981; Loescher *et al.* 1982). All are 'sink' tissues. Negm (1981, 1982) has also reported the presence of NAD-dependent galactitol and mannitol 1-phosphate dehydrogenases in leaves of *Euonymus* and tissue cultures of ash respectively. Several of these papers are only abstracts but Negm & Loescher (1979*a*) have characterized the polyol dehydrogenase from apple callus in some detail. Oxidation of glucitol to fructose has a pH optimum between 9 and 10 and a K_m for glucitol of 86 mM. The enzyme oxidized xylitol and L-threitol at rates in excess of 80% of that of glucitol but oxidized mannitol at only 8% of that of its natural substrate. It was completely inactive with NADP. In the reverse direction, the pH optimum for reduction of fructose was between 5.7 and 7.0, the K_m for fructose was 1.5 M and NADPH could substitute to some extent for NADH. These data are consistent with the enzyme acting as one degrading glucitol *in vivo* but the pH optimum seems higher than would be found in the cytoplasm and the paper does not state whether extracted activity could account for the rate of degradation of glucitol in intact callus. It is to be hoped that this and other studies from W. H. Loescher's laboratory will catalyse further studies of the enzymology of utilization (and synthesis) of alditols.

Storage of reducing power

As alditols are more reduced than sugars, they represent convenient compounds for the short- or long-term storage of reducing power as well as of carbon. In vascular plants, the only alditol which is known to turn over rapidly is allitol in *Itea*. The mechanism appears to be a reversible interconversion of allitol and allulose, comparable to one between sedoheptitol and sedoheptulose in certain leafy liverworts (A. Christie & D. H. Lewis, unpublished). In these leafy liverworts, the regeneration of metabolic reducing power by the conversion of alditol to ketose appears to fuel an unusual dark fixation of carbon dioxide into carbohydrates, the significance of which can only be guessed at present. Whether a similar enhanced rate of dark fixation occurs in *Itea* remains to be seen. As far as other alditols are concerned, their turnover is not fast enough for them to act as rapid modulators of NAD(P)/NAD(P)H ratios.

The chemical advantage of polyols over sucrose as stores of energy may be gauged as follows. In order to synthesize sucrose from glucose and fructose, three 'high-energy' phosphates are consumed (two to phosphorylate the hexoses and one to form UDPG – see Chapter 2). In order to convert a hexose to an alditol via a hexose phosphate, four 'high energy' phosphate equivalents are consumed (one to phosphorylate the hexose and three equivalents as NADPH). In the utilization of sucrose, either none or only one of these 'high energy' phosphates is conserved depending on whether degradation is via invertase or sucrose synthetase. However, three equivalents are recoverable when an alditol is converted to a sugar via an NAD-linked polyol dehydrogenase. On these grounds, alditols are thermodynamically a better bet than sucrose as stores of energy in addition to that which degradation of hexose via glycolysis and the tricarboxylic acid cycle releases.

As most plants do not make use of this extra kind of energy storage, it may be that possession of alditols incurs penalties in other areas of physiology and metabolism. One of these in vascular plants may be differentiation of xylem. Lewis (1980) suggested that sucrose had been selected as a universal metabolite in vascular plants rather than alditols because of its lack of complexing with borate. This permitted boron to exert its role as an essential element for vascular plants, probably via an involvement in phenolic metabolism, especially lignification. By complexing with borate, alditols would adversely interfere with this role in differentiation. Clearly, alditol-containing plants differentiate perfectly well. Perhaps they do so by rigorous control of polyol concentration at sites where boron exerts its effects. The price of this control may be a heavy one, outweighing for most plants the advantages of employing alditols as storage compounds.

Osmo-regulation

It was mentioned above that concentrations of alditols in tissues were frequently higher in winter than in summer. These observations have led to the suggestion that, via lowering of freezing points, they act as cryo-protective agents preventing, or at least minimizing, frost damage. Lewis & Smith (1967) had a brief discussion of this possibility and further suggested that the abundance of polyols in many angiospermous parasites (see also Smith, Muscatine & Lewis, 1969) could facilitate the passage of water from their hosts. In both these situations, alditols are postulated as acting in a biophysical rather than a biochemical role. Such a biophysical role has been supported by the observations of Sellmair & Kandler (1970) for hamamelitol, Hansen & Grausland (1973) for glucitol and Kull & Baitinger-Haardt (1977) for allitol. More experimental studies to complement the observational correlations are required.

Lewis & Smith (1967) also suggested that, during development of low internal osmotic potentials, 'it would presumably raise fewer physiological difficulties if the high concentration of solutes were to consist mainly of relatively inert, non-polar solutes such as polyols'. This conclusion is now embraced within the concept of 'compatible solutes' (Brown, 1978; Hellebust, 1980), i.e. compounds which, when present in substantial concentrations, can effectively substitute for water in maintenance of physiological activity (e.g. in enzyme activity and membrane integrity) by taking the place of ions inhibitory to cell metabolism. Although the concept of compatible solutes developed from studies of micro-organisms, it has been effectively extended to vascular plants in which alditols represent one such class of compounds. Figures 7.3*a–c* from Ahmad, Larher & Stewart (1979) show the effects of ionic osmotic stress on concentration of glucitol in *Plantago maritima* (see also Stewart, Larher, Ahmed & Lee, 1979; Jeffries, Davy & Rudmik, 1979; Jeffries, Rudmik & Dillon, 1979; Briens & Larher, 1983; Gorham, Hughes & Wyn Jones, 1981; and Königshofer, 1983). Table 7.4 demonstrates the strikingly different effect of 0.5 M glucitol and NaCl on the activities of various enzymes from this species. However, Jeffries (1980) provides a word of caution about observations of this kind (i.e. inhibitory effects of single salts) and mentions that, by concurrent adjustment of concentrations of other ions (K^+, Ca^{2+}, Mg^{2+}), enzyme activity can be maintained even in the presence of 1 M NaCl.

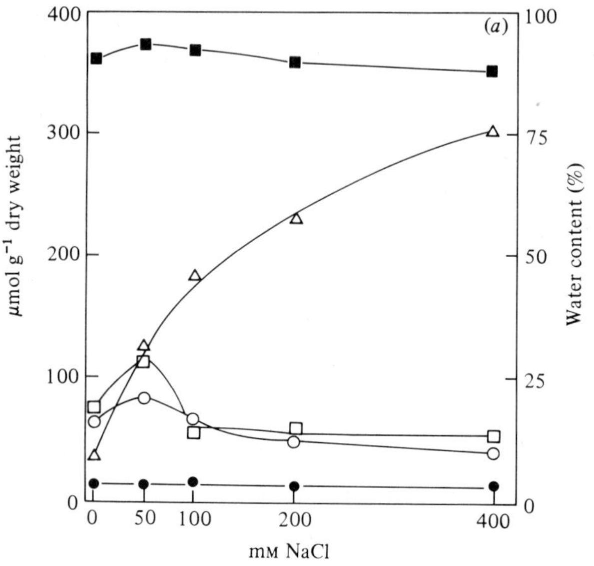

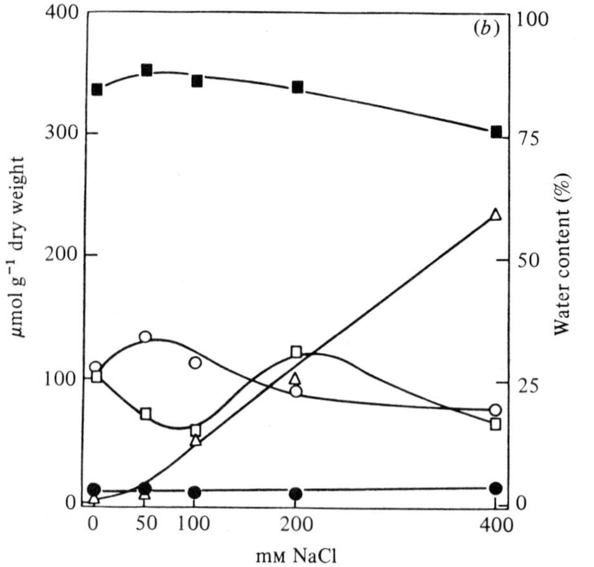

Fig. 7.3 Effect of exogenous sodium chloride on content of water and organic solutes in (*a*) shoots and (*b*) roots of *Plantago maritima*; (*c*) relationship between endogenous concentrations of glucitol (sorbitol) and of sodium and chloride ions. (*a*) and (*b*): △, glucitol; □, reducing sugars; ○, amino acids; ●, methylated onium compounds; ■, water. (Amino acids and methylated onium compounds act as 'compatible solutes' in other plants.) (*c*): ●, root; ■, shoot. Data plotted on basis of water content. Seedlings were grown in a complete nutrient solution for 5 weeks. Treatment with salt involved daily additions of 50 mM NaCl to give final concentrations shown on *x*-axis of (*a*)

Lambers *et al.* (1981) have suggested that, in *P. coronopus*, there is a causal relationship between synthesis of glucitol in response to increased external ionic concentration and the concomitant inhibition of the 'alternative' respiratory pathway [i.e. the cyanide-resistant but SHAM (salicylhydroxamic acid)-sensitive pathway by which metabolic reducing power (NADH) can be oxidized]. In their experiments, concentration of glucitol rapidly increased in

Table 7.4 *Influence of glucitol or sodium chloride (both 500 mM) on enzymes extracted from the halophyte,* Plantago maritima *(data from Ahmad* et al. *1979)*

	Activity as % control	
Enzyme	Glucitol	Sodium chloride
Acid phosphatase	103	92
Glutamine synthetase	98	62
Glutamate dehydrogenase	101	39
G-6-P dehydrogenase	97	46
Malate dehydrogenase	104	23
Nitrite reductase	98	49

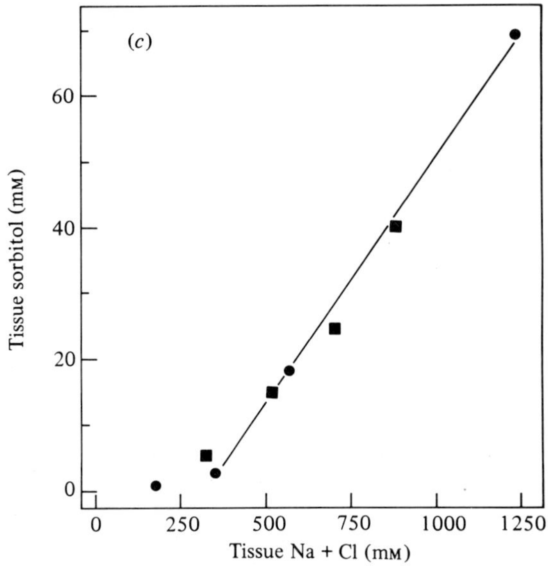

Caption for fig. 7.3 (*cont.*)
and (*b*). Plants were analysed 7 days after the salt series was completed. (Data reproduced from Ahmad *et al.* 1979, with permission of the *New Phytologist Trust.*)

roots before shoots, an unexpected observation in view of the discussion above about synthesis in 'sources' (i.e. shoots) and consumption in 'sinks' (especially root tips). The metabolic processes underlying changes in concentrations of alditol in response to osmotic stress require further investigation. The studies of Kauss (1979) of these comparable metabolic responses in the wall-less alga, *Poteriochromonas* (= *Ochromonas*) *malhamensis*, provide a fascinating model on which to base testable hypotheses concerning similar responses in vascular plants.

Conclusions

The chapter began by indicating the differences between acyclic and cyclic polyols. The prevalence of cyclitols in particular sea-grasses and the sluggish nature of their metabolism was emphasized in Chapter 6. A similar metabolic quiescence and tendency to accumulate under ionic stress are features of acyclic polyols, especially in particular tissues of certain species. Under these circumstances, they are probably compounds of biophysical importance rather than of potential biochemical significance. With these situations in mind, the answer to the question, 'when is a storage carbohydrate not a storage carbohydrate?' is, 'when it is a polyol'.

References

Ahmad, T., Larher, F. & Stewart, G. R. (1979). Sorbitol, a compatible osmotic solute in *Plantago maritima*. *New Phytologist*, **82**, 671–8.

Anderson, J. D., Andrews, P. & Hough, L. (1961). The biosynthesis and metabolism of polyols. Sorbitol (D-glucitol) of plum leaves. *Biochemical Journal*, **81**, 149–54.

Anderson, J. D., Andrews, P. & Hough, L. (1962). The biosynthesis and metabolism of polyols. II. The metabolism of ^{14}C-labelled glucose, D-glucuronic acid and D-glucitol (sorbitol) by plum leaves. *Biochemical Journal*, **84**, 140–6.

Arnold, W. N. (1968). The selection of sucrose as the translocate in higher plants. *Journal of Theoretical Biology*, **21**, 13–20.

Asai, T. (1937). Untersuchungen über die Bedeutung des Mannits im Stoffwechsel einiger höheren Pflanzen. *Japanese Journal of Botany*, **8**, 343–66.

Bieleski, R. L. (1969). Accumulation and translocation of sorbitol in apple phloem. *Australian Journal of Biological Sciences*, **22**, 611–20.

Bieleski, R. L. (1977). Accumulation of sorbitol and glucose by leaf slices of Rosaceae. *Australian Journal of Plant Physiology*, **4**, 11–24.

Bieleski, R. L. (1982). Sugar alcohols. In *Plant Carbohydrates I. Intracellular Carbohydrates*, ed. F. A. Loewus & W. Tanner, *Encyclopedia of Plant Physiology*, New Series, vol. 13A, pp. 158–92. Berlin: Springer-Verlag.

Bieleski, R. L. & Redgwell, R. J. (1977). Synthesis of sorbitol in apricot leaves. *Australian Journal of Plant Physiology*, **4**, 1–10.

Bieleski, R. L. & Redgwell, R. J. (1980). Sorbitol metabolism in nectaries from flowers of Rosaceae. *Australian Journal of Plant Physiology*, **7**, 15–26.

Bird, I. F., Cornelius, M. J., Keys, A. J. & Whittingham, C. P. (1974). Intracellular site of sucrose synthesis. *Phytochemistry*, **13**, 59–64.

Bliss, C. A., Hamon, N. W. & Lukaszewski, T. P. (1972*a*). Metabolism of dulcitol in *Euonymus japonica*. *Phytochemistry*, **11**, 1695–703.

Bliss, C. A., Hamon, N. W. & Lukaszewski, T. P. (1972*b*). Biosynthesis of dulcitol in *Euonymus japonica*. *Phytochemistry*, **11**, 1705–11.

Briens, M. & Larher, F. (1983). Sorbitol accumulation in Plantaginaceae; further evidence for a function in stress tolerance. *Zeitschrift für Pflanzenphysiologie*, **110**, 447–58.

Brown, A. D. (1978). Compatible solutes and extreme water stress in eukaryotic micro-organisms. *Advances in Microbial Physiology*, **17**, 181–242.

Chen-She, S-H., Lewis, D. H. & Walker, D. A. (1975). Stimulation of photosynthetic starch formation by sequestration of cytoplasmic orthophosphate. *New Phytologist*, **74**, 783–92.

Chong, C. (1971). Study of the seasonal and daily distribution of sorbitol and related carbohydrates within apple seedlings by analysis of selected tissues and organs. *Canadian Journal of Plant Science*, **51**, 519–25.

Chong, C. & Taper, C. D. (1971). Daily variation of sorbitol and related carbohydrates in *Malus* leaves. *Canadian Journal of Botany*, **49**, 173–7.

Chong, C. & Taper, C. D. (1972). *Malus* tissue cultures. I. Sorbitol (D-glucitol) as a carbon source for callus initiation and growth. *Canadian Journal of Botany*, **50**, 1399–404.

Coombe, B. G. (1976). The development of fleshy fruits. *Annual Review of Plant Physiology*, **27**, 507–28.

De Villiers, O. T. (1978). Incorporation of $^{14}CO_2$ into sorbitol and other metabolites in chloroplasts of *Prunus salicina* Lindl. leaves. *South African Journal of Science*, **74**, 742–3.

Donen, I. (1939). The role of sorbitol in the metabolism of the Kelsey plum. I. Changes in chemical composition during growth and storage. *Biochemical Journal*, **33**, 1611–20.

Edson, N. L. (1953). The metabolism of sugar alcohols. *Report of 29th Meeting of the Association of the Australian and New Zealand Association for the Advancement of Science*, pp. 281–99.

Gorham, J., Hughes, Ll. & Wyn Jones, R. G. (1981). Low-molecular weight carbohydrates in some salt-stressed plants. *Physiologia Plantarum*, **53**, 27–33.

Gorrod, A. R. N. (1961). Sorbitol metabolism in apple fruits. *Nature*, **190**, 190.

Grant, C. R. & ap Rees, T. (1981). Sorbitol metabolism by apple seedlings. *Phytochemistry*, **20**, 1505–11.

Greenland, A. J. & Lewis, D. H. (1981*a*). The acid invertases of the developing third leaf of oat. I. Changes in activity of invertase and concentrations of ethanol-soluble carbohydrates. *New Phytologist*, **88**, 265–77.

Greenland, A. J. & Lewis, D. H. (1981*b*). The acid invertases of the developing third leaf of oat. II. Changes in the activities of soluble isoenzymes. *New Phytologist*, **88**, 279–88.

Hansen, P. (1967). ^{14}C-studies on apple trees. III. The influence of season on storage and mobilization of labelled compounds. *Physiologia Plantarum*, **20**, 1103–11.

Hansen, P. (1970). ^{14}C-Studies on apple trees. V. Translocation of labelled compounds from leaves to fruit and their conversion within the fruit. *Physiologia Plantarum*, **23**, 564–73.

Hansen, P. (1979). ^{14}C-studies on apple trees. IX. Seasonal changes in the formation of fruit constituents and their subsequent conversions. *Physiologia Plantarum*, **47**, 190–4.

Hansen, P. & Grausland, J. (1973). ^{14}C-studies on apple trees. VIII. The seasonal variation and nature of reserves. *Physiologia Plantarum*, **28**, 24–32.

Hansen, P. & Ryugo, K. (1979). Translocation and metabolism of carbohydrate fraction of ^{14}C-photosynthates in 'French' prune, *Prunus domestica* L. *Journal of the American Society for Horticultural Science*, **104**, 622–5.

Hellebust, J. A. (1980). Reactions to water and salt stress in plants. In *Plant Membrane Transport: Current Conceptional Issues*, ed. R. M. Spanswick, W. J. Lucas & J. Dainty, pp. 147–56. Amsterdam: Elsevier/North-Holland Biomedical Press.

Herold, A. & Lewis, D. H. (1977). Mannose and green plants: occurrence, physiology and metabolism, and use as a tool to study the role of orthophosphate. *New Phytologist*, **79**, 1–40.

Herold, A., Lewis, D. H. & Walker, D. A. (1976). Sequestration of cytoplasmic orthophosphate by mannose and its differential effect on photosynthetic starch synthesis in C_3 and C_4 species. *New Phytologist*, **76**, 397–407.

Herold, A., McGee, E. E. M. & Lewis, D. H. (1980). The effect of orthophosphate concentration and exogenously supplied sugars on the distribution of newly fixed carbon in sugar beet leaf discs. *New Phytologist*, **85**, 1–13.

Hirai, M. (1979). Sorbitol-6-phosphate dehydrogenase from loquat fruit. *Plant Physiology*, **63**, 715–17.

Hodgson, J. F. (1973). *Aspects of the Carbon Nutrition of Angiospermous Parasites*. Ph.D. Thesis, University of Sheffield, U.K.

Hough, L., Shankar Iyer, P. N. & Stacey, B. E. (1973). Metabolism of ^{14}C-labelled D-glucose, D-fructose and allitol by *Itea* plants. *Phytochemistry*, **12**, 573–82.

Hough, L. & Stacey, B. E. (1966). Biosynthesis and metabolism of allitol and D-allulose in *Itea* plants; incorporation of $^{14}CO_2$. *Phytochemistry*, **5**, 215–22.

Jeffries, R. L. (1980). Discussion to paper by J. A. Hellebust above. In *Plant Membrane Transport: Current Conceptional Issues*, ed. R. M. Spanswick, W. J. Lucas & J. Dainty, pp. 157–8. Amsterdam: Elsevier/North-Holland Biomedical Press.

Jeffries, R. L., Davy, A. J. & Rudmik, T. (1979). The growth strategies of coastal halophytes. In *Ecological Processes in Coastal Environments*, ed. R. L. Jeffries & A. J. Davy, pp. 243–68. Oxford: Blackwell Scientific Publications.

Jeffries, R. L., Rudmik, T. & Dillon, E. M. (1979). Responses of halophytes to high salinities and low water potential. *Plant Physiology*, **64**, 989–94.

Jukes, C. & Lewis, D. H. (1974). Planteose, the major soluble carbohydrate of seeds of *Fraxinus excelsior*. *Phytochemistry*, **13**, 1519–21.

Kauss, H. (1979). Osmotic regulation in algae. *Progress in Phytochemistry*, **5**, 1–27.

Kidd, F., West, C., Griffiths, D. E. & Potter, N. A. (1940). An investigation of the changes in chemical composition and respiration during the ripening and storage of Conference pears. *Annals of Botany*, **4**, 1–30.

Kocourek, J., Tichá, M. & Kŏstíř, J. (1964). Formation of ribulose in plants fed L-arabitol. *Archives of Biochemistry and Biophysics*, **108**, 349–51.

Königshofer, H. (1983). Changes in ion compositoion and hexitol content of different *Plantago* species under the influence of salt stress. *Plant and Soil*, **72**, 289–96.

Kull, U. & Baitinger-Haardt, C. (1977). Physiology of ribohexulose (D-allulose) and allitol in *Itea* plants. *Zeitschrift für Pflanzenphysiologie*, **82**, 301–9.

Lambers, H., Blacquière, T. & Stuiver, B. C. E. E. (1981). Interactions between osmoregulation and the alternative respiratory pathway in *Plantago coronopus* as affected by salinity. *Physiologia Plantarum*, **51**, 63–8.

Lewis, D. H. (1980). Boron, lignification and the origin of vascular plants – a unified hypothesis. *New Phytologist*, **84**, 209–27.

Lewis, D. H. & Smith, D. C. (1967). Sugar alcohols (polyols) in fungi and green plants. I. Distribution, physiology, and metabolism. *New Phytologist*, **66**, 143–84.

Loescher, W. H., Marlow, G. C. & Kennedy, R. A. (1982). Sorbitol metabolism and sink-source interconversions in developing apple leaves. *Plant Physiology*, **70**, 335–9.

Loescher, W. H., Redgwell, R. & Bieleski, R. (1982). Mannitol biosynthesis in higher plants: detection and characterization of a NADPH-dependent mannose 6-phosphate reductase in mannitol-synthesizing plants. *Plant Physiology*, **69**, Supplement 51.

McCorkindale, J. (1953). Polyol dehydrogenase activity in *Sorbus* berries. *Proceedings of the University of Otago Medical School*, **31**, 7–8.

Marlow, G. C. & Loescher, W. H. (1980). Sorbitol dehydrogenase, sorbitol and watercore in apple fruit tissues. *HortScience*, **15**, 400 (Abstract.)

Martin, W. E. (1937). Chemical study of the ripening process of Bosc pears. *Botanical Gazette*, **99**, 42–68.

Negm, F. B. (1981). Characterization and partial purification of aldose reductase (alditol-$NADP^+$ oxidoreductase) from *Euonymus* sp. leaves. *Plant Physiology*, **67**, Supplement 14.

Negm, F. B. (1982). Mannitol-1-phosphate dehydrogenase in *Fraxinus pennsylvanica*. *Plant Physiology*, **69**, Supplement 51.

Negm, F. B. & Loescher, W. H. (1979*a*). Detection and characterization of sorbitol dehydrogenase from apple callus tissue. *Plant Physiology*, **64**, 69–73.

Negm, F. B. & Loescher, W. H. (1979*b*). Purification of sorbitol dehydrogenase from pear callus tissues. *Plant Physiology*, **63**, Supplement 601.

Negm, F. B. & Loescher, W. H. (1981). Characterization of aldose 6-phosphate reductase (alditol 6-phosphate: NADP l-oxidoreductase) from apple leaves. *Plant Physiology*, **67**, 139–42.

Nuccorini, R. (1930). The function of mannitol in the olive. *Annali di chimica applicata*, **40**, 535–46.

Nuccorini, R. & Bartoli, O. (1932). Chemical investigations on fruits. VII. Experiments on the ripening of sorb apples. *Bollettino dell'Istuto superiore agrario di Pisa*, **8**, 277–90.

Obaton, F. (1929). Evolution de la mannite (Mannitol) chez les végétaux. *Revue générale de Botanique*, **41**, 282–4 and 622–33.

Okonkwo, S. N. C. (1964). *In vitro* culture of seedlings of *Striga senegalensis* Benth. *Nature*, **204**, 1108.

Plouvier, V. (1963). Distribution of aliphatic polyols and cyclitols. In *Chemical Plant Taxonomy*, ed. T. Swain, pp. 313–36. New York: Academic Press.

Redgwell, R. J. & Bieleski, R. L. (1978). Sorbitol-1-phosphate and sorbitol-6-phosphate in apricot leaves. *Phytochemistry*, **17**, 407–9.

Rumpho, M. E., Edwards, G. E. & Loescher, W. H. (1983). A pathway for photosynthetic carbon flow to mannitol in celery leaves. *Plant Physiology*, **73**, 869–73.

Schradie, J. (1966). Ph.D. Thesis, University of Southern California, USA.

Sellmair, J., Beck, E. & Kandler, O. (1969). Zur Physiologie der Clusianose. *Zeitschrift für Pflanzenphysiologie*, **61**, 338–42.

Sellmair, J. & Kandler, O. (1970). Zur Physiologie von Hamamelose und Hamamelit in *Primula clusiana* Tausch. *Zeitschrift für Pflanzenphysiologie*, **63**, 65–83.

Smith, D., Muscatine, L. & Lewis, D. (1969). Carbohydrate movement from autotrophs to heterotrophs in parasitic and mutualistic symbiosis. *Biological Reviews*, **44**, 17–90.

Stacey, B. E. (1974). Plant polyols. In *Plant Carbohydrate Biochemistry*, ed. J. B. Pridham, pp. 47–59. New York: Academic Press.

Stewart, G. R., Larher, F., Ahmed, I. & Lee, J. A. (1979). Nitrogen metabolism and salt-tolerance in higher plant halophytes. In *Ecological Processes in Coastal Environments*, ed. R. L. Jeffries & A. J. Davy, pp. 211–27. Oxford: Blackwell Scientific Publications.

Stokes, D. M. & Walker, D. A. (1972). Photosynthesis by isolated chloroplasts. Inhibition by DL-glyceraldehyde of carbon dioxide assimilation. *Biochemical Journal*, **128**, 1147–57.

Trip, P., Krotkov, G. & Nelson, C. D. (1963). Biosynthesis of mannitol-C^{14} from $C^{14}O_2$ by detached leaves of white ash and lilac. C^{14}-mannitol of high specific activity isolated from *Syringa vulgaris* and *Fraxinus americana*. *Canadian Journal of Botany*, **41**, 1005–10.

Trip, P., Krotkov, G. & Nelson, C. D. (1964). Metabolism of mannitol in higher plants. *American Journal of Botany*, **51**, 825–35.

Trip, P., Nelson, C. D. & Krotkov, G. (1965). Selective and preferential translocation of C^{14}-labeled sugars in white ash and lilac. *Plant Physiology*, **40**, 470–7.

Wallaart, R. A. M. (1980). Distribution of sorbitol in Rosaceae. *Phytochemistry*, **19**, 2603–10.

Webb, K. L. & Burley, J. W. A. (1962). Sorbitol translocation in apple. *Science*, **137**, 766.

Whetter, J. M. & Taper, C. D. (1963). Note on seasonal occurrence of sorbitol (D-glucitol) in buds and leaves of *Malus*. *Canadian Journal of Botany*, **41**, 175–7.

Whetter, J. M. & Taper, C. D. (1966). Occurrence of sorbitol (D-glucitol) and certain related sugars in germinating seeds and developing seedlings of *Malus*. *Canadian Journal of Botany*, **42**, 51–5.

Whiting, G. C. (1970). Sugars. In *The Biochemistry of Fruits and Their Products*, ed. A. C. Hulme, pp. 1–31. New York: Academic Press.

Zeigler, H. (1975). Nature of transported substances. In *Transport in Plants, I. Phloem Transport*, ed. M. H. Zimmerman & J. A. Milburn, *Encyclopedia of Plant Physiology*, New Series, vol. 1, pp. 59–136. Berlin: Springer-Verlag.

Zimmermann, M. H. & Zeigler, H. (1975). List of sugar alcohols in sieve tube exudates. In *Transport in Plants, I. Phloem Transport*, ed. M. H. Zimmermann & J. A. Milburn, *Encyclopedia of Plant Physiology*, New Series, vol. 1, pp. 480–503. Berlin: Springer-Verlag.

A. HEROLD

8 Biochemistry and physiology of synthesis of starch in leaves: autotrophic and heterotrophic chloroplasts

Introduction

Starch occurs in leaves, both as a product of photosynthesis and as a result of the metabolism of imported sugars. Starch content varies both within and between species at different stages of life cycles but why such differences occur and how formation of starch is regulated have still to be explained. This article presents a historical survey to emphasize the problems and proceeds to discuss certain aspects of the regulation of starch synthesis which have emerged from recent research and which contribute to solving these long-standing problems. Furthermore, because synthesis of starch is inevitably governed, in part, by availability of substrate, the investigations described also give information about the regulation of fluxes of sugar from one cellular compartment to another, especially between chloroplasts and cytoplasm. All observations exclude the special case of starch in stomatal guard cells (see Chapter 1). A full account of biosynthetic pathways has not been included since these have been more than adequately detailed in several excellent articles which are cited below and include Chapter 10 of this volume on the metabolism of storage starch.

Historical background

Sachs (1862, 1864) reported the relationship between photosynthetic assimilation of CO_2 and synthesis of starch, and produced 'starch prints' to demonstrate the requirement for light. (See Fig. 8.1.) He observed the diurnal/nocturnal fluctuations of starch in leaves and proposed that starch represented a temporary storage product, synthesized during the day and mobilized at night. Brown & Morris (1893) provided a comprehensive survey of the early literature. Interspecific differences with respect to starch synthesis were noted early and have been discussed in Chapter 1. Parkin (1899) made a valuable contribution by studying in detail three families, the Liliaceae, Amaryllidaceae and Iridaceae. Sampling over a period of months revealed changes which occur within a single species during the life cycle. Thus, in the tiger lily, there was abundant starch prior to flowering but, following anthesis, it was confined to the leaf base and even there was very sparse. Eckerson (1909) also noted that only the very young leaves of *Primula obconica*, *P. sinensis* and *Cineraria cruenta* could be 'emptied of starch'. Young plants of geranium lost starch after about 40 h in the dark but older plants still contained the polysaccharide even after 4 days (cf. Komura, Yano & Hattori, 1966). It was suggested that the leaves may act as storage organs in mature plants and this view has been expressed by later workers (e.g. Tollenaar, 1925; Matheson & Wheatley, 1963; Badenhuizen, 1969; see Herold, 1977, for other references). If the starch of mature leaves does serve a storage function, the distinction between the 'assimilatory' starch of leaves and the 'storage' starch of seeds and vegetative organs is no longer valid.

Early workers also proposed that there was a relationship between accumulation of starch in leaves and the rate of translocation of sugar from the leaf, as influenced by distribution of vascular tissue, or by the presence and activity of non-photosynthetic 'sinks'. Schimper (1885) suggested that the low starch content of leaves of monocotyledons was due to the leaf shape and pattern of venation which favours rapid translocation. Altering the ratio of photosynthesizing leaves to non-photosynthetic sinks influences translocation rate and the starch content of leaves. Saposchnikoff (1890) found that detaching 12 of the 14 leaves of a sunflower plant, i.e. decreasing the amount of photosynthetic tissue, caused the rate of loss of starch from the remaining two leaves to increase by a factor of 3.5 (cf. Porter & Bird, 1962;

Fig. 8.1 Starch print produced by illuminating a starch-free leaf through a photographic negative. Starch was detected with iodine solution.

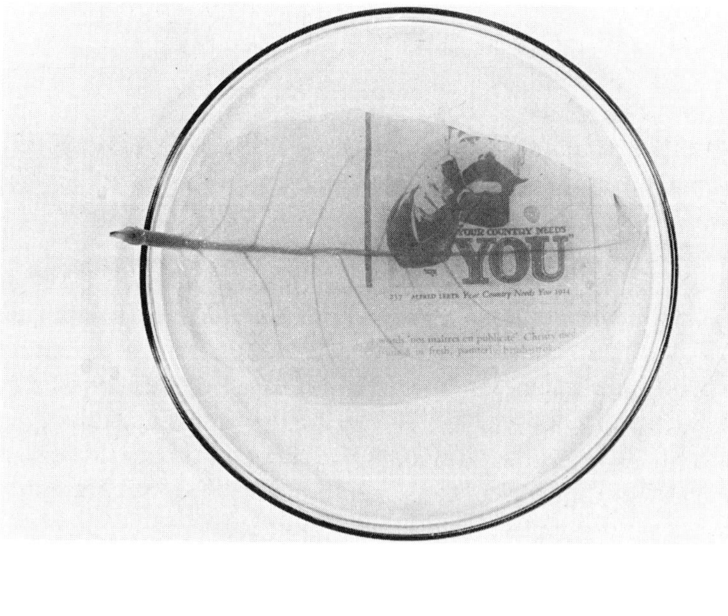

De Fekete & Vieweg, 1974; Herold & Walker, 1979). Conversely, absence of storage organs in certain aquatic plants, i.e. an increased ratio of photosynthetic source to sink, was thought to account for the marked accumulation of starch in the leaves (Schimper, 1885). More recent work using sugar cane has shown that a partial inhibition of translocation induced synthesis of starch in mesophyll cells, which do not normally contain this polysaccharide (Laetsch, 1969).

The terms 'high-starch' and 'low-starch' plants are still applied today but just why some species synthesize more starch than others and why the rate of mobilization is so varied remains as much of a mystery today as it was to botanists of the nineteenth century. Similarly, although changes in the starch content of leaves during the life cycle are accepted as a feature of growth and maturation, the controls which regulate the apportioning of photosynthetic product between starch, translocatory sucrose and metabolites for immediate use in the leaf are only just beginning to be understood (see also Chapter 9).

Observations of formation of starch from exogenous sugar were first reported at the end of the last century. While some, normally 'starch-free', species have been induced to form starch by supplying abnormally high concentrations of exogenous sugar over a period of days (see below), this was not possible in various species of onion (Boehm, 1883). Boehm also investigated the metabolism of exogenous sucrose and glucose in the dark and concluded that photosynthetically-produced sugars must be translocated and used by non-green areas of plants. Meyer (1886) fed a variety of sugars and sugar alcohols and noted that those sugars which occur naturally in plants are utilized for starch synthesis (see Herold, 1978). All species investigated could metabolize sucrose in this way. Phillis & Mason (1937) recorded the requirement for either light or oxygen during synthesis of starch from exogenous sugars and this was subsequently related to the source of ATP (Maclachlan & Porter, 1959). Feeding sugars to leaf pieces or disks provides an easily manipulated system, assumed to mimic natural accumulation of metabolites, although exogenous sugars may enter compartments not accessible to translocatory sugar or enter at rates much greater than is normal for endogenous sugars (Thomas, Ranson & Richardson, 1973).

Sachs (1862) located the chloroplast as the site of starch synthesis and later work with isolated chloroplasts indicated that synthesis was independent of the surrounding cytoplasm. However, the importance of continuous exchange of metabolites between chloroplast and cytoplasm is now acknowledged (see reviews of Walker & Herold, 1977; Herold & Walker, 1979). As currently understood, synthesis of starch in leaves is achieved by the same enzymes as described first for storage tissue (Chapter 10). A degradative role is generally ascribed to phosphorylase although this enzyme catalyses formation of starch *in vitro* and some workers still feel that it is involved in synthesis *in vivo* (Porter, 1962; Badenhuizen, 1965, 1969; Baxter & Duffus, 1973) (see Chapter 9). Leloir, De Fekete & Cardini (1961) established a synthetic pathway involving nucleotide diphosphate sugars, in which glucose units are transferred from adenosine diphosphate glucose (ADPG) to an existing glucose chain (Chapter 10). It now seems likely that the transferase enzyme (starch synthase) occurs in several forms some of which do not require primer (see e.g. Manners, 1974). Subsequently, chloroplasts from leaves of spinach have been shown to contain the enzymes associated with this pathway (see Preiss & Levi, 1979, for details). ADPG pyrophosphorylase is the major regulatory enzyme (e.g. Preiss, Ghosh & Wittkop, 1967). Triose phosphate (Fig. 8.2) is the primary product of CO_2 fixation and synthesis of assimilatory starch is an alternative to its transport from the chloroplast or its use to replenish intermediates of the Benson-Calvin Cycle (Walker & Herold, 1977). Regulation of the consumption of TP is discussed in a subsequent section. Where imported or exogenously supplied sugars provide the substrate for formation of starch it has been assumed, because of the limited direct transport of other sugars and sugar phosphates (Fig. 8.3), that degradation to TP occurs in the cytoplasm prior to entry into the chloroplast. However, the possibility that alternative metabolites in the TP→starch pathway (Fig. 8.2) may traverse the envelope of immature chloroplasts is proposed in a later section.

Although forms of starch synthase which have no requirement for primer have been identified, starch, in most cases, is synthesized only when a short chain glucan is provided. The source of this primer *in vivo* is uncertain.

Fig. 8.2 The Benson-Calvin cycle of photosynthesis, showing diversion of its intermediates into the biosynthesis of starch.

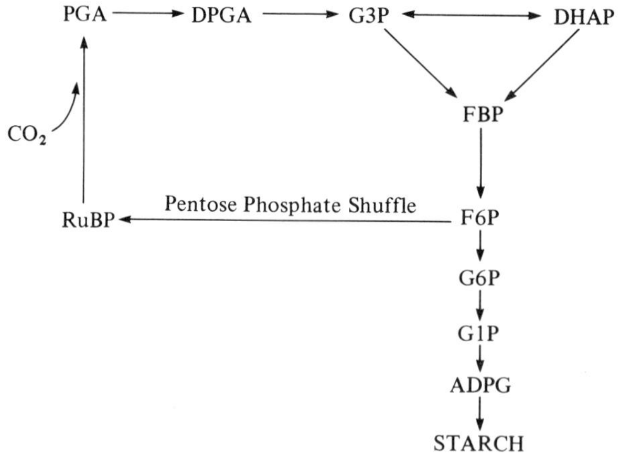

Similarly unexplained is the role, if any, of maltose in the synthesis of starch (De Fekete & Vieweg, 1974; Herold, McGee & Lewis, 1980, and see Chapters 5 and 9).

Detection and measurement of starch in leaves

Starch grains can be located microscopically (Fig. 1.1 in Chapter 1) or be simply detected with a solution of iodine. The usual procedure is to drop the leaf or leaf pieces into boiling 80% ethanol to extract chlorophyll, leaving a colourless tissue. The leaf is then washed and immersed in a dilute solution of iodine and potassium iodide which forms a blue-black complex with the amylose component of starch. Details of this technique as used in the preparation of starch prints (to be discussed below) are described fully by Walker (1979). There are several chemical methods for the quantitative measurement of starch (e.g. MacRae, 1971) but these are being superseded by assay of the free glucose released during incubation of an ethanol-insoluble leaf fraction with amyloglucosidase, an enzyme which cleaves the glucose chain (see Chen-She, Lewis & Walker, 1975). However, it should be noted that true 'starch' (by definition molecules of more than 300 glucose units – see Chapter 1) is probably overestimated by 20–40% (Porter, Martin & Bird, 1959; Kakie, 1969) since other glucans of shorter chain length will also yield

Fig. 8.3 Flux of sugars and sugar phosphates across the chloroplast envelope. Thickness of arrow indicates the relative ease of permeation. The envelope is virtually impermeable to sucrose (Heldt & Sauer, 1971) and sugar bisphosphates, and has increasing permeability to hexose monophosphates, free hexoses and pentoses (Schafer *et al.* 1977) and pentose monophosphates (see Heber, 1974; Walker, 1974 for reviews). DHPA and PGA together with Pi are the major metabolites transported (Heldt, 1976).

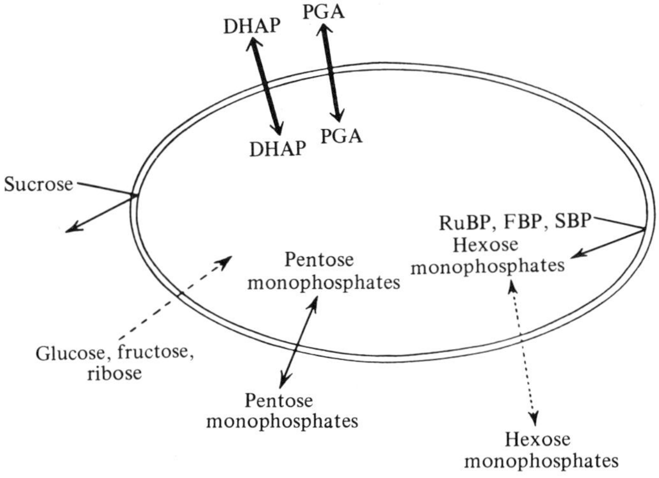

glucose. Conversely, an extraction procedure which is designed to detect a range of carbohydrates could underestimate starch unless appropriate steps are taken to recover its water-soluble component (Herold *et al.* 1980). Reports should state precisely the methods used for extraction and assay. Care should also be exercised in expressing results. The comparative starch content of different species varies markedly depending on the way in which results are calculated (see comments on Fig. 8.4 in the text below). Results are more meaningful when expressed in as many ways as possible, for example on chlorophyll, leaf area, protein, wet weight and dry weight bases.

The regulation of assimilatory (autotrophic) starch synthesis

In intact plants, starch is synthesized during the day and mobilized by night and can account for up to 30% of the total CO_2 incorporated (Walker & Herold, 1977). Figure 8.4 shows the starch content of leaves of three species grown in the glasshouse and assayed at hourly intervals on a June day. Expressed on a leaf area basis, tobacco, with less chlorophyll per unit leaf area, would have considerably higher starch content than the other two species. In isolated chloroplasts, starch can be identified as an early product

Fig. 8.4 The changing starch content of leaves of three species on a day in June. Plants were grown under glass with supplementary lighting between 8 a.m. and 5 p.m. and with supplementary CO_2. At the times shown, a disk was cut from each of 10 mature leaves and bulked for determination of starch. Chlorophyll was extracted in 80% ethanol and estimated according to Arnon (1949). Starch was measured enzymatically (Chen-She, Lewis & Walker, 1975).

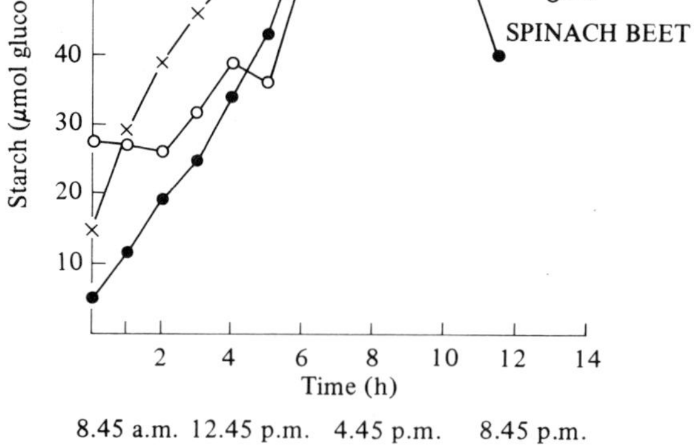

of photosynthesis (Walker & Herold, 1977; Herold & Walker, 1979, for discussion) and forms a relatively constant fraction of the total CO_2 incorporated. As such it can be regarded as a direct product from TP as an alternative to accumulation of intermediates of the Benson-Calvin cycle (Fig. 8.2) or export of TP from the chloroplast. When phosphorylated sugars accumulate in the chloroplast, the concentration of orthophosphate is lowered and synthesis of starch can be regarded as a mechanism for recycling Pi. Thus, although differences in content and activity of those enzymes associated with synthesis of starch could account for both interspecific and intraspecific variation in starch content, availability of substrate TP, governed by those processes which regulate its diversion through the alternative channels, is likely to be equally important. This has, in fact, been indicated in recent work. Increase or decline in starch content has been ascribed jointly to a direct effect of [Pi] in the chloroplast on certain enzymes and the influence of external [Pi] on availability of substrate TP within the chloroplast. This latter effect is an outcome of the activity of the 'phosphate translocator' (Heldt, 1976), a mechanism for the stoichiometric exchange of TP, Pi and phosphoglycerate across the chloroplast envelope. During photosynthesis, efflux of TP coupled with influx of Pi is the normal situation and the external [Pi], i.e. that in the cytoplasm, influences this exchange.

Regulation by Pi in isolated chloroplasts, protoplasts and leaf disks

[Pi] in the medium of isolated chloroplasts affects synthesis of starch in a manner which is unrelated to total CO_2 fixation (Preiss & Levi, 1979). Low external [Pi] favours formation of starch in spite of decreased CO_2 fixation and this is a result of decreased export of TP via the phosphate translocator. High external [Pi] is associated with decline in starch content and this is explained by enhanced export of TP to the detriment of both starch formation and the maintenance of intermediates of the Calvin cycle. Figure 8.5 shows the three alternative channels for TP in relation to the effect of external [Pi]. Enzyme activities are also affected. Low [Pi] creates an increased PGA/Pi ratio inside the chloroplast stimulating the activity of ADPG pyrophosphorylase (Preiss *et al.* 1967). At the same time, degradation of starch, where phosphorylase is involved, is likely to be impeded. In this way, effects on both enzymes and on the phosphate translocator contribute to synthesis and maintenance of starch. High [Pi] in the medium or cytoplasm will have the opposite effect on enzyme activities (see Preiss & Levi, 1979, for discussion) and will enhance export of TP. Light–dark regulation of starch metabolism *in vivo* is likely to be achieved by the effect of changes in concentration of metabolites on synthesis of ADPG (e.g. Kaiser & Bassham, 1979) and on the activity of phosphorylase.

In certain leaf disks, e.g. from spinach, lowering [Pi] by supply of exogenous mannose and consequent irreversible formation of mannose phosphate also causes increased starch formation, again in spite of a marked lowering in CO_2 fixation (see Herold & Lewis, 1977, for a review of effects due to mannose). Mannose can also lower [Pi] of spinach protoplasts. When a combination of sorbitol and mannose are used in the enzyme digestion medium rather than

Fig. 8.5 The effect of external [Pi] on formation of starch in the chloroplast. External [Pi] regulates the internal concentration of TP, the substrate for synthesis of starch, and the PGA/Pi ratio (see text). In 'A', the normal situation, 60–80% of the fixed carbon is exported from the chloroplast (e.g. Hofstra & Nelson, 1969; Ho, 1976) and a relatively constant fraction of 10–30%, dependent on species, age etc. is diverted into starch. In 'B', low external [Pi] restricts efflux of TP and conditions favour synthesis of starch. Conversely, in 'C', a high external [Pi] stimulates efflux of TP, so restricting synthesis of starch.

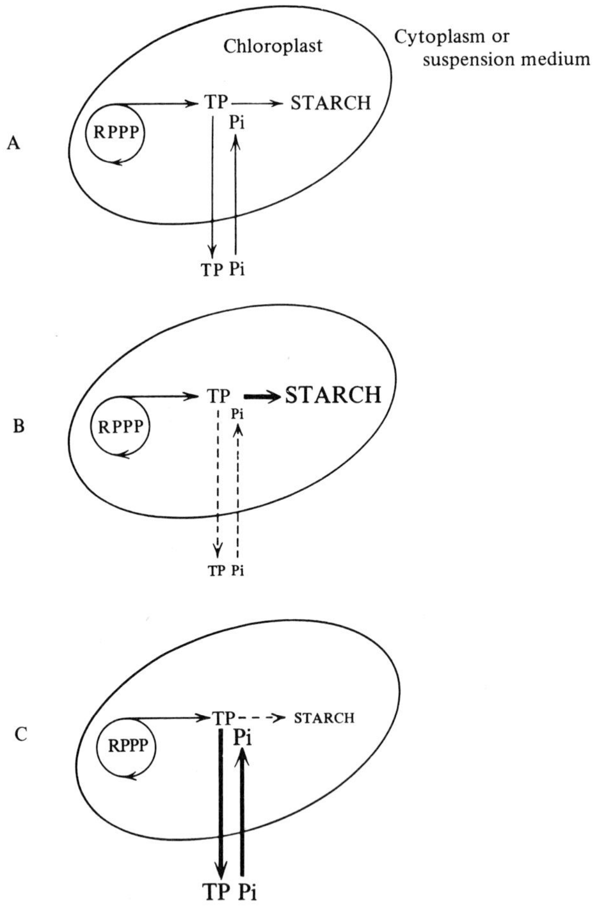

sorbitol alone (Herold, Leegood, McNeil & Robinson, 1981), subsequent photosynthesis is lowered but there is an increase in the fraction of CO_2 incorporated into starch (but not necessarily an increase in total starch content). This effect increased with increasing mannose concentration up to 100 mM (Table 8.1). The importance of the cytoplasm in regulating chloroplastic metabolism is emphasized by the fact that chloroplasts, released by rupturing control and mannose-pretreated protoplasts (Edwards, Robinson, Tyler & Walker, 1978), were similar in their photosynthetic activity

Table 8.1. *The effect of mannose on photosynthesis and incorporation of ^{14}C from $^{14}CO_2$ into starch in spinach protoplasts and chloroplasts released from the protoplasts (Herold & McNeil, unpublished)*

	Photosynthesis (μmol mg^{-1} chl h^{-1})		
	O_2 evolution (maximum)	$^{14}CO_2$ uptake (based on initial 12 min)	^{14}C starch as % total ^{14}C
A. Protoplasts			
Control	66.0	37.8	16.2
Mannose-pretreated (mM)			
50	35.0	23.8	23.6
100	27.0	16.4	26.0
250	22.0	15.4	19.7
500	16.0	9.1	19.4
B. Chloroplasts			
Control	71.0	28.4	8.0
Mannose-pretreated (mM)			
50	78.0	26.9	8.2
100	70.0	28.0	7.0
250	70.0	44.3	6.9
500	65.0	35.2	6.6

Protoplasts were isolated using the method of Edwards *et al.* (1978). Pretreatment with mannose involved incubation in an enzyme medium containing a range of mannose concentrations in addition to sorbitol to give an overall concentration of 500 mM. Protoplasts (50 g chl ml^{-1}) were illuminated for 12 min in a medium containing 500 mM sorbitol, 25 mM Tricine, pH 7.6, 1 mM $CaCl_2$ and 5 mM $NaH^{14}CO_3$ (20 Ci/mol) and photosynthesis measured by oxygen evolution and $^{14}CO_2$ uptake. The reaction was terminated by the addition of 4 ml boiling absolute ethanol and the protoplast pellet re-extracted in a further 2 ml 80% ethanol. The radioactivity of the ethanol-soluble and the ethanol-insoluble extract following digestion with amyloglucosidase was determined from planchets. Chloroplasts were obtained from the protoplasts (Edwards *et al.* 1978). For photosynthesis, the assay medium contained 330 mM sorbitol, 20 mM HEPES, pH 7.6, 10 mM EDTA, 0.2 mM Pi and 5 mM $NaH^{14}CO_3$ (20 Ci/mol).

and starch formation, i.e. mannose did not appear to have any direct effect on the chloroplasts. Also, as regularly observed, the starch content of isolated chloroplasts was lower than that in intact protoplasts, implying that the medium bathing chloroplasts *in vitro* does not match the cytoplasm in its regulatory properties.

Plants grown in phosphate-deficient conditions frequently have a high starch content (see Herold & Lewis, 1977; Herold & Walker, 1979) and, although other factors are involved at the whole-plant level, a direct effect of low [Pi] on enzyme activities and the export of TP from the chloroplast is likely to be a factor in determining the high starch content. In more recent work, the effect of enhanced phosphate concentration in intact soybean plants was examined. Not only was there a decreased starch/sucrose ratio during photosynthesis (i.e. the opposite of P-deficient conditions) but there was increased translocation of sucrose to the developing pod (R. Giaquinta, personal communication, see Chapter 3). Changes in content or availability of Pi in leaves during the life cycle of plants may contribute to the regulation of starch metabolism *in vivo* (see e.g. Hopkinson, 1964; Thorne & Koller, 1974).

Accumulation of starch in the leaves of intact plants: sink–source interactions

Ageing leaves frequently have a high starch content, for example those of tobacco (e.g. Matheson & Wheatley, 1963). This may reflect a lowered demand for sugars by sink tissues subsequent to flowering, seed production etc. This phenomenon and other examples of accumulation of starch and sucrose associated with decreased sink activity have been reviewed by Neales & Incoll (1968), Herold (1980) and others. The question of a correlation, or 'link' (Herold, 1980), between increased carbohydrate content and the photosynthetic decline which is also a feature of reduced sink activity (e.g. Geiger, 1976) will not be considered here but rather the way in which increased synthesis of starch in the chloroplast may be initiated and regulated by distant sinks. Greater understanding of the extreme situation may then help to appreciate the day to day and minute to minute regulation of synthesis of starch.

Decreased consumption of translocated sucrose by distant 'secondary' (Herold, 1980) sinks causes its subsequent accumulation in the leaf (see Geiger, 1979). Sucrose cannot permeate the membrane of the mature chloroplast (Heldt & Sauer, 1971) and cannot, therefore, have a direct effect on chloroplastic metabolism but its concentration is likely to affect levels of certain other metabolites and, thereby, favour starch formation. Figure 8.6 shows the reactions involved in synthesis of sucrose from triose phosphate in the cytoplasm (see also Chapter 2). Sucrose inhibits sucrose phosphate synthetase (Salerno & Pontis, 1978) and sucrose phosphate phosphatase (Hawker, 1967) and the possible significance of these inhibitions in relation to likely changes in Pi and TP concentrations has already been discussed at length (Herold & Walker, 1979; Herold *et al.* 1980; Herold, 1980) and is mentioned in Chapter 2. In these circumstances, an increase in starch would be induced by lowered [Pi] (see previous section) or raised [TP], conditions initiated by the increased concentration of sucrose. That inhibition of sucrose phosphate synthetase is likely to cause accumulation of intermediates in the TP → sucrose pathway has recently been confirmed. When leaf disks of sugar beet were floated on sucrose prior to and during photosynthesis, concentrations of hexose phosphates were significantly increased over those in control disks floated on water (Table 8.2). Although, in the same disks, the exogenous sucrose was metabolized to glucose and fructose, there was no measurable change in hexose phosphates until CO_2 fixation commenced. This supports

Fig. 8.6 Intermediates of the pathway synthesizing sucrose. End-product inhibition has been demonstrated for sucrose phosphate phosphatase (1). Sucrose also inhibits sucrose phosphate synthetase (2).

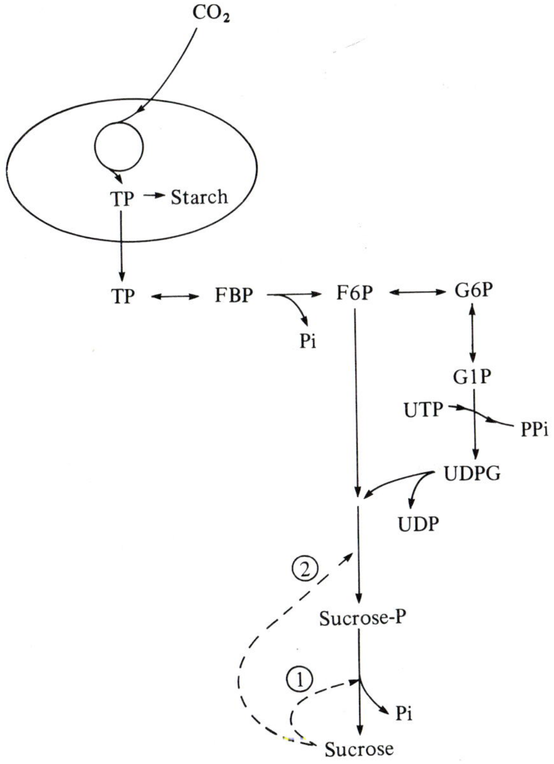

Table 8.2 *The effect of pretreatment with sucrose on the incorporation of CO_2 into hexose phosphates in sugar beet leaf disks (A. Herold, unpublished)*

	Hexose phosphate (nmol mg^{-1} chl)	
Pretreatment[a]	Glc-6-P	Fru-6-P
Experiment 1		
Water	128 ± 7	30 ± 1
Sucrose	261 ± 17	61 ± 3
Experiment 2		
Water	99 ± 3	24 ± 2
Sucrose	170 ± 13	43 ± 2

[a] Sets of 40 disks (50 mm in diameter) were floated overnight on water or 50 mM sucrose solution in low light.
Leaf extracts were prepared by grinding the disks to a powder in liquid nitrogen, adding 2 ml 10% $HClO_4$ and neutralizing with K_2CO_3. Metabolites were measured enzymatically (Lowry & Passonneau, 1972). The increased hexose phosphate concentrations of sucrose-pretreated disks were only observed following a period of illumination. Each result is the mean of three (Expt 1) or four (Expt 2) sets of disks ± standard deviation.

Table 8.3 *Photosynthesis and starch content of pot-bound tobacco plants prior to and following 'potting-on' (from Herold & McNeil, 1979)*

	Photosynthesis (mg CO_2 dm^{-2} h^{-1})	Starch (mg dm^{-2})
A. Prior to potting-on		
14 November	14.4	2.4
21	12.3	6.6
28	10.3	17.6
5 December	5.1	60.7
B. Following potting-on		
6 December	9.2	63.5
7	10.2	52.5
9	12.6	31.4
12	15.7	7.4
15	18.2[a]	8.0

[a] Values ranged from 15.2 in the oldest leaves to 22.0 in leaf 1, the youngest from which disks could be taken.
Each result is the mean value of six leaves taken from two plants. Seedlings were transferred to 6.5-cm pots on 31 October 1977 and re-potted to 9.0-cm pots on 5 December.

the view that accumulation of sucrose affects the distribution of newly fixed carbon. The significance of this in relation to accumulation of starch has still to be explained.

In some circumstances, accumulation of starch in leaves may be an important feature of the life cycle comparable to its temporary storage in the stems of some plants, for example wheat (see Herold, 1980, for review). This view, which has been proposed for tobacco on several occasions (see Introduction), is further supported by recent work – again with tobacco. In pot-bound plants, starch in leaves increased several-fold, although there was a marked decline in photosynthesis and content of RuBP carboxylase (Herold & McNeil, 1979). When plants were then transferred to large pots there was a rapid loss of starch and an equally rapid restoration of photosynthesis (Table 8.3). Following complete recovery, such plants are frequently more robust and have higher rates of photosynthesis than those which have been grown in large pots throughout. It was proposed that, in this situation, mobilization of the starch which accumulates during a period of low sink activity provides substrate for an accelerated synthesis of new plant materials. In the same way, starch which accumulates in mature leaves of some species may be important in providing substrate for processes associated with ageing at a later stage (cf. Starch – plan or accident, Chapter 9). The possibility that the chloroplasts of old leaves may function heterotrophically as amyloplasts (see 'Heterotrophic chloroplasts' below) and metabolize, for synthesis of starch, sugars imported from other leaves has still to be examined (cf. 'Starch in infected plants' below).

Heterotrophic chloroplasts

Accumulation of starch is a feature of the early development of chloroplasts (Whatley, 1974) and here it is a product of substrate imported from photosynthetically active parts of the plant and from storage organs (Joy, 1967; Dickson & Larson, 1975). Similarly, when starch is formed from exogenously-supplied sugars in the dark, the chloroplasts are entirely dependent on the external substrate and are, in this situation, heterotrophic organelles, functioning as amyloplasts. The supplied or imported sugar, or some product of the sugar, enters the chloroplast and is metabolized to starch. However, availability of sugar to darkened leaf disks or pieces, or to intact leaves does not necessarily result in the synthesis of starch. Likewise, for 'starch prints' to be possible, darkened leaf cells fail to form starch from translocatory sucrose. For heterotrophic chloroplasts, regulation of import of sugar is likely to be the major factor in the control of starch synthesis.

Synthesis of starch from exogenous sugars in leaf disks of tobacco

When disks from leaves of tobacco of different age were floated on sugars in the dark, abundant starch was formed in the youngest leaves but there was a marked decline with leaf age (Herold, 1978) (Fig. 8.7). This contrasted with autotrophic starch formation where disks had a similar content, irrespective of leaf age. Uptake of exogenous sugar was comparable in the differently aged leaves. Although other factors are involved, such as the increasing retention of sugar in the vascular tissue with age (Fellows & Geiger, 1974), it was concluded that metabolism of exogenous sucrose or glucose to starch was likely to be influenced both by a decline in the rate of respiratory degradation of sugars to TP and by changes in permeability of the chloroplastic envelope.

As discussed above, the envelope of the mature chloroplast is virtually impermeable to sucrose (Fig. 8.3), and glucose enters only very slowly (Schafer, Heber & Heldt, 1977). It is assumed, therefore, that, for the mature chloroplast, degradation to TP is a prerequisite to entry of sugar. However,

Fig. 8.7 Synthesis of starch from exogenous sugar as a function of leaf age. Disks were cut from successive leaves of a starch-free tobacco plant and floated on 0.5 M sucrose solution in the dark for 24 h. Leaf 3 was the youngest from which the required number of disks could be taken. Starch was measured enzymatically (Herold, 1978).

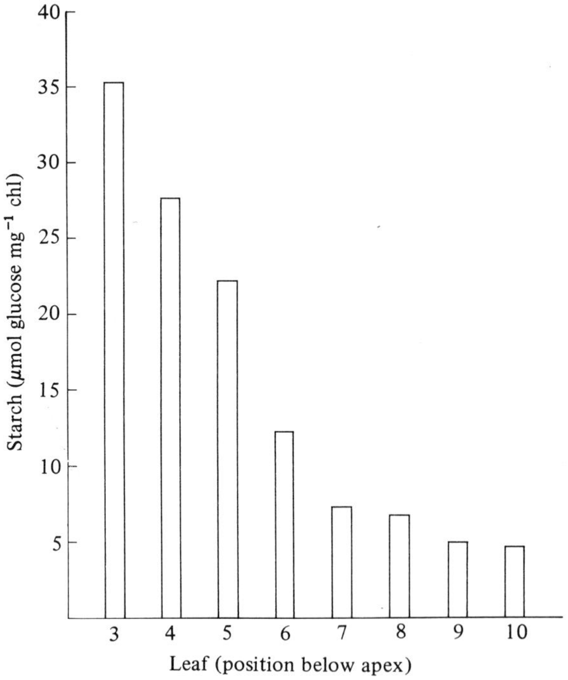

for young leaves which are still at the importing stage, the permeability of chloroplasts may differ as also may those of amyloplasts of non-photosynthetic tissues (Chapter 10). Changes in the selective permeability of the inner membrane during development of chloroplasts have been recognized (Hampp & Wellburn, 1976) and these are likely to be linked with developmental changes in the structure and biochemistry of the envelope (Kirk, 1970; Whatley, 1974; Bradbeer, 1976; Leech, 1976). It follows that there may be direct entry of a range of metabolites, including free hexoses and sucrose, hexose phosphates and sugar nucleotides into immature chloroplasts (Fig. 8.8 and see Herold, 1978), thereby providing increased substrate for synthesis of starch. Direct entry of hexoses would explain the incomplete randomization of ^{14}C in the glucose released from starch which was synthesized from exogenously-supplied [^{14}C]glucose (Kandler & Gibbs, 1959; Maclachlan & Porter, 1959, 1961; Marker & Whittingham, 1966; Raven, 1976). If degradation to TP was always obligatory, complete randomization of label would be expected. Likewise, greater permeability to ATP in immature as compared to mature chloroplasts would explain how a high rate of starch formation,

Fig. 8.8 Possible mechanisms for the synthesis of starch from exogenously-supplied sucrose and glucose in immature chloroplasts. In addition to the major flux of carbohydrate across the envelope (1), greater permeability to hexose phosphates (2), sucrose, (3), nucleotide diphosphates (4), synthesized in a reaction catalysed by sucrose synthetase (see Chapter 2), and ATP (5) may contribute to the enhanced synthesis of starch in young leaves.

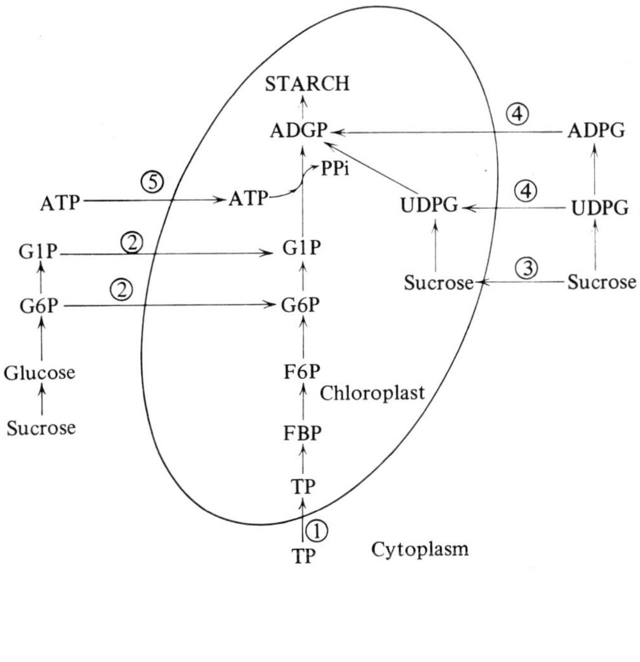

a process which consumes ATP, could be maintained in the absence of photophosphorylation.

Starch prints

In 1887, Sachs observed that 'this experiment (i.e. starch print formation) demonstrates yet another fact, viz. that the action of light is strictly local and is not transferred to neighbouring shaded places'. For the layman, starch prints still provide a very convincing demonstration of the light requirement in photosynthesis but, for the present-day plant physiologist, they pose problems concerning the regulation of the synthesis of starch (Walker, 1974, 1979). Sucrose is the major translocatory sugar in higher plants (Arnold, 1968; see Chapter 3). Rapid translocation of sucrose out of leaves during photosynthesis (e.g. Trip, 1969) and the demonstration of starch formation in the dark from exogenously supplied sucrose (above) must be reconciled with a complete absence of starch in darkened areas of the leaf (Fig. 8.1). This problem is most likely one of regulation of transport of sucrose and its availability, presumably via degradation to TP, for synthesis of starch in the chloroplasts of neighbouring darkened cells.

There is ample documentation of the restricted movement of assimilatory sucrose in leaves. In 1953, Porter observed that movement of assimilates across the leaf lamina is very slow and subsequent work with isotopes (Aronoff, 1955; Jones & Eagles, 1962; Porter, 1966; Penny & Nelson, 1970) indicates that, when $^{14}CO_2$ is supplied to a limited area of a leaf, ^{14}C is not detected in either illuminated (Aronoff, 1955) or darkened (Jones & Eagles, 1962) neighbouring cells. This restricted transport can be demonstrated visually by producing autoradiographs of leaves which have been illuminated through a photographic negative and exposed to $^{14}CO_2$, so explaining the 'coarse' regulation associated with starch print formation (Herold, 1977). Nevertheless, the clarity of definition which can be achieved in starch prints (Walker, 1979) implies a cut-off of starch formation at the cellular, or possibly chloroplastic, level and this requires further explanation regarding 'fine' regulation. Photosynthetically-produced sucrose crosses at least a few cells before entering the minor veins, and vein-spacing dimensions are such (see e.g. Esau, 1953; Fellows & Geiger, 1974) that slight loss of clarity would be observed if starch was synthesized from translocatory sucrose in these cells. For example, Wylie (cited by Esau, 1953) measured minimum distances of 85–124 μm between the minor veins of five different species. The route for short-distance transport of sucrose is still uncertain but current opinion favours a route which is partly symplastic and partly apoplastic. At the phloem complex, sucrose is actively loaded from the apoplast by a transport mechanism which may involve co-transport of protons (see Chapter 3). If, as is proposed, this transport has a symplastic component, it must be assumed that some form of compartmentalization exists, preventing either degradation of sucrose or entry of its products of degradation into the chloroplast, the site of formation of starch.

Starch in the leaves of infected plants

Abnormal accumulation of starch in leaves infected by fungal and other pathogens is well documented (see Whipps & Lewis, 1981, for a comprehensive list of references). They consider the mechanisms involved in this altered pattern of starch synthesis and discuss the likely possibility that the biotrophic pathogens cause a decline in [Pi] with consequent stimulation of photosynthetic starch formation (see 'Regulation by Pi', above). Decreased [Pi] in such infected leaves has, in fact, been reported on one occasion (MacDonald & Strobel, 1970) although in other studies results have been inconclusive. Changes in the pattern of translocation of sucrose occur in infected plants and mature leaves assume an importing status. It is reasonable to speculate, therefore, that imported sucrose is also utilized for formation of starch – i.e. the chloroplast functions concurrently as an autotrophic and heterotrophic organelle – and regulation of the sucrose to starch pathway is relevant (see 'Heterotrophic chloroplasts', above). In addition to induced high activity of certain cytoplasmic enzymes, including invertase (Whipps & Lewis, 1981) and those associated with respiratory degradation of glucose (see Montalbini & Buchanan, 1974, for a review), there is again the possibility that permeability of the chloroplast envelope is altered in the presence of the pathogen.

A miscellany of effectors

The foregoing sections here show that, in many circumstances, the regulation of synthesis of starch is imposed by availability of substrate and that the selective permeability of the chloroplast envelope is a major factor. However, a wide range of chemicals can influence formation of starch (though not always that in leaves) and, whilst it is tempting to suggest that modification of the chloroplastic envelope and its selective properties may occur, the possibility of direct effects on enzymes of the synthetic pathway must also be considered. There is a continued turnover of starch (Tollenaar, 1925; Jones, Martin & Porter, 1959; Chan & Bird, 1960; Kakie, 1969; De Fekete & Vieweg, 1974) and changes in the activity of degradative enzymes, for example the diurnal variation in amylolytic activity in spinach chloroplasts (Pongratz & Beck, 1978), may also contribute to measured levels of starch.

Metabolism of starch may be influenced by certain growth factors. For example, McLaren & Smith (1977) found that the starch content of duckweed was increased when abscisic acid was included in the incubation medium, in spite of a decreased rate of CO_2 uptake. They concluded, from pulse-chase experiments, that there was a decreased turnover of starch. For stomatal guard cells also, abscisic acid caused increased starch content (Mansfield & Jones, 1971). Conversely, Thorpe & Meier (1974), working with illuminated tobacco callus tissue, found that gibberellic acid caused an *increased* turnover of starch, likely to be associated with an increase in activity of phosphorylase and a lowered content of starch. Again, when bean leaves were sprayed with gibberellic acid, amounts of starch decreased and those of soluble sugars increased as compared to unsprayed leaves (Marcelle *et al.* 1974). Hawker, Marschner & Downton (1974) have studied the effect of sodium and potassium ions on synthesis of starch in the leaves of several species. For sugar beet, bean and *Atriplex nummularia*, K^+, but not Na^+, appeared to be essential for its formation by stimulating the activity of starch granule-bound starch synthase activity (see Fig. 8.2). On one occasion the starch content of sugar beet leaves was increased 9-fold by adjusting the $[K^+]$ of the growth medium. Accumulation of starch is likely to occur also when low levels of nitrogen, phosphorus or magnesium limit metabolism of photosynthetic sugars for protein synthesis, thereby increasing substrate available for formation of starch (Herold & Walker, 1979). Just what significance, if any, these substances have in the regulation of formation of starch *in vivo* has still to be established.

Finally, environmental factors, in particular temperature, irradiance and concentration of CO_2, are likely to affect metabolism of starch independently of their effect on photosynthesis (e.g. Forde, Whitehead & Rowley, 1975). High $[CO_2]$ in some way increases formation of starch (Madsen, 1968; Hofstra & Hesketh, 1974; Nafziger & Koller, 1976). Temperature has a marked effect on the rate at which sugars are converted to a range of storage products, including starch (Spiertz, 1974) and changes in membrane properties may be a factor here (Downton & Hawker, 1974).

Concluding comments

For many years, the significance of synthesis of starch in leaves has been questioned. Does assimilatory starch consume a genetically determined fraction of the CO_2 incorporated during photosynthesis, thereby making an essential contribution to growth processes during the dark period or has it evolved as a safety mechanism so that variable synthesis prevents osmotic disruption of chloroplasts when synthesis of chloroplastic sugars occurs more rapidly than consumption? Also, do the mature leaves of some species have a definite storage function comparable to that of conventional storage organs, metabolizing to starch sugars imported from other parts of the plant? What structural and biochemical changes occur such that healthy and infected mature leaves sometimes can reassume an importing status and metabolize translocatory sucrose to starch? Until more is known about the regulation of both autotrophic and heterotrophic starch metabolism, until the effects of a range of chemical and environmental factors are better understood, and until the relationship, if any, between the starch content of leaves and growth processes throughout the plant is determined, we cannot hope to provide an answer to these questions. Leaf starch frequently accounts for up to 20% of the total fixed carbon and its ultimate fate should be of major interest at a time when the need to increase crop productivity is so urgent.

I wish to thank Professors D. A. Walker and D. H. Lewis who first interested me in the problems of the metabolism of starch. The Science and Engineering Research Council and Imperial Chemical Industries are gratefully acknowledged for financial support during postgraduate and postdoctoral studies.

References

Arnold, W. N. (1968). The selection of sucrose as the translocate of higher plants. *Journal of Theoretical Biology*, **21**, 13–20.

Arnon, D. I. (1949). Copper enzymes in isolated chloroplasts. Polyphenoloxidase in *Beta vulgaris. Plant Physiology, Lancaster*, **24**, 1–15.

Aronoff, S. (1955). Translocation from soybean leaves. II. *Plant Physiology, Lancaster*, **30**, 184–5.

Badenhuizen, N. P. (1965). Occurrence and development of starch in plants. In *Starch: Chemistry and Technology*, vol. 1, ed. R. L. Whistler & E. F. Paschall, pp. 65–103. London, New York: Academic Press.

Badenhuizen, N. P. (1969). *The Biogenesis of Starch Granules in Higher Plants*. New York: Appleton-Century-Crofts.

Baxter, E. D. & Duffus, C. M. (1973). Phosphorylase activity in relation to starch synthesis in developing *Hordeum distichum* grain. *Phytochemistry*, **12**, 2321–30.

Boehm, J. A. (1883). Ueber Starkebildung aus Zucker. *Botanische Zeitung*, **41**, 33–54.

Bradbeer, J. W. (1976). Chloroplast development in greening leaves. In *Perspectives in Experimental Botany*, vol. 2, *Botany*, ed. N. Sunderland, pp. 131–43. Oxford: Pergamon Press.

Brown, H. T. & Morris, G. H. (1893). Chemistry and physiology of foliage leaves. *Journal of the Chemical Society*, **63**, 604–77.

Chan, T. & Bird, I. F. (1960). Starch dissolution in tobacco leaves in the light. *Journal of Experimental Botany*, **11**, 335–40.

Chen-She, S., Lewis, D. H. & Walker, D. A. (1975). Stimulation of photosynthetic starch formation by sequestration of cytoplasmic orthophosphate. *New Phytologist*, **74**, 383–92.

De Fekete, M. A. R. & Vieweg, G. (1974). Starch metabolism: synthesis versus degradation pathways. In *Plant Carbohydrate Biochemistry*, ed. J. B. Pridham, pp. 127–44. London, New York: Academic Press.

Dickson, R. E. & Larson, P. R. (1975). Incorporation of ^{14}C-photosynthate into major chemical fractions of source and sink leaves of cottonwood. *Plant Physiology, Lancaster*, **56**, 185–93.

Downton, W. J. S. & Hawker, J. S. (1974). Response of starch synthesis to temperature in chilling-sensitive plants. In *Environmental and Biological Control of Photosynthesis*, ed. R. Marcelle, pp. 81–8. The Hague: W. Junk.

Eckerson, S. (1909). On the demonstration of the formation of starch in leaves. *Botanical Gazette*, **48**, 224–8.

Edwards, G. E., Robinson, S. P., Tyler, N. J. C. & Walker, D. A. (1978). Photosynthesis by isolated protoplasts, protoplast extracts and chloroplasts of wheat. Influence of orthophosphate, pyrophosphate and adenylates. *Plant Physiology, Lancaster*, **62**, 313–19.

Esau, K. (1953). *Plant Anatomy*. New York: John Wiley; London: Chapman & Hall.

Fellows, R. J. & Geiger, D. R. (1974). Structural and physiological changes in sugar beet leaves during sink to source conversions. *Plant Physiology, Lancaster*, **54**, 877–85.

Forde, B. J., Whitehead, H. C. M. & Rowley, J. A. (1975). Effect of light intensity and temperature on photosynthetic rate, leaf starch content and ultrastructure of *Paspalum dilatatum*. *Australian Journal of Plant Physiology*, **2**, 185–95.

Geiger, D. R. (1976). Effects of translocation and assimilate demand on photosynthesis. *Canadian Journal of Botany*, **54**, 2337–45.

Geiger, D. R. (1979). Control of partitioning and export of carbon in leaves of higher plants. *Botanical Gazette*, **140**, 241–8.

Hampp, R. & Wellburn, A. R. (1976). Early changes in the envelope permeability of developing chloroplasts. *Journal of Experimental Botany*, **27**, 778–84.

Hawker, J. S. (1967). Inhibition of sucrose phosphatase by sucrose. *Biochemical Journal*, **102**, 401–6.

Hawker, J. S., Marschner, H. & Downton, W. J. S. (1974). Effects of sodium and potassium on starch synthesis in leaves. *Australian Journal of Plant Physiology*, **1**, 491–500.

Heber, U. (1974). Metabolite exchange between chloroplast and cytoplasm. *Annual Review of Plant Physiology*, **25**, 393–421.

Heldt, H. W. (1976). Metabolite carriers of chloroplasts. In *Encyclopedia of Plant Physiology*, vol. 3, ed. C. R. Stocking & U. Heber, pp. 137–43. London, Heidelberg, New York: Springer-Verlag.

Heldt, H. W. & Sauer, F. (1971). The inner membrane of the chloroplast envelope as the site of specific metabolite transport. *Biochimica et Biophysica Acta*, **234**, 83–91.

Herold, A. (1977). *Some Aspects of Carbon Metabolism in Leaves*. Ph.D. Thesis, University of Sheffield.

Herold, A. (1978). Starch synthesis from exogenous sugars in tobacco leaf discs. *Journal of Experimental Botany*, **29**, 1391–1401.

Herold, A. (1980). Regulation of photosynthesis by sink activity – the missing link. *New Phytologist*, **86**, 131–44.

Herold, A. & Lewis, D. H. (1977). Mannose and green plants: occurrence, physiology and metabolism, and use as a tool to study the role of orthophosphate. *New Phytologist*, **79**, 1–40.

Herold, A., Leegood, R. C., McNeil, P. H. & Robinson, S. P. (1981). Accumulation of maltose in protoplasts isolated from spinach leaves treated with mannose. *Plant Physiology, Lancaster*, **67**, 85–8.

Herold, A., McGee, E. E. M. & Lewis, D. H. (1980). The effect of orthophosphate concentration and exogenously supplied sugars on the distribution of newly fixed carbon in sugar beet leaf discs. *New Phytologist*, **85**, 1–13.

Herold, A. & McNeil, P. H. (1979). Restoration of photosynthesis in pot-bound tobacco plants. *Journal of Experimental Botany*, **30**, 1187–94.

Herold, A. & Walker, D. A. (1979). Transport across chloroplast envelopes – the role of phosphate. In *Handbook of Transport Across Membranes*, ed. G. Giebisch, D. C. Tosteson & H. H. Ussing, pp. 411–39. Heidelberg, New York: Springer-Verlag.

Ho, L. C. (1976). The relationship between the rates of carbon transport and of photosynthesis in tomato leaves. *Journal of Experimental Botany*, **27**, 87–97.

Hofstra, G. & Hesketh, J. D. (1974). The effects of temperature and CO_2 enrichment on photosynthesis in soybean. In *Environmental and Biological Control of Photosynthesis*, ed. R. Marcelle, pp. 71–80. The Hague: W. Junk.

Hofstra, G. & Nelson, C. D. (1969). The translocation of photosynthetically assimilated ^{14}C in corn. *Canadian Journal of Botany*, **47**, 1435–42.

Hopkinson, J. M. (1964). Studies on the expansion of the leaf surface. IV. The carbon and phosphorus economy of a leaf. *Journal of Experimental Botany*, **15**, 125–37.

Jones, H. & Eagles, J. E. (1962). Translocation of ^{14}carbon within and between leaves. *Annals of Botany*, **26**, 505–10.

Jones, H., Martin, R. V. & Porter, H. K. (1959). Translocation of ^{14}carbon in tobacco following assimilation of ^{14}carbon dioxide by a single leaf. *Annals of Botany*, **23**, 493–508.

Joy, K. W. (1967). Carbon and nitrogen sources for protein synthesis. *Journal of Experimental Botany*, **18**, 140–50.

Kaiser, W. M. & Bassham, J. A. (1979). Light–dark regulation of starch metabolism in chloroplasts. II. Effect of chloroplastic metabolite levels on the formation of ADP-glucose by chloroplast extracts. *Plant Physiology, Lancaster*, **63**, 109–13.

Kakie, T. (1969). Effect of phosphorus deficiency on the photosynthetic carbon dioxide fixation products in tobacco leaves. *Soil Science and Plant Nutrition*, **15**, 245–51.

Kandler, O. & Gibbs, M. (1959). Untersuchungen uber den Einfluss Photosynthese auf die Austauschvorgange Innenhalbe des Hexose Molekuls. *Zeitschrift für Naturforschung*, **B14**, 8–13.

Kirk, J. T. O. (1970). Biochemical aspects of chloroplast development. *Annual Review of Plant Physiology*, **21**, 11–42.

Komuro, T., Yano, K. & Hattori, S. (1966). The changes in the starch content of tobacco leaves by dark treatment. *Botanical Magazine, Tokyo*, **79**, 376–80.

Laetsch, W. M. (1969). Relationship between chloroplast structure and photosynthetic carbon-fixation pathways. *Science Progress, Oxford*, **57**, 323–51.

Leech, R. M. (1976). Plastid development in isolated etiochloroplasts and isolated etioplasts. In *Perspectives in Experimental Botany*, vol. 2, ed. N. Sunderland, pp. 145–62. Oxford: Pergamon Press.

Leloir, L. F., De Fekete, M. A. R. & Cardini, C. E. (1961). Starch and oligosaccharide synthesis from uridine diphosphate glucose. *Journal of Biological Chemistry*, **236**, 636–41.

Lowry, O. H. & Passoneau, J. U. (1972). *A Flexible System of Enzymatic Analysis*. London, New York: Academic Press.

MacDonald, P. W. & Strobel, G. A. (1970). Adenosine diphosphate glucose pyrophosphorylase control of starch accumulation in rust infected wheat leaves. *Plant Physiology, Lancaster*, **46**, 126–35.

Maclachlan, G. A. & Porter, H. K. (1959). Replacement of oxidation by light as the energy source for glucose metabolism in tobacco leaf. *Proceedings of the Royal Society*, **B150**, 460–73.

Maclachlan, G. A. & Porter, H. K. (1961). Metabolism of glucose carbon atoms by tobacco leaf discs. *Biochimica et Biophysica Acta*, **46**, 244–58.

McLaren, J. S. & Smith, H. (1977). Effect of abscisic acid on photosynthetic products of *Lemna minor*. *Phytochemistry*, **16**, 219–21.

MacRae, J. C. (1971). Quantitative measurement of starch in very small amounts of leaf tissue. *Planta*, **96**, 101–8.

Madsen, E. (1968). Effect of CO_2-concentration on the accumulation of starch and sugar in tomato leaves. *Physiologia Plantarum*, **21**, 168–75.

Manners, D. J. (1974). The structure and metabolism of starch. In *Essays in Biochemistry*, vol. 10, ed. P. N. Campbell & F. Dickens, pp. 37–71. London, New York, San Francisco: Academic Press.

Mansfield, T. A. & Jones, R. J. (1971). Effects of abscisic acid on potassium uptake and starch content of stomatal guard cells. *Planta*, **101**, 147–58.

Marcelle, R., Clijsters, H., Oben, G., Bronchart, R. & Michel, J. M. (1974). Effects of CCC and GA in photosynthesis of primary bean leaves. In *Plant Growth Substances*, ed. Y. Sumiki & O. Hirokawa, pp. 1169–74. Tokyo.

Marker, A. F. H. & Whittingham, C. P. (1966). The photoassimilation of glucose in *Chlorella* with reference to the role of glycollic acid. *Proceedings of the Royal Society*, **B165**, 473–85.

Matheson, N. K. & Wheatley, J. M. (1963). Diurnal–nocturnal changes in the starch of tobacco leaves. *Australian Journal of Biological Science*, **16**, 70–6.

Meyer, A. (1886). Bildung der Starkekorner in den Laubblattern aus Zuckerarten, Mannit und Glycerin. *Botanische Zeitung*, **44**, 81–160.

Montalbini, P. & Buchanan, B. B. (1974). Effect of a rust infection on photophosphorylation by isolated chloroplasts. *Physiological Plant Pathology*, **4**, 191–6.

Nafziger, E. D. & Koller, H. R. (1976). Influence of leaf starch concentration on CO_2 assimilation in soybean. *Plant Physiology, Lancaster*, **57**, 560–3.
Neales, T. F. & Incoll, L. D. (1968). The control of leaf photosynthetic rate by the level of assimilate concentration in the leaf. A review of the hypothesis. *Botanical Reviews*, **34** 107–25.
Parkin, J. (1899). Starch synthesis from endogenous sugars. II. Contributions to our knowledge of the formation, storage and depletion of carbohydrates in monocotyledons. *Royal Society of London Philosophical Transactions, Series B*, **191**, 35–79.
Penny, P. & Nelson, C. D. (1970). Movement within leaves and plants of ^{14}C applied as $^{14}CO_2$. *Canadian Journal of Botany*, **48**, 1033–7.
Phillis, E. & Mason, T. G. (1937). On the effects of light and of oxygen on the uptake of sugar by the foliage leaf. *Annals of Botany*, **1**, 231–7.
Pongratz, P. & Beck, E. (1978). Diurnal oscillation of amylolytic activity in spinach chloroplasts. *Plant Physiology, Lancaster*, **62**, 687–9.
Porter, H. K. (1953). Starch synthesis and degradation *in vivo*. In *Biological Transformations of Starch and Cellulose, Symposium of the Biochemical Society*, vol. 11, ed. R. T. Williams, pp. 27–41. Cambridge: Cambridge University Press.
Porter, H. K. (1962). Synthesis of polysaccharides of higher plants. *Annual Review of Plant Physiology*, **13**, 303–28.
Porter, H. K. (1966). Leaves as collecting and distributing agents of carbon. *Australian Journal of Science*, **29**, 31–40.
Porter, H. K. & Bird, I. F. (1962). Assimilation and respiration by tobacco leaf tissue. A quantitative study using ^{14}C. *Indian Journal of Plant Physiology*, **5**, 5–32.
Porter, H. K., Martin, R. V. & Bird, I. F. (1959). Synthesis and dissolution of starch labelled with ^{14}carbon in tobacco leaf tissue. *Journal of Experimental Botany*, **10**, 264–76.
Preiss, J., Ghosh, H. P. & Wittkop, J. (1967). Regulation of the biosynthesis of starch in spinach leaf chloroplasts. In *Biochemistry of Chloroplasts*, vol. 2, ed. T. W. Goodwin, pp. 131–43. London, New York: Academic Press.
Preiss, J. & Levi, C. (1979). Metabolism of primary products of photosynthesis. 23. Metabolism of starch in leaves. In *Encyclopedia of Plant Physiology*, vol. 6, *Photosynthesis II, Photosynthetic Carbon Metabolism and Related Processes*, ed. M. Gibbs & E. Latzko, pp. 282–312. Berlin, Heidelberg, New York: Springer-Verlag.
Raven, J. A. (1976). Glucose metabolism in *Hydrodictyon africanum* in relation to cell energetics. *New Phytologist*, **76**, 195–204.
Sachs, J. (1862). Ueber den Einfluss des Lichtes auf die Bildung des Amylums in den Chlorophyllkornen. *Botanische Zeitung*, **20**, 365–76.
Sachs, J. (1864). Ueber die Auflosung und Wiederbildung des Amylums in den Chlorophyllkornen bei wechselnder Beleuchtung. *Botanische Zeitung*, **22**, 289–96.
Sachs, J. (1887). *Lectures on the Physiology of Plants*. London: Clarendon Press.
Salerno, G. L. & Pontis, M. G. (1978). Studies on sucrose phosphate synthetase. The inhibitory action of sucrose. *FEBS Letters*, **86**, 263–7.

Saposchnikoff, W. (1890). Bildung und Wanderung der Kohlenhydrate in dem Laubblattern. *Berichte Deutsche Botanische Gesellschaft*, **8**, 233–42.

Schafer, G., Heber, U. & Heldt, H. W. (1977). Glucose transport into spinach chloroplasts. *Plant Physiology, Lancaster*, **60**, 286–9.

Schimper, A. F. W. (1885). Ueber Bildung und Wanderung der Kohlenhydrate in dem Laubblattern. *Botanische Zeitung*, **43**, 737–87.

Spiertz, J. H. J. (1974). Grain growth and distribution of dry matter in the wheat plant as influenced by temperature, light energy and ear size. *Netherlands Journal of Agricultural Science*, **22**, 207–20.

Thomas, M., Ranson, S. L. & Richardson, J. A. (1973). *Plant Physiology*. London: Longman.

Thorne, J. H. & Koller, H. R. (1974). Influence of assimilate demand on photosynthesis, diffusive resistance, translocation, and carbohydrate levels of soybean leaves. *Plant Physiology, Lancaster*, **54**, 201–7.

Thorpe, T. A. & Meier, D. D. (1974). Enzymes of starch metabolism in *Nicotiana tabacum* callus. *Phytochemistry*, **13**, 1329–34.

Tollenaar, D. (1925). *Formation of Starch in Tobacco*. Dissertation, Wageningen.

Trip, P. (1969). Sugar transport in conducting elements of sugar beet leaves. *Plant Physiology, Lancaster*, **44**, 717–25.

Walker, D. A. (1974). Chloroplast and cell concerning the movement of certain key metabolites etc. across the chloroplast envelope. In *Medical and Technical Publications of the International Review of Science and Biochemistry, Series 1*, vol. 11, ed. D. H. Northcote, pp. 1–49. London: Butterworths.

Walker, D. A. (1979). *Energy, Plants and Man*. London: Packard.

Walker, D. A. & Herold, A. (1977). Can the chloroplast support photosynthesis unaided? In *Photosynthetic Organelles: Structure and Function*, a special issue of *Plant and Cell Physiology*, ed. S. Miyachi, S. Katoh, Y. Fujita & K. Shibata, pp. 295–310. Tokyo.

Whatley, J. M. (1974). Chloroplast development in primary leaves of *Phaseolus vulgaris*. *New Phytologist*, **73**, 1097–110.

Whipps, J. & Lewis, D. H. (1981). Patterns of translocation, storage and interconversion of carbohydrates. In *Effects of Disease on the Physiology of the Growing Plant*, ed. P. G. Ayres, pp. 47–83. Cambridge: Cambridge University Press.

MARK STITT

9 Degradation of starch in chloroplasts: a buffer to sucrose metabolism

Introduction

In the light, chloroplasts convert water, CO_2 and Pi into TP. Most of this TP is exported to the cytosol where it is converted into sucrose and the esterified Pi is released so that it can re-enter the chloroplast and be used for further photosynthesis. However, some TP can be retained in the chloroplast and converted to starch, again releasing Pi but this time directly into the stroma (see Chapter 8). By recycling the stromal Pi, synthesis of starch allows photosynthesis to occur at high rates even when the external Pi is limiting. Accordingly, synthesis of starch is highest in isolated chloroplasts when Pi in the medium is low (Steup, Peavey & Gibbs, 1978; Heldt *et al.* 1977; Preiss & Levi, 1980). In leaves and protoplasts, a lowering of the cytosolic Pi is thought to lead to increased synthesis of starch (see Chapter 8). It is suggested that, under conditions when the rate of photosynthesis exceeds that of synthesis of sucrose, Pi falls and an increasing proportion of the fixed carbon is then retained temporarily in the chloroplast as starch. Later, this starch is remobilized. In this view, starch is seen as a buffer for sucrose metabolism. Two questions arise – by what route is the starch remobilized and how is this process controlled? Recently, it has become possible to study breakdown of starch using isolated chloroplasts. They provide a simpler system than leaves and one in which the enzymes and pathways can be investigated and aspects of the control studied. However, it is also important to consider the physiological significance of results obtained using isolated organelles. In this chapter, I shall discuss recent work carried out on breakdown of starch in isolated chloroplasts and then try to relate this to some features of its metabolism in whole leaves.

Products of mobilization of starch in isolated chloroplasts

What are the products of mobilization of starch in isolated chloroplasts? By establishing what these are, it becomes possible to see which enzymes and pathways are needed. Experiments with isolated spinach chloroplasts by Peavey, Steup & Gibbs (1977) and Heldt *et al.* (1977)

indicated that they degrade starch into TP, 3PGA, maltose and glucose, but that the rates of breakdown were far lower than those in spinach leaves (Stitt & Heldt, 1981*a*; A. Herold personal communication). When the rates are not physiological, it is possible that some pathways are not operating fully and that some products may not be formed. Demonstration of fully physiological rates of breakdown in isolated chloroplasts requires that intact chloroplasts with a high starch content are prepared. Such chloroplasts can be prepared from pre-illuminated leaves, which have been left in the light until they have a high starch content. For species with small starch grains, like pea, conventional methods of chloroplast preparation can still be used (Stitt, Bulpin & ap Rees, 1978). However, in species with larger starch grains, like spinach, the fragile chloroplasts are easily disrupted by the starch grains and intact starch-loaded chloroplasts are best isolated via protoplasts (Stitt & Heldt, 1981*a*), so avoiding the necessity for mechanical homogenization and rapid centrifugation.

After isolating chloroplasts, methods are needed by which the rate of breakdown can be accurately measured and the products identified. This can be done by a combination of radioactive analysis and enzymatic analysis of substrates, the former to identify the products and the latter to measure their rates of accumulation (Stitt & ap Rees, 1980*a*). Both of these approaches on their own are limited but they complement each other. By establishing a stoichiometric balance between the fall of ^{14}C in pre-labelled starch and the increase in the measured products, it can be shown that at least the major products have all been identified. The flow of ^{14}C also provides an estimate of the relative fluxes into the various products but it cannot be used directly to calculate absolute rates because some unlabelled starch is broken down together with the radioactive starch so that the specific activity of the degraded starch is not known. However, once the products have been identified, it is possible to measure the rate at which they accumulate, and this will give absolute rates. Used on its own, enzymatic analysis would have the disadvantages that it can neither show directly that the increase in a compound is actually due to degradation of starch nor that all the products have been identified.

A technical problem arises over how to pre-label the starch. When either ^{14}C-bicarbonate (Peavey, Steupp & Gibbs, 1977; Heldt *et al.* 1977) or [^{14}C]glucose (Stitt & ap Rees, 1980*a*) are supplied to isolated chloroplasts in the light, only a minute proportion of the radioactivity is metabolized in 10–15 min. The remaining radioactivity will make it impossible to detect any formation of CO_2, or glucose, respectively, from starch. In these experiments, the chloroplasts were washed by repeated centrifugation to remove most of the unincorporated radioactivity. Such centrifugation not only may damage the chloroplasts but also is not fully effective. Enough unmetabolized [^{14}C]glucose remained to complicate the measurement of subsequent formation of glucose. The amount of $^{14}CO_2$ remaining after washing was not reported but later investigations (M. Stitt, unpublished) suggested spinach chloroplasts must be washed about 5 times before enough of the supplied ^{14}C-bicarbonate is removed to allow evolution of $^{14}CO_2$ in the dark to be detected.

A method has recently been developed to avoid this problem (Stitt & Heldt, 1981*a*) and involves supplying 20–50 μM ^{14}C-bicarbonate to protoplasts in the light at pH 6 to pre-label the starch. In these conditions, the bicarbonate is present almost wholly as CO_2, the active species for photosynthesis, so that low concentrations of bicarbonate support high rates of photosynthesis. Within 5–10 min almost all the supplied ^{14}C-bicarbonate is fixed. $^{14}CO_2$-free chloroplasts are isolated by passing the protoplasts through a net (Edwards, Robinson, Tyler & Walker, 1978) and then gently centrifuging. As shown in Table 9.1, the flux of ^{14}C from starch into CO_2, glucose, maltose and phosphorylated intermediates can then be measured. The absolute fluxes into the major products – glucose and the phosphorylated intermediates – are then directly determined by enzymatic substrate analysis. For minor products,

Table 9.1 *Estimation of breakdown of starch in isolated spinach chloroplasts*

Source of data	Relative quantities of breakdown products as percentages of pre-labelled starch degraded	Rate of accumulation of product (μmol mg^{-1} chl h^{-1}): Measured by enzymatic analysis	Rate of accumulation of product (μmol mg^{-1} chl h^{-1}): Extrapolated from measured absolute and relative rates
Total	100	–	10.3
Neutral compounds	56 ± 12	–	5.2
Glucose	37 ± 9	3.3 ± 0.7	–
Maltose	19 ± 5	–	1.9
Anions	35 ± 11	4.6 ± 2.8	–
CO_2	9 ± 2.1	–	0.9

Protoplasts, prepared from starch-loaded leaves, were labelled with 50 μM [^{14}C]HCO_3 at pH 6.0 and chloroplasts were then isolated and the movement of ^{14}C out of starch into CO_2, anions and neutral compounds measured. CO_2 was collected in KOH, starch was separated by centrifugation, anions and neutral compounds were separated by ion-exchange resin, and glucose and maltose by thin layer chromatography. The accumulation of carbon in glucose, HP, TP, fructosebisP and 3PGA was measured photometrically. The last four were summed as 'anions'. The results are the mean ±S.E. of seven radioactive and four enzymatic experiments, respectively.

like CO_2 and maltose which are not easily measured, the absolute rates can be extrapolated (Table 9.1).

Physiological rates of breakdown of starch have been demonstrated in chloroplasts from pea (Stitt & ap Rees 1980*a*) and spinach (Stitt & Heldt, 1981*a*). Those from pea degraded starch at 3.3 μatom C mg^{-1} chl h^{-1} compared with rates of 2–4 in leaves. Spinach chloroplasts and leaves degraded starch at over 10, and 12, μatom C mg^{-1} chl h^{-1}, respectively. In spinach, the main products were TP and 3PGA, CO_2, glucose and maltose (Table 9.1). In intact pea chloroplasts, the main products were TP, 3PGA and CO_2, plus maltose. As discussed above, the methods used with pea would not have detected formation of small amounts of glucose, because some of the [^{14}C]glucose supplied to label starch was still present during the period of degradation. Preliminary experiments indicate that small amounts of glucose are indeed produced by pea chloroplasts (M. Stitt, unpublished). On the other hand, pea chloroplast extracts produced little glucose from added amylopectin (Levi & Preiss, 1977). Further experiments are needed to clarify this point.

Enzymes which mobilize starch in chloroplasts

The next question which arises is how starch is converted into these products in chloroplasts. What are the relative contributions of phosphorolysis and hydrolysis to starch breakdown, and what contributions do glycolysis and the oxidative pentose phosphate pathway make to the formation of TP and 3PGA? One approach to these questions is to determine which enzymes are present in high enough activity for them to make a significant contribution. This involves measuring the enzyme activity accurately and showing that it is in the chloroplast.

Measurement of starch-degrading enzymes poses special problems since they have similar substrates and products (ap Rees, 1974). Enzymes are distinguished by using conditions under which only one enzyme can be active, e.g. using ionic conditions, precise structure of substrate or pre-treatment of the extract to inhibit other enzymes. This pre-supposes an adequate classification of the enzymes likely to be present in a tissue. A system has been developed for non-photosynthetic plant tissues, based on mobilization of seed starches. (Manners, 1974; ap Rees, 1974; Preiss & Levi, 1980). Starch can be broken down by a phosphorolytic or an amylolytic attack. The latter involves at least two enzymes which attack α-(1–4) glucans. α-Amylase is an endoenzyme, attacking α-(1–4) bonds in the middle of the glucan and releasing shorter glucans and, eventually, maltose. The α-amylose contains Ca^{2+} and is heat stable. β-Amylase, an exo-amylase releasing maltose from the end of glucans, is neither heat stable nor does it contain Ca^{2+}. However, this classification may not be adequate for the enzymes of leaves. Indeed, an exception has already been found in the non-photosynthetic spadices of *Arum maculatum*, where starch is degraded by an endoamylase which produces products similar to α-amylase but which is not heat stable and does not contain Ca^{2+}. Therefore, particular attention is needed when selecting assays to measure degrading enzymes in leaves.

Before assigning an enzyme to the chloroplast, marker enzymes ought to be measured to check that the activity is not due to cross-contamination from other cellular fractions. Often it is also advisable to use sucrose gradients to purify chloroplasts more fully than is possible by differential centrifugation (Stitt *et al.* 1978; Okita, Greenberg, Kuhn & Preiss, 1979). During cell fractionation and in sucrose gradients, enzymes may lose their activity, and this can lead to incorrect estimates of the distribution of enzymes or of cross-contamination. Checking that the total enzyme activity recovered in all the separated fractions is the same as that in the starting material allows such errors to be detected. One problem in studying compartmentation of enzymes in plants has always been the difficulty of obtaining high yields of intact organelles, owing to the forces needed to break the plant cell walls. However, it is now possible to isolate protoplasts from a wide range of leaves and they provide a favourable starting material from which intact chloroplasts can be isolated in high yield (Okita *et al.* 1979). This classical approach to compartmentation can be supplemented in two ways. Establishing that an enzyme is latent, i.e. that it is only detected after the chloroplasts are ruptured, provides independent evidence that it is not present due to cross-contamination and, further, that it is really present within the stroma, and not attached somehow on the outside of the chloroplast envelope (Kelly & Latzko, 1977*a*; Stitt *et al.* 1978; Pongratz & Beck, 1978; Steup & Latzko, 1979). Another important and elegant approach is to study the isoenzyme pattern of a leaf and to assign the isoenzymes to separate compartments (Kelly & Latzko, 1977*a*; Steup & Latzko, 1979). This involves separation of proteins, for example by gel chromatography. In view of the difficulty involved in constructing specific assays for many of the enzymes which degrade starch, it is probably useful to accompany all cell fractionation procedures with a partial resolution on gels of the different enzymes measured (Okita *et al.* 1979). The location of enzymes which degrade starch in chloroplasts has recently been reviewed (Preiss & Levi, 1980; Latzko, Steup & Schaechtele, 1981) and I shall concentrate on recent work using intact chloroplasts.

Hydrolysis

Spinach leaves contain a debranching enzyme, which can hydrolyse the α-(1–6)branch-points in amylopectin, and about half of this enzyme is chloroplastic (Okita *et al.* 1979). A variety of enzymes can attach the α-(1–4) links of starch. Of these, α-amylase is widely distributed in leaves (Gates & Simpson, 1968) but has not yet been conclusively identified in chloroplasts and is absent from those of pea (Stitt *et al.* 1978). α-Amylase has recently been reported to be found on starch grains in chloroplasts from cotton (Chang, 1980) and this could represent a chloroplastic α-amylase. Leaves can also have high β-amylase activity, but, at least in pea, it is not chloroplastic (Levi & Preiss, 1977; Stitt *et al.* 1978). The only hydrolytic enzyme degrading α(1–4) glucosidic chains as yet unambiguously shown to be in chloroplasts is a heat-labile Ca^{2+}-independent endoamylase in spinach (Okita *et al.* 1979) with an action spectrum like α-amylase. Its activity is high enough to play a role in degradation. Pongratz & Beck (1978) also found a latent amylolytic activity in spinach chloroplasts which was not fully characterized but included an endo-amylolytic component.

How widespread these heat-labile endoamylases are is not known. In spinach, activity could be detected because there is no high β-amylase activity. In the most frequently used assay for α-amylase (Briggs, 1967), the extract is first heated in the presence of Ca^{2+} to destroy β-amylase. The residual amylolytic activity represents the heat-stable α-amylase and any heat-unstable endoamylase would not be detected. However, if high β-amylase activity is present, omitting the heating step will still not reveal a heat-unstable endoamylase, because it will now be masked by β-amylase activity. For this reason, reports that pea chloroplasts contain no amylolytic enzymes (Levi & Preiss, 1977; Stitt *et al.*, 1978) are not conclusive and need confirmation using methods capable of resolving a heat-unstable endoamylase in the presence of the high β-amylase activity found in pea shoots. This could be done by careful choice of substrate and conditions, or by analysing chloroplast extracts using gels to resolve different amylolytic activities (Okita *et al.* 1979).

The endoamylase releases dextrins and maltose, but hardly any glucose (Okita *et al.* 1979). While maltose is released from starch in chloroplasts, glucose is also a major product (Peavy *et al.* 1977, and Table 9.1). Presumably, much of the dextrin and maltase is further metabolized by other enzymes. In seeds, this occurs via maltase but significant levels of maltase cannot be found in chloroplasts from spinach (Okita *et al.* 1979) or pea (Levi & Preiss, 1977). Spinach leaves contain α-glucan transferase (Linden, Tanner & Kandler, 1974) and α-glucosyl transferase, enzymes which rearrange glucose units between glucose, maltose and glucans. On the one hand, glucose can be exchanged into maltose which, with other short-chain glucans, can then be lengthened by transfer of glucose on to them from long-chain glucan. Such activity would convert short glucans back into a suitable substrate for enzymes which are active on longer chain glucans, for example phosphorylase (see below). On the other hand, α-glucosyl transferase releases glucose from maltotriose by virtue of an associated exoamylolytic activity. It is not clear whether, *in vivo*, the enzymes will catalyse a lengthening of short-chain glucans or a release of glucose. Possibly, it might depend upon whether the amounts of glucan are high or low, and ultimately upon whether there is a sink for long-chain glucans or glucose. When first described, the location and exact activities of these enzymes was not further investigated. Later Okita *et al.* (1979), assaying D-enzyme (see Chapter 10 for details of the action of D-enzyme) by release of glucose from maltotriose at pH 6, found some activity associated with chloroplasts, although the majority was extrachloroplastic. A similar activity at the optimum pH of 7.2 appears exclusively chloroplastic in spinach protoplasts (M. Stitt, unpublished). There is negligible activity with amylopectin, unhydrolysed potato amylose or partially hydrolysed amylose suggesting that this activity really represents D-enzyme rather than an unspecific exoamylase. The activity at 1mM maltotriose would account for the observed rate of glucose formation in these chloroplasts but, until the concentrations *in situ* of maltotriose and of other glucans are known, the role of these enzymes cannot be fully established.

The glucose formed can be exported from chloroplasts and metabolized in the cytosol, or metabolized in the chloroplasts. Leaves contain hexokinase (Sutton, 1975) but there is still no conclusive evidence that chloroplasts themselves contain significant activity. Up to a third of the hexokinase in pea leaves is found in preparations of chloroplast (Levi & Preiss, 1977; Stitt *et al.* 1978) but this activity was not latent inside chloroplasts (Stitt *et al.* 1978). The maximal activity which could be ascribed to the chloroplast (0.3 μmol hexose mg^{-1} chl h^{-1}) was probably still an overestimate and is already lower than the rate of breakdown of starch in these pea chloroplasts (0.55 μmol hexose mg^{-1} chl h^{-1} (Stitt & ap Rees, 1980*a*). A recent preliminary report (Chang, 1980) noted that hexokinase is associated with starch grains from cotton chloroplasts. Provided this association is not an artefact resulting during fractionation, it shows hexokinase can be present in at least some chloroplasts.

Phosphorolysis

There is ample evidence that chloroplasts from pea (Levi & Preiss, 1977; Stitt *et al.* 1978; Steup & Latzko, 1979) and from spinach (Okita *et al.* 1979; Steup & Latzko, 1979) have enough α-glucan phosphorylase to account for the observed rates of breakdown of starch, making it possible that a considerable part of its mobilization is phosphorolytic. The α-glucan phosphorylase is most active on long-chain glucans but, in pea chloroplasts, a maltose phosphorylase has also been reported (Levi & Preiss, 1977) which could provide another means of using maltose, releasing glucose-1-P and free glucose. Whether it could synthesize maltose from the *in vivo* concentrations of Pi and glucose 1-P is not clear. In spinach chloroplasts, no maltose phosphorylase has been detected (M. Steup, personal communication; M. Stitt, unpublished). Further work is needed to clarify this difference between pea and spinach. The relation between maltose and glucose 1-P is further complicated by the recent discovery of maltose synthetase, which synthesizes maltose from two glucose 1-P molecules (see Chapter 4). This enzyme is thought to be important in forming primers during synthesis of starch but whether it is also active during its degradation is unknown.

Considerable 3PGA and TP are formed during mobilization of starch, involving either glycolysis or the OPP cycle. Although most of the enzymes required are members of the Calvin cycle, some are unique to respiratory metabolism. The unique glycolytic enzyme, phosphofructokinase, is present in chloroplasts from spinach (Kelly & Latzko, 1975, 1977*a*, *b*), pea (Stitt & ap Rees, 1979) and other species (Kelly & Latzko, 1977*a*). In pea, a third of the total activity is chloroplastic and the activity exceeds the rates of breakdown of starch (Stitt & ap Rees, 1979). The cytosolic and chloroplastic activities represent different isoenzymes which have been partially purified and their regulatory properties studied (Kelly & Latzko, 1977*a*, *b*). Of the enzymes of the OPP cycle, three are absent from the Calvin cycle; glucose 6-phosphate dehydrogenase, 6-phosphogluconate dehydrogenase and transaldolase. Activity of transaldolase in the leaf is low and, although it has been reported in chloroplasts (Latzko & Gibbs, 1968), this enzyme needs more attention. The glucose 6-phosphate dehydrogenase and 6-phosphogluconate dehydrogenase are present in purified chloroplasts (Miflin & Beevers, 1974) and are latent in chloroplasts (Lendzian & Ziegler, 1970; Stitt & ap Rees, 1979); the cytosolic and chloroplastic activities represent different isoenzymes (Anderson, Ng & Park, 1974; Herbert, Burkhard & Schnarrenberger, 1979).

Degradative enzymes – summary

Spinach chloroplasts which have been most intensively studied contain R-enzyme, endoamylase, α-glucan phosphorylase and, probably, transglucosylase allowing a joint hydrolytic and phosphorolytic attack. Maltase is absent. Pea chloroplasts contain ample α-glucan phosphorylase but also a maltose phosphorylase, and may lack amylolytic enzymes, although the latter point requires confirmation. Chloroplasts contain ample activities of the enzymes of glycolysis and the OPP cycle. An unexpected finding has been the high activities of starch-degrading enzymes present outside the chloroplasts, including an endoamylase in spinach (Okita *et al.* 1979), α-amylase and β-amylases in pea (Levi & Preiss, 1977; Stitt *et al.* 1978), and α-glucan phosphorylase in both pea and spinach (Stitt *et al.* 1978; Steup

& Latzko, 1979; Okita *et al.* 1979). The extrachloroplastic and chloroplastic activities represent distinct isoenzymes (Steup *et al.* 1979, 1980) and are found even when intact photosynthesizing chloroplasts are prepared which still contain their starch (Stitt *et al.* 1978). The chloroplastic and extrachloroplastic enzymes can be separated by DEAE chromatography (Preiss, Okita & Greenberg, 1980; Okita & Preiss, 1980) or by elegant affinity chromatography with bound glucan (Steup *et al.* 1980). For both the amylase and phosphorylase, affinity for substrates differs between the chloroplastic and extrachloroplastic enzymes (Preiss *et al.* 1980; Okita *et al.* 1979; Steup & Schaechtele, 1981). At least for the phosphorylase, the two forms have different molecular weights (Steup *et al.* 1979, 1980). While the extrachloroplastic enzymes are well authenticated, their function is unknown. Possibilities include the metabolism of an extrachloroplastic dextrin or that they are in non-photosynthetic cells in the leaf, or in stomata. An implication of these results is that measuring enzymes in whole leaves may not provide information about chloroplastic metabolism of starch since much of the activity observed may be extrachloroplastic.

Fluxes during mobilization of starch in isolated chloroplasts

Since both phosphorolytic and hydrolytic enzymes are found in chloroplasts, the question arises as to the role of these enzymes. Phosphorylated intermediates are a major product of mobilization of starch in both pea and spinach chloroplasts. They could be produced either by phosphorolysis, or by a hydrolytic release and a phosphorylation of glucose. The major route in pea and spinach appears to be phosphorolysis, which could account for up to half of the starch mobilized. The first line of evidence is that conversion of starch to phosphorylated intermediates is Pi-dependent (Peavey *et al.* 1977; Levi & Preiss, 1977; Heldt *et al.* 1977; Stitt & ap Rees, 1980*a*) and 3PGA will not substitute for Pi. This suggests phosphorolysis but it not conclusive because accumulation of intermediates in chloroplasts in the absence of Pi might indirectly inhibit other reactions. The second argument is that the activity of hexokinase in pea chloroplasts (Stitt *et al.* 1978) and the rate of phosphorylation of glucose supplied to spinach chloroplasts (Stitt & Heldt, 1981*a*) are both lower than the flux from starch into phosphorylated intermediates, while chloroplasts contain ample α-glucan phosphorylase (see above). Also, phosphorolysis was clearly the major route in converting amylopectin to hexose P in a pea chloroplast stromal extract (Levi & Preiss, 1977).

However, hydrolysis is also required in spinach chloroplasts since glucose is a major product. The nature of the hydrolytic pathway is unclear since the endoamylase releases maltose, α-limit dextrin and higher α(1–4) glucan as products, but hardly any glucose (Okita *et al.* 1979; Peavey *et al.* 1977 and Table 9.1). The origin of the glucose produced in chloroplasts is not clear. At present the best hypothesis, based on the enzymatic capacities of spinach chloroplasts, is that glucose is released during rearrangement of short-chain glucans. Whether hydrolysis occurs as well as phosphorolysis in pea is not yet clear. A recent report suggests that hydrolysis is important in cotton chloroplasts (Chang, 1980).

Which enzymes are responsible for the primary attack on starch grains? In non-photosynthetic tissues, it is generally accepted that the initial attack is by an endoamylase (Preiss & Levi, 1980) although the earlier view that α-amylase is always responsible must be modified because no α-amylase was present in the club of *Arum maculatum* (Bulpin & ap Rees, 1978). In leaves, also, dissolution of starch occurs without α-amylase in pea and spinach chloroplasts (Stitt *et al.* 1978; Okita *et al.* 1979), but at least in spinach, α-amylase could be replaced by a heat-labile endoamylase (Okita *et al.* 1979). Possibly α-amylase is a particularly stable endoamylase functioning especially during extracellular breakdown of starch, and can be replaced by other endoamylases during breakdown of starch within a fully functional cell where the compartmentation is unimpaired. At the moment, it is not clear whether the primary attack on leaf starch is always endoamylolytic (see Preiss & Levi, 1980). Both Beck, Pongratz & Reuter (1981) and Okita *et al.* (1979) propose that the spinach starch is first attacked by amylases and that phosphorylase acts on released glucans. Although starch grains are degraded by phosphorylase, this process is much faster when the grains have been predigested by the chloroplastic amylases (Beck *et al.* 1981). In cotton leaves, the starch grains could only be attacked by phosphorylase after pretreatment with the cotton α-amylase (Chang, 1980). If phosphorylase can only act on free glucans, it should be possible to establish the presence of a pool of glucans during breakdown. Earlier work showed that glucans did not accumulate at significant levels in leaves during this process (Matheson & Wheatley, 1962). Only traces of glucans may be needed but it has not yet been established whether the concentrations of free glucans are high enough for them to provide a better substrate for phosphorylase than would be provided by the starch grains. On the other hand, an observation that phosphorylase can degrade isolated starch grains does not provide sound support for the ability of phosphorylase to attack starch grains *in situ*, because the isolation of the starch grain may have altered its structure (Preiss & Levi, 1980).

Metabolism of hexose phosphate in isolated chloroplasts

The distribution of enzymes suggests that the conversion of HP to TP and 3PGA in chloroplasts occurs by a combination of glycolysis and the

OPP cycle. This view has been confirmed by flux measurements in pea (Stitt & ap Rees, 1980*a*, *b*) and spinach (Stitt & Heldt, 1981*a*). In general, the results show that the OPP pathway operates as a cycle, interacting closely with glycolysis. When glucose 6-P enters the cycle, it is first converted to pentoseP, and the carbon-1 is released as CO_2. A preferential release of CO_2 from the first carbon of glucose, as observed in chloroplasts (Stitt & ap Rees, 1980*a*), is diagnostic for the OPP cycle. The pentoseP does not accumulate but is converted to fructose6P and TP (Stitt & Heldt, 1981*a*) which can then either be cycled around the OPP pathway or metabolized further via glycolysis. Cycling of fructose6P is demonstrated by the release of $^{14}CO_2$ from [2-^{14}C]glucose. Since during conversion of pentoseP to HP, the ^{14}C is brought into the C1 position of the rearranged fructose6P and is released when it re-enters the cycle. Cycling of TP is indicated by release of $^{14}CO_2$ from [6-^{14}C]glucose. This release cannot occur via transketolase and transaldolase, since these enzymes are not able to remove Pi from the C6 of HP, and bring the C6 into the C1 position. Stitt & ap Rees (1980*a*) proposed that TP, randomized by TP-isomerase, are then reformed into fructosebisP, which is converted to HP by fructosebisphosphatase. In this HP, the C1 and C6 positions will be 50% exchanged. A similar scheme could also occur for sedoheptulosebisP and sedoheptulosebisphosphatase. Alternatively, as discussed in Stitt & Heldt (1981*a*), the release of $^{14}CO_2$ from [6-^{14}C]glucose could be due to the presence of the L-path of the OPP cycle (Williamson, 1980) in chloroplasts. Operation of both glycolysis and the OPP pathway is also shown in experiments with reconstituted spinach (Kaiser & Bassham, 1979; Lendzian, 1980*b*) and pea chloroplasts (Furbank & Lilley, 1981). In such systems, dilution of the stromal contents may prevent cycling (Kaiser & Bassham, 1979) or make analysis more complicated (Lendzian, 1980*b*) but the operation of both pathways can be clearly demonstrated.

The relative contribution of glycolysis and the OPP cycle to HP metabolism has been estimated (Stitt & ap Rees, 1980*b*) both by the relative release of $^{14}CO_2$ from [1-^{14}C]- and [2-^{14}C]glucose, and from the distribution of ^{14}C between CO_2 and 3PGA plus TP during metabolism of [U-^{14}C]glucose or ^{14}C-starch. In pea and spinach, 15–30% of the hexose is converted to TP and CO_2 in the OPP cycle (Stitt & ap Rees, 1980*b*; Stitt & Heldt, 1981*a*) and the remainder is metabolized via glycolysis. This means that about half the hexose will pass through glucose 6-phosphate dehydrogenase and half through phosphofructokinase (Stitt & ap Rees, 1980*b*). In the absence of Pi when export from the chloroplast is restricted, the activity of the OPP cycle rises and cycling of HP and TP increases. Usually, phosphorolytic mobilization of starch is a Pi starch-consuming process. The Pi is obtained when TP moves out of the chloroplast in exchange for Pi from the surroundings. If cycling of TP is occurring, however, some of the esterified Pi would be actually released again within the chloroplast, allowing recycling of the stromal Pi.

The last four sections are summarized in Fig. 9.1, a tentative scheme for metabolism in spinach chloroplasts. Starch is hydrolysed to dextrins and maltose by R-enzyme and the endoamylase. Phosphorolysis is an important pathway, leading initially to HP. The exact substrate for phosphorolysis is not clear but may be glucans released by the endoamylase. Glucose is produced, possibly during rearrangement of the dextrins. The interconversions of dextrins, maltose, glucose and HP need further attention. The HP is converted to TP by an interaction of glycolysis and the OPP cycle and the TP is exported in exchange for Pi either directly or after conversion to 3PGA. The Pi is then available for further phosphorolytic mobilization of starch. In pea chloroplasts, the relations between phosphorolysis, the OPP cycle and glycolysis resemble those in spinach. There may be differences in the initial

Fig. 9.1 A tentative scheme for mobilization of starch in spinach chloroplasts.

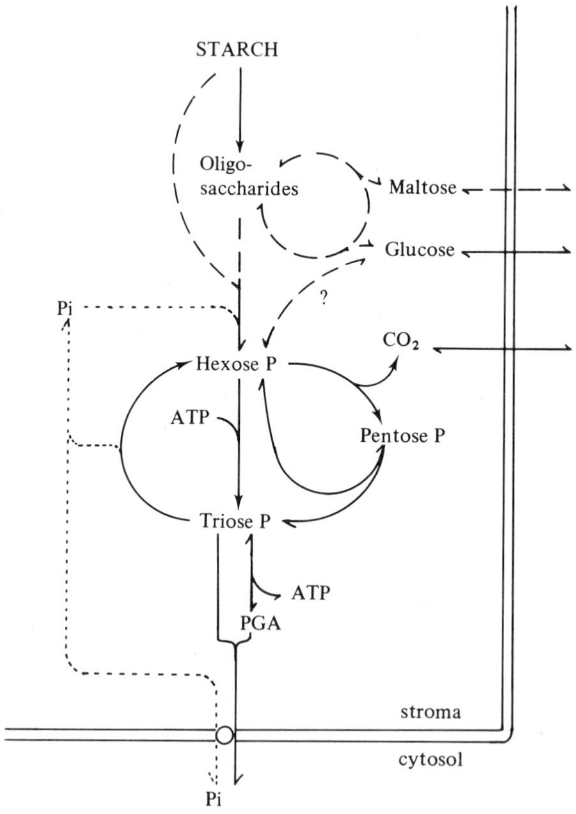

mechanism of degradation but further work is required firmly to establish this point. In cotton, preliminary work suggests that hydrolysis is more important than phosphorolysis (Chang, 1980).

The use of starch

Since both hydrolysis and phosphorolysis occur in spinach and the available data suggest that their importance may vary between different plants, the question arises as to why both routes can be utilized. No definite answer can be given but, by considering the uses to which starch is put in the leaf, it may be possible to propose a hypothesis. Moreover, it is useful to consider briefly how starch is used, before discussing control of its degradation.

Some starch will be used as a substrate for chloroplast metabolism. The conversion of phosphorolytically derived HP through to TP and 3PGA in the chloroplast is significant compared to the metabolism of the whole leaf in the dark. A large proportion of the enzymes of carbohydrate oxidation are present in chloroplasts (Kelly & Latzko, 1977*a*; Stitt & ap Rees, 1979) and the rate at which pea chloroplasts metabolize starch via glycolysis and the OPP cycle (Stitt & ap Rees, 1980*a*) is high compared with the total rate of respiration in pea leaves (Smillie, 1966). Production of NADPH within the chloroplast may be necessary because the enzymes which shuttle NADP(H) between cytosol and chloroplast are inactivated in the dark (Buchanan, 1980). The OPP cycle appears more active in pea chloroplasts than in pea leaves as a whole (Stitt & ap Rees, 1980*b*). Chloroplasts contain 10–20 nmol NADP(H) (Krause & Heber, 1976), which is reduced by about 10% in the dark. When HP is produced at 0.6–1.0 μmol mg^{-1} chl h^{-1} and 20% is then metabolized in the OPP cycle, NADPH will be formed at 10–16 nmol mg^{-1} chl min^{-1}. In this way, the NADPH pool turns over in 4–12 s but it is not known how the NADPH is being used although recent workers have suggested that it may be oxidized by ferredoxin (Kow, Erbes & Gibbs, 1982) and NADPH can also be oxidized by glutathione (see Halliwell, 1981).

Most of the products of breakdown of starch are exported from the chloroplast as TP, PGA and sugars. *In situ*, these will be available for metabolism in the cytosol and could be respired or used for synthesis of sucrose. The extent to which starch is used for respiration has not been directly demonstrated but there is good evidence that starch can be converted to sucrose and exported in the dark (Sharkey & Pate, 1976; Geiger & Batey, 1967). When leaves are pulse-chased with $^{14}CO_2$ in the light and then darkened, a pulse of ^{14}C appears in sucrose in the phloem at the same time as the ^{14}C moves out of starch in the dark. Carbohydrate is retained in leaves in the day and mobilized to maintain a continued, if somewhat lower (Sharkey & Pate, 1976), export of sucrose during the night. Similar conclusions have been drawn elsewhere although the origin of the sucrose in the dark is not always starch (Bailey, 1958; Porter, Martin & Bird, 1959; Melvin, 1965; Harrt & Kortschack, 1967; Taylor, 1968; Pearson, 1973) but may be a remobilization of stored sucrose.

The route of mobilization of starch will affect the relation between sucrose and respiration in the cytosol (Fig. 9.2). Normally during breakdown of starch, phosphorolysis saves 1 ATP/HP mobilized, compared with hydrolysis. However, the chloroplast envelope separates starch from the cytosolic sinks and modifies the energy balance because HP cannot cross the envelope directly (Fliege, Fluegge, Werdan & Heldt, 1978). When chloroplastic starch is being used for respiration in the cytosol, phosphorolysis is more economical than hydrolysis. However, if starch is being converted to sucrose in the cytosol, the energy demands of hydrolysis and phosphorolysis are the same because the phosphorolytic route involves the conversion of HP into TP in the chloroplast, using 1 ATP/HP, and then the reconversion of TP to HP in the cytosol, with the loss of the newly esterified Pi. During conversion of starch to sucrose, hydrolysis appears a simpler route than phosphorolysis, in terms of the implications for enzyme regulation. The phosphorolytic route requires the cytosolic fructosebisphosphatase to be active in the dark while the cytosolic phosphofructokinase is inactive; at the same time, an active phosphofructokinase and inactive fructose bisphosphatase are needed in the chloroplasts. It is not known how these enzymes are regulated but it is

Fig. 9.2 The phosphorolytic and hydrolytic routes from starch to sucrose and respiration.

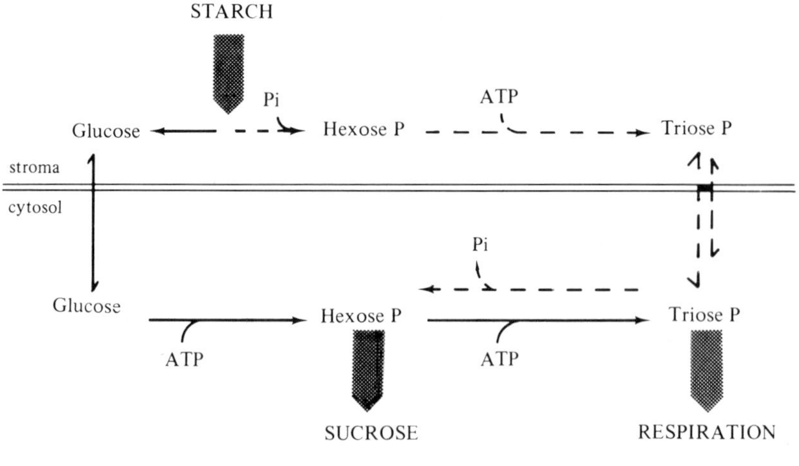

tempting to speculate that phosphorolysis usually provides substrates for respiration while hydrolysis provides substrates for synthesis of sucrose. If this is so, hydrolysis should become relatively more important in species whose leaves are very rich in starch, when an increasing proportion of the starch is being converted to sucrose.

Control of mobilization of starch in the dark

In general, two features of control of degradation of starch might be distinguished. The first is the extent to which it is modulated in the dark to reflect demand for respiratory substrate or synthesis and export of sucrose. The second is how far it is inhibited in the light during periods of synthesis and accumulation of starch. There is evidence in leaves that breakdown can be curtailed in the dark by accumulation of substrates. If some of the vascular strands are nicked in a petiole of grape leaves, breakdown is decreased in those areas of leaf supplying them but not in other areas (Czapek, 1905). Breakdown also decreases when sinks are removed from a plant (Mason & Phillis, 1934, and see later). Apparently, the rate of its mobilization falls when export of sucrose is restricted. A high concentration of sucrose in leaves of barley (Gordon, Ryle, Powell & Mitchell, 1980*a*; Gordon, Ryle & Webb, 1980*b*) was directly correlated with an inhibition of breakdown of starch. Plants were illuminated at varying irradiances so that different amounts of sucrose accumulated in the leaf in the day. This sucrose was exported at night. When leaves contained more sucrose, there was a longer lag before breakdown of starch started and this began when the concentration of sucrose fell to a particular level. It is not clear exactly how this inhibition is produced since it is not yet possible to compare rates of breakdown with routes and properties of this process for the same species. There is a need for such co-ordinated experiments on the biochemistry and physiology of starch. Nevertheless, based on work in pea and spinach, some methods of control can be proposed.

If starch is being degraded hydrolytically, it is not possible to put forward any control mechanism. No control properties were found in the enzymes (Okita & Preiss, 1980). In spinach chloroplasts, no inhibition of the formation of glucose and maltose from starch was produced by adding as much as 25 mM glucose, fructose or sucrose, or 1 mM TP, ribose5P or 3PGA. Low Pi actually stimulated this conversion (Stitt & Heldt, 1981*a*). On the other hand, phosphorolysis is stimulated by Pi and inhibited by 3PGA (Peavey *et al.* 1977; Heldt *et al.* 1977). A stimulation of phosphorolysis is accompanied by increased glycolysis and lower activity of the OPP cycle (Stitt & ap Rees, 1980*a*). Although including Pi greatly stimulates accumulation of phosphorylated products, less carbon flows into CO_2 and neutral products, so that total mobilization is only increased by 25–30% (Stitt & Heldt 1981*a*). The response to Pi occurs gradually as the Pi level in the medium rises from 10 μM to 2 mM and is accompanied by a gradual fall of stromal fructosebisP, TP and 3PGA, which are exported in exchange for Pi, and also by a marked fall in the stromal HP level (M. Stitt & H. W. Heldt, unpublished). *In vivo*, a decline in concentration of cytosolic substrate, accompanied by a rise in cytosolic Pi, could stimulate phosphorolysis to provide more substrates for the cytosol.

How is this stimulation of phosphorolysis regulated? The fall in HP which accompanies the stimulation of phosphorolysis and glycolysis shows that the chloroplastic phosphofructokinase catalyses an important step for control of phosphorolysis. Phosphofructokinase shows numerous potential regulatory properties (Kelly & Latzko, 1977*b*) and, in the present context, it is of interest that its activity is inhibited by 3PGA which in turn is overcome by rising Pi. It is not clear whether phosphorolysis is controlled solely through changes in the glucose 1-P/Pi-ratio resulting from alterations in the use of HP or if phosphorylase itself is also regulated. On the one hand, no regulatory properties have yet been found for the chloroplastic phosphorylase itself (Preiss *et al.* 1980) and, as phosphorylase is not a strongly irreversible enzyme, it may not be a good site for regulation. On the other hand, the view that the glucose 1-P/Pi ratio is never high enough *in vivo* to catalyse a net synthesis of starch (see Preiss & Levi, 1980) implies that phosphorylase could never be completely inhibited by changes in the levels of its substrates. More work is needed, both on the properties of the chloroplastic phosphorylase and on the extent to which carbon can move from HP back into glucans and starch, via phosphorylase *in vivo*.

Control of breakdown of starch in the light

Some control of breakdown of starch may be expected in the light, especially when it is being actively accumulated. However, it is important to establish whether light and the operation of photosynthesis alone suffice to shut off degradation or whether the control of degradation in the light is related more to its function as a buffer for synthesis of sucrose. One approach is to ask whether the enzymes of degradation are controlled so that they would be inactive during illumination. There is no evidence at present to suggest that the catalytic activity of the enzymes is greatly lowered by the conditions found during photosynthesis. In the light, the stroma becomes more alkaline (Heldt *et al.* 1973) but both the endoamylase and phosphorylase from spinach chloroplasts have a broad pH optimum and would only be about 50% inhibited. No inhibition was produced by a wide range of intermediates either (Preiss *et al.* 1980; Okita & Preiss, 1980). Another way to control the enzymes would be by altering their state of activation. This seems to occur to a limited extent. When chloroplasts were isolated from spinach leaves in the night, they showed about 2–3 times as much amylolytic activity as in those taken from leaves during the day (Pongratz & Beck, 1978). The exact role of these alterations of activity is unclear, because the maximal rates of breakdown were reached at the start of the night before the activity of amylase had risen much above that found in the light. Since the enzyme activity found in the light can support maximal rates of breakdown during this period at the start of the night, it would appear that the oscillations in activity of amylase on their own would not prevent similar rates occurring in the light too so that other factors must also be involved. Activity of amylase rose slowly through the night and peaked at the end, when starch was almost exhausted. Possibly the high activity allows rapid breakdown to occur even when little starch is left, thus ensuring its efficient turnover. The cause of the variation in activity is unknown; it is not a direct response to light since the oscillation of activity of amylase was maintained even when leaves were transferred into continuous light or continuous darkness. In studies with several Calvin cycle enzymes, thioredoxin has been proposed as a system for modifing enzyme activities between the light and dark, and dithiothreitol is widely used to stimulate these effects (see Buchanan, 1980). However, neither the endoamylase nor the phosphorylase was inhibited by dithiothreitol or thioredoxin (Okita & Preiss, 1980; Preiss *et al.* 1980). Diurnal changes of amylolytic activity have also been found in leaves of *Kolanchoe* (Vieweg & Fekete, 1977) but not in those of pea (Stitt *et al.* 1978). However, the significance of measurements in whole leaves is unclear because the enzymes may be extrachloroplastic (see above).

A second approach is to investigate whether a flux of carbon out of starch occurs in the light. At least sometimes degradation continues in the light (Jones, Martin & Porter, 1959; Dickson & Larwood, 1973). Leaves labelled with $^{14}CO_2$ will remobilize ^{14}C from starch during a subsequent light chase, even though the starch continues to accumulate. Sometimes, even the total amount of starch decreases in the light. When peas are transferred to continuous light, the starch content begins to fall after about 20 h (Bulpin, 1977). Similar results have been found for spinach (Pongratz & Beck, 1978) and when barley was grown in summer sunlight in Munich, the amount of starch started to decrease by 3 p.m. (M. Stitt, unpublished). However, none of these experiments allow a quantitative comparison of the rates of synthesis and breakdown in light in order to assess the significance of the continued mobilization. Recently, we have measured the rates of synthesis and breakdown of starch in the light, and the rate of breakdown in the dark in isolated starch-loaded spinach chloroplasts (Stitt & Heldt, 1981*b*). Breakdown occurred in the light. At high Pi(2 mM) the rate in light was scarcely inhibited, compared with the dark. A larger inhibition was found at lower Pi which supported maximal rates of allowed photosynthesis. This inhibition was due mainly to an inhibition of phosphorolysis; the formation of neutral products was less strongly inhibited. At all concentrations of Pi, rates of degradation in light were significant compared with the rates of synthesis of starch. Even at concentrations of Pi allowing optimal synthesis, a significant flux of carbon out of starch continued and degradation exceeded synthesis at the optimal Pi for CO_2 fixation. These results imply that the rate of accumulation of starch depends upon control of degradation as well as control of synthesis and that the balance of synthesis and degradation is very sensitive to Pi. In contrast to degradation of starch, release of CO_2 was strongly inhibited in light. This is in agreement with earlier work showing that the isolated G6P dehydrogenase is likely to be inactivated by the NADPH/NADP levels found in the stroma in light (Lendzian, 1980*a*) and also with observations that this enzyme is directly inactivated by light (Buchanan, 1980). These experiments showing rapid degradation of starch in light were carried out with starch-loaded chloroplasts, typical of those found at the end of the day. It would be interesting to study the interaction of synthesis and degradation in chloroplasts with lower starch contents, representative of those found earlier in the day when rates of accumulation are maximal. Towards the end of a light period, the accumulation of starch often becomes slower (Upmeyer & Kroller, 1973; Stitt *et al.* 1978; Pongratz & Beck, 1978) and this might reflect a mounting rate of degradation which balances and can eventually exceed synthesis after about 20 h.

That degradation of starch can occur in light and that the relation between synthesis and degradation in the light depends upon Pi can be understood in terms of the view that starch is primarily an overflow pool from sucrose. Then, the extent of both accumulation and remobilization would depend upon concentration of intermediates and Pi in the cytosol in light. Direct control of breakdown by pH, Mg^{2+} and thioredoxin would not be needed. This contrasts with regulation of the Calvin cycle itself, where the enzymes are very tightly regulated by pH, Mg^{2+} and thioredoxin. This ensures that they are not active in the dark, otherwise they would then seriously disrupt the metabolism of the leaf in darkness (Laing, Stitt & Heldt, 1981).

Starch – plan or accident

If starch is viewed as a buffer for sucrose metabolism, it may be asked how far its accumulation and subsequent degradation are just a response of the plant to uncontrolled changes in its environment, or whether the amount of starch accumulated and the degree of partitioning of carbon between sucrose and starch are controlled processes. There are several arguments showing that the extent of accumulation is indeed in some way selected. For example, different species in given environmental conditions will accumulate greatly different amounts (Chapters 1 and 8) and there may well be some rationale behind these differing strategies in the use of photosynthate. Another argument is that the percentage of photosynthate retained in starch actually rises when many species are grown over 3–4 days in lower irradiances or in a shorter daylength (Silvius, Chatterton & Kremer, 1979; Chatterton & Silvius, 1980). This would not occur if starch were simply an overflow product since less and not more starch should then be made. The author suggested that starch levels are programmed and that plants make optimal use of the phloem by retaining carbon in the leaf in order to export it at night. Observations that metabolism of starch in leaves is affected by the status of the plant, over a timescale of days, also indicate that it is in some way a controlled process. In general, amounts of starch in leaves are increased by processes which decrease demand for exported sucrose, such as slow growth (McCarty, 1935; Hilliard & West, 1970) or removal (Hall & Milthorpe, 1978) or absence (Chatterton, Carlson, Hungerford & Lee, 1972) of large sinks. On the other hand, increasing demand for sucrose from leaves caused by removing (Hall & Milthorpe, 1978) or shading (Thorne & Koller, 1974) other leaves will lower the starch content in the remaining leaves. Although part of this response may involve changes in concentrations of intermediates in the leaf, the slow time response, and the fact that the characteristics of gas exchange and activity of ribulosebisphosphate carboxylase alter (Thorne & Koller, 1974) suggests that changes in enzyme capacities may also be involved.

Variations in metabolism of starch occur within a single leaf as it develops and then ages, and these changes are also best understood when such metabolism is seen as a controlled process. In old leaves, the diurnal turnover of starch becomes modified so that starch accumulates to high levels which do not fall during the night (see Chapter 8). When young and old cottonwood leaves were pulsed with $^{14}CO_2$, and then given a light chase, radioactivity disappeared from starch in young leaves, but continued to rise in starch in old leaves (Dickson & Larwood, 1973). Moreover, ^{14}C appeared in starch of old leaves which had not been supplied initially with $^{14}CO_2$. Photosynthate is not only retained in old leaves as starch but is also imported as its precursor. Recently Giaquinta, Quebedaux & Wittenbach (1981) have shown that starch, accumulated in old soybean leaves, is then used to fill the bean pods. Young leaves showed a diurnal turnover of starch but, in old leaves, this gradually disappeared and the overall amount of starch increased. Pod-fill occurred as photosynthesis declined and was accompanied by remobilization of the starch. The implication is, that in older leaves, starch changes from a transient to a more long-term storage product.

Exactly how metabolism of starch is varied between different leaves is not known. Presumably, its general nature will depend upon coarse control, such as changes in the activities of the enzymes of degradation and synthesis. Within these limits, rates of turnover will vary, depending upon fine control, such as concentrations of Pi and 3PGA (Heldt *et al.* 1977; Preiss & Levi, 1980). The presence of such fine control is essential if starch is to act as a flexible buffer to sucrose metabolism, allowing the plant, for example, to spread the synthesis and export of sucrose over a longer period when photosynthetic rates are above average. It is of interest that the control of synthesis (Heldt *et al.* 1977) and degradation (see above) of starch by the concentrations of Pi and 3PGA is a gradual process permitting fine tuning of the short-term flow of carbon into, and out of, starch. The fact that synthesis may occur in the dark (see Chapter 8) and that degradation can occur in light (see above) adds to the flexibility with which the metabolism of starch and sucrose metabolism can be co-ordinated.

The conclusions and suggestions in this chapter are based on chloroplasts from leaves of vegetatively growing C3 plants. It is possible that pathways and mechanisms of control may differ in older leaves, in specialized leaves of C4 and CAM plants and in leaves with particularly large amounts of starch. Methods are, however, now available which will allow these questions to be investigated in a wide range of plants. In particular, this means, on the one hand, using protoplasts as a source of material to study activities of enzymes in chloroplasts, and as a source of chloroplasts to study degradation of starch. On the other hand, it would be fruitful simultaneously to study metabolism of starch in leaves on the plant and to investigate changes in the activities of starch-metabolizing enzymes in relation to starch, sucrose and photosynthetic metabolism. In this way, the fine details of starch metabolism could be integrated with an understanding of the contribution of starch to the balance of metabolism in the leaf and the plant.

I am indebted to Professor H. W. Heldt for his stimulating co-operation in experimental work and for constructive criticism of this article and to Frau U. Kaiser for photographing the figures.

References

ap Rees, T. (1974). Pathways of carbohydrate metabolism in higher plants. In *International Review of Science*, Series One, vol. **11**, ed. D. H. Northcote, pp. 89–129. Medical and Technical Publications.

Anderson, L. E., Ng, T. C. L. & Park, K. E. Y. (1974). Inactivation of pea leaf chloroplastic and cytoplasmic glucose 6-phosphate dehydrogenases by light and dithiothreitol. *Plant Physiology*, **53**, 835–9.

Bailey, R. W. (1958). Carbohydrates in pasture species I. The starch content of clover and rye grass. *Journal of the Science of Food and Agriculture*, **9**, 743–7.

Beck, E., Pongratz, P. & Reuter, I. (1981). In *Proceedings of the 5th International Congress on Photosynthesis*, vol. 4, ed. G. Akoyunoglou, pp. 529–37.

Briggs, D. E. (1967). A modified asaay for α-amylase in germinating barley. *Journal of the Institute of Brewing*, **73**, 361–70.

Buchanan, B. B. (1980). Role of light in the regulation of chloroplast enzymes. *Annual Review of Plant Physiology*, **31**, 341–74.

Bulpin, P. V. (1977). Starch metabolism in *Arum maculatum* spaddica and Pea leaves. Ph.D. Thesis, University of Cambridge.

Bulpin, P. V. & ap Rees, T. (1978). Starch breakdown in the spadix of *Arum maculatum*. *Phytochemistry*, **17**, 391–6.

Chang, C. W. (1980). The partition of starch metabolising enzymes in chloroplasts and their functional dependencies. *Plant Physiology*, **65**, (suppl.) 114.

Charles, S. A. & Halliwell, B. (1980). Properties of freshly purified and thiol-treated spinach chloroplast fructosebisphosphatase. *Biochemical Journal*, **185**, 689–93.

Chatterton, N. J., Carlson, G. E., Hungerford, W. E. & Lee, D. R. (1972). Effect of tillering and cool nights on photosynthesis and chloroplast starch in *Pangola*. *Crop Science*, **12**, 206–18.

Chatterton, N. J. & Silvius, J. E. (1980). Photosynthate partitioning into leaf starch as affected by daily photosynthetic period duration in several species. *Physiologia Plantarum*, **49**, 141–4.

Czapek, F. (1905). *Biochemie der Pflanze*. Jena: Fischer.

Dickson, R. E. & Larwood, P. R. (1973). Incorporation of ^{14}C photosynthate into major chemical fractions of source and sink leaves in cottonwood. *Plant Physiology*, **57**, 185–93.

Edwards, G. E., Robinson, S. P., Tyler, N. J. C. & Walker, D. A. (1978). Photosynthesis by isolated protoplasts, protoplast extracts and chloroplasts of wheat. *Plant Physiology*, **62**, 313–19.

Fliege, R., Flugge, U.-I., Werdan, K. & Heldt, H. W. (1978). Specific transport of inorganic phosphate, 3-phosphoglycerate and triose phosphates across the inner membrane of the envelope in spinach chloroplasts. *Biochimica et Biophysica Acta*, **502**, 232–47.

Furbank, R. & Lilley, R. McC. (1981). Reductive pentose phosphate cycle and oxidative carbohydrate metabolic activities in pea chloroplast stroma extracts. *Plant Physiology*, **67**, 1036–41.

Gates, J. W. & Simpson, G. M. (1968). The presence of starch and alpha-amylase in the leaves of plants. *Canadian Journal of Botany*, **46**, 1459–62.

Geiger, D. R. & Batey, J. W. (1967). Translocation of ^{14}C sucrose in sugar beet during darkness. *Plant Physiology*, **42**, 1743–9.

Giaquinta, R. T., Quebedaux, B. & Wittenbach, B. (1981). Partitioning and translocation of photosynthate in exporting soybean leaves during seed filling. In *Proceedings of the 5th International Congress on Photosynthesis*, vol. 4, ed. G. Akoyunoglou, pp. 549–50.

Gordon, A. J., Ryle, G. J. A., Powell, D. & Mitchell, D. (1980*a*). Export, mobilisation and respiration of assimilates in uniculm barley during light and darkness. *Journal of Experimental Botany*, **131**, 461–73.

Gordon, A. J., Ryle, G. J. A. & Webb, G. (1980*b*). The relationship between sucrose and starch during 'dark' export from leaves of uniculm barley. *Journal of Experimental Botany*, **131**, 845–50.

Hall, A. J. & Milthorpe, F. L. (1978). Assimilate source-sink relationship in *Capsicum annuum* L. III. The effects of fruit excision on photosynthesis, and leaf and stem carbohydrates. *Australian Journal of Plant Physiology*, **5**, 1–13.

Halliwell, B. (1981). *Chloroplast Metabolism*, pp. 183–6. Oxford: Clarendon Press.

Harrt, C. E. & Kortschak, H. P. (1967). Translocation of ^{14}C in sugar cane plants during the day and night. *Plant Physiology*, **42**, 89–94.

Heldt, H. W., Chon, C. J., Maronde, D., Herold, A., Stankovic, Z. S., Walker, D. A., Kraminer, A., Kirk, M. R. & Heber, U. (1977). Role of orthophosphate and other factors in the regulation of starch formation in leaves and isolated chloroplasts. *Plant Physiology*, **59**, 1146–55.

Heldt, H. W., Laing, W., Lorimer, G. H., Stitt, M. & Wirtz, W. (1981). On the regulation of CO_2 fixation by light. In *Proceedings of the 5th International Congress on Photosynthesis*, vol. 4, ed. G. Akoyunoglou, pp. 213–26.

Heldt, H. W., Werdan, K., Milovancev, M. & Geller, G. (1973). Alkalisation of the chloroplast stroma caused by light-dependent proton flux into the thylakoid space. *Biochimica et Biophysica Acta*, **314**, 224–41.

Herbert, M., Burkhard, Ch. & Schnarrenberger, C. (1979). A survey for isoenzymes of glucose phosphate isomerase, phosphoglucomutase, glucose 6-phosphate dehydrogenase and 6-phosphogluconate dehydrogenase in C3, C4- and Crassulacean-acid-metabolism in plants, and green algae. *Planta*, **145**, 96–104.

Hilliard, J. H. & West, S. H. (1970). Starch accumulation associated with growth reduction at low temperatures in a tropical plant. *Science*, **198**, 494–6.

Jones, H., Martin, R. V. & Porter, H. K. (1959). Translocation of ^{14}C carbon in tobacco following assimilation of ^{14}C carbon dioxide by a single leaf. *Annals of Botany*, N.S., **23**, 493–508.

Kaiser, W. M. & Bassham, J. A. (1979). Carbon metabolism of chloroplasts in the dark: oxidative pentose phosphate pathway versus glycolytic pathway. *Planta*, **144**, 193–200.

Kelly, G. J. & Latzka, E. (1975). Evidence for phosphofructokinase in chloroplasts. *Nature*, **256**, 429–30.

Kelly, G. J. & Latzko, E. (1977*a*). Chloroplast phosphofructokinase. I Proof of phosphofructokinase activity in chloroplasts. *Plant Physiology*, **60**, 290–4.

Kelly, G. J. & Latzko, E. (1977*b*). Chloroplast phosphofructokinase. II Partial purification, kinetic and regulatory properties. *Plant Physiology*, **60**, 295–9.

Kow, Y. W., Erbes, D. L. & Gibbs, M. (1982). Chloroplast respiration. A means of supplying oxidized pyridine nucleotide for dark chloroplastic metabolism. *Plant Physiology*, **69**, 442–7.

Krause, G. H. & Heber, U. (1976). Energetics of intact chloroplasts. In *The Intact Chloroplast*, ed. J. Barber, pp. 171–214. Amsterdam: Elsevier.

Jones, H., Martin, R. V. & Porter, H. K. (1959). Translocation of ^{14}C carbon in tobacco following assimilation of ^{14}carbon dioxide by a single leaf. *Annals of Botany*, N. S., **23**, 493–508.

Laing, W. A. L., Stitt, M. & Heldt, H. W. (1981). Control of CO_2 fixation. Changes in the activity of ribulose phosphate kinase and fructose- and sedoheptulose-bisphosphatase in chloroplasts. *Biochimica et Biophysica Acta*, **638**, 348–59.

Latzko, E. & Gibbs, M. (1968). Distribution and activity of enzymes of the reductive pentose phosphate cycle in spinach leaves and chloroplasts isolated by different methods. *Zeitschrift für Pflanzenphysiologie*, **59**, 184–94.

Latzko, E., Steup, M. & Schaechtele, Chr. (1981). Starch synthesis and degradation in photosynthetic tissues. In *Proceedings of the 5th International Congress on Photosynthesis*, vol. 4, ed. G. Akoyunoglou, pp. 517–28.

Lendzian, K. J. (1980*a*). Modulation of glucose 6-phosphate dehydrogenase by NADPH, NADP and dithiothreitol at various NADPH/NADP ratios in an illuminated reconstituted spinach chloroplast system. *Planta*, **148**, 1–6.

Lendzian, K. J. (1980*b*). Metabolism of specifically labelled glucose, glucose 1-phosphate and glucose 6-phosphate via the oxidative pentose phosphate cycle in a reconstituted spinach chloroplast system in darkness and in light. *Plant Physiology*, **66**, 8–12.

Lendzian, K. J. & Ziegler, H. (1970). Ueber die Regulation der Glucose 6-phosphatdehydrogenase in Spinatchloroplasten durch Licht. *Planta*, **94**, 27–36.

Levi, C. & Preiss, J. (1977). Amylopectin degradation in pea chloroplast extracts. *Plant Physiology*, **61**, 218–20.

Linden, J. C., Tanner, W. & Kandler, O. (1974). Properties of glucosyltransferase and glucan transferase from spinach. *Plant Physiology*, **54**, 752–7.

McCarty, E. (1935). Seasonal march of carbohydrates in *Elymus ambiguus* and *Muhlenbergia gracilis* and their reaction under moderate grazing use. *Plant Physiology*, **10**, 727–38.

Manners, D. (1974). Some aspects of the enzymic degradation of starch. In *Plant Carbohydrate Biochemistry*, ed. J. P. Pridham, pp. 109–25. London: Academic Press.

Mason, T. G. & Phillis, E. (1934). Studies on the transport of nitrogenous substances in the cotton plant. *Annals of Botany*, **48**, 315–33.

Matheson, N. K. & Wheatley, J. M. (1962). Starch changes in developing and senescing tobacco leaves. *Australian Journal of Biological Science*, **15**, 445–58.

Melvin, J. F. (1965). Variation in the carbohydrate content of lucerne and the effect of ensilage. *Australian Journal of Agricultural Research*, **16**, 951–9.

Miflin, B. & Beevers, H. (1974). Isolation of intact plastids from a range of plant tissues. *Plant Physiology*, **53**, 870–4.

Okita, T. W., Greenberg, E., Kuhn, D. N. & Preiss, J. (1979). Subcellular localisation of the starch degradative and biosynthetic enzymes of spinach leaves. *Plant Physiology*, **64**, 187–92.

Okita, T. W. & Preiss, J. (1980). Starch degradation in spinach leaves: isolation and characterisation of the amylases and R-enzyme of spinach leaves. *Plant Physiology*, **66**, 870–6.

Pearson, C. J. (1973). Daily changes in stomatal aperture and in carbohydrates and metabolites in epidermis and mesophyll of leaves of *Corralina cyanea* and *Vicia faba*. *Australian Journal of Biological Science*, **26**, 1035–44.

Peavey, D. G., Steup, M. & Gibbs, M. (1977). Characterisation of starch breakdown in the intact spinach chloroplast. *Plant Physiology*, **60**, 305–8.

Pongratz, P. & Beck, E. (1978). Diurnal oscillation of amylolytic activity in spinach chloroplasts. *Plant Physiology*, **62**, 687–9.

Porter, H. K., Martin, R. V. & Bird, I. F. (1959). Synthesis and degradation of starch labelled with ^{14}C carbon in tobacco leaf tissue. *Journal of Experimental Botany*, **10**, 264–76.

Preiss, J. & Levi, C. (1980). Starch synthesis and degradation. In *The Biochemistry of Plants*, vol. 3, ed. P. Stumpf, pp. 371–424. London, New York: Academic Press.

Preiss, J., Okita, T. W. & Greenberg, E. (1980). Characterisation of the spinach leaf phosphorylases. *Plant Physiology*, **66**, 864–9.

Schaefer, G. & Heber, U. (1977). Glucose transport into spinach chloroplasts. *Plant Physiology*, **60**, 286–9.

Sharkey, P. J. & Pate, J. S. (1976). Translocation from leaves to fruits of legume, studied by a phloem bleeding technique: diurnal changes and effect of continual darkness. *Planta*, **128**, 63–72.

Silvius, J. E., Chatterton, N. J. & Kremer, D. F. (1979). Photosynthate partitioning in soybean leaves at two irradiance levels. Comparative response of acclimated and unacclimated leaves. *Plant Physiology*, **64**, 872–5.

Silvius, J. E., Kremer, D. F. & Lee, D. R. (1978). Carbon assimilation and translocation in soybean leaves at different stages of development. *Plant Physiology*, **62**, 54–8.

Smillie, R. (1966). Photosynthetic and respiratory activities of growing pea leaves. *Plant Physiology*, **37**, 716–21.

Steup, M. & Latzko, E. (1979). Intracellular localisation of phosphorylases in spinach and pea leaves. *Planta*, **145**, 69–75.

Steup, M., Peavey, D. G. & Gibbs, M. (1978). The regulation of starch metabolism by inorganic phosphate. *Biochemical and Biophysical Research Communications*, **72**, 1554–62.

Steup, M., Schaechtele, Chr. & Latzko, E. (1979). Separation and partial characterisation of chloroplast and non-chloroplast α-glucan phosphorylase from spinach leaves. *Zeitschrift für Pflanzenphysiologie*, **96**, 365–74.

Steup, M., Schaechtele, Chr. & Latzko, E. (1980). Purification of a non-chloroplastic a-glucan phosphorylase from spinach leaves. *Planta*, **148**, 168–73.

Steup, M. & Schaechtele, Chr. (1981). Mode of glucan degradation by purified phosphorylase forms from spinach leaves. *Planta*, **153**, 351–61.

Stitt, M., Bulpin, P. V. & ap Rees, T. (1978). Pathway of starch breakdown in photosynthetic tissues of *Pisum sativum. Biochimica et Biophysica Acta*, **544**, 200–14.

Stitt, M. & ap Rees, T. (1979). Capacities of pea chloroplasts to catalyse the oxidative pentose phosphate pathway and glycolysis. *Phytochemistry*, **18**, 1905–11.

Stitt, M. & ap Rees, T. (1980*a*). Carbohydrate breakdown by chloroplasts of *Pisum sativum. Biochimica et Biophysica Acta*, **627**, 131–43.

Stitt, M. & ap Rees, T. (1980*b*). Estimation of the activity of the oxidative pentose phosphate pathway in pea chloroplasts. *Phytochemistry*, **19**, 1583–5.

Stitt, M. & Heldt, H. W. (1981*a*). Physiological rates of starch breakdown in isolated intact spinach chloroplasts. *Plant Physiology*, **68**, 755–61.

Stitt, M. & Heldt, H. W. (1981*b*). Simultaneous synthesis and degradation of starch in spinach chloroplasts in the light. *Biochimica et Biophysica Acta*, **638**, 1–11.

Sutton, B. G. (1975). Glycolysis in CAM planta. *Australian Journal of Plant Physiology*, **2**, 389–402.

Taylor, A. O. (1968). The influence of temperature and time of day on the level of readily fermentable carbohydrate in white clover leaves. *Journal of the New Zealand Agricultural Society*, **2**, 172–3.

Thorne, J. H. & Koller, H. R. (1974). Influence of assimilate demand on photosynthesis, diffusion resistance, translocation levels and carbohydrate levels of soybean leaves. *Plant Physiology*, **54**, 201–7.

Upmeyer, D. T. & Koller, H. R. (1973). Diurnal trends in net photosynthetic rate and carbohydrate levels of soybean leaves. *Plant Physiology*, **51**, 871–4.

Vieweg, G. H. & Fekete, M. A. R. de (1977). Diurnal fluctuations of amylase activity in *Kalanchoe diagramontiana* leaves. *Zeitschrift für Pflanzenphysiolgie*, **81**, 74–9.

Williamson, J. F. (1980). A critical examination of the evidence for the reactions of the pentose pathway in animal tissues. *Trends in Biochemical Sciences*, **60**, 315–19.

C. M. DUFFUS

10 Metabolism of reserve starch

Introduction

Starch is a major component of agricultural production. For example, cereal grains, which contain between 60 and 70% starch are the dominating influence in world food production. Around 1500 million tonnes of cereals were produced in 1979 and, on a world basis, these account for over half of our dietary energy supplies. The main source for industrial use is maize although, in some countries, other plants, e.g. potato, sago, tapioca and wheat, may also make a contribution. In the food industry, starch is used directly in a range of 'instant' or cold water pastes. Other applications include the production of syrups from starch by acid or enzymatic hydrolysis. These are used as sweeteners or as cereal substitutes in the fermentation industries. Starch products also find extensive use in the pharmaceutical, textile and paper industries as well as in the offshore oil industry where they are incorporated into the artificial muds used during the drilling of oil wells. In the face of declining oil supplies, it is likely that starch may have a new future as a fermentable substrate in the alcohol-for-fuel industry which may yet overtake its present role in the alcohol-for-drinks industry. Thus, there is considerable interest both in the mechanisms controlling the biosynthesis of starch as well as its degradation. The objective, as far as biosynthesis is concerned, is to identify the biological limitations to deposition of starch in plants with the hope of modifying them to increase the amount produced. Interestingly, there is probably even greater current interest in the degradation of starch, probably as a direct result of its role as a source of alcohol and sugars. This interest extends not only to degradation of starch *in vivo*, particularly in relation to the brewing and distilling industries, but also *in vitro* to the use of added enzymes from other sources such as micro-organisms, for the industrial production of fermentation products. The purpose of this account is, however, to describe the metabolism of starch in plants and industrial processes will not be described further.

Biosynthesis of starch

Morphological changes accompanying deposition of starch

The biological system most commonly used for studies on the biosynthesis of starch is the developing cereal grain. A major reason for this is the enormous economic importance of cereals, which can justify a relatively large programme of research into improving both quality and yield of grain. The system has also a number of experimental attractions: firstly, the seeds mature over a relatively long period, facilitating convenience of sampling and assay; secondly, a grain or seed is a well defined organ of uniform size and hence results can be expressed conveniently on a per grain basis; thirdly, by careful selection, it is possible to breed varieties of cereal having a range of unusual reserve polysaccharides which are ideal material for both biochemical and genetic investigations. To some extent, these arguments apply to other seeds also, including those of legumes, and there is an increasing body of literature relating to metabolism of starch in peas and beans. Oddly, rather less work has been done with starch-storing tubers such as the potato or yam.

The events leading up to fertilization are similar for most developing angiospermous seeds. The pollen germinates on the stigma forming a pollen tube which penetrates the style and carries two male nuclei into the embryo sac. One male nucleus fuses with the egg nucleus to form the diploid zygote. The second nucleus fuses with the two polar nuclei forming a triploid primary endosperm nucleus from which the endosperm is derived (Brink & Cooper, 1947; Luxová, 1967). The subsequent development of zygote and endosperm differs considerably in different groups of plants. For example, in cereals, the endosperm grows rapidly to become the major storage organ of the mature grain. In legumes, the embryo grows to fill the seed almost completely at maturity, its cotyledons becoming the major storage organs. Cereal endosperm is generally a starchy tissue and embryos are normally reservoirs of lipid and protein, hence legume seeds have a much higher protein-to-starch ratio than cereals.

Cereal grains. The endosperm nucleus divides rapidly and, by about two or three days after fertilization when cell-wall formation commences, there might be 200–500 free nuclei within the embryo sac in barley (Thompson & Johnston, 1945). Mitochondria, proplastids, endoplasmic reticulum and Golgi apparatus are distinguishable from the earliest stages of development of the endosperm (Buttrose, 1963). Deposition of starch within its amyloplasts commences early in development, as does the formation of small spherical protein bodies in vacuoles.

Forming part of the endosperm, but distinct in structure from the cells storing starch, is the aleurone layer. This tissue differentiates later and the cells are initially cuboidal with thin walls and large nuclei (Morrison, Kuo & O'Brien, 1975). At the crease region (Fig. 10.1) where nutrients enter the endosperm, the aleurone cells are considerably modified with thick irregular walls and an extensive membrane system (Cochrane & Duffus, 1980). They are most prolific in barley, forming a layer 3–5 cells thick at maturity. Since these cells are the site of synthesis of α-amylase during germination, the success of barley in the degradation of starch during malting is often attributed to its well-developed aleurone layer. The starch granules grow steadily throughout development, reaching a maximum diameter in wheat of around 25 μm. There is some evidence for a secondary formation of small starch granules 2–3 weeks after fertilization (May & Buttrose, 1959; Williams & Duffus, 1977*a*). The suggestion of Buttrose (1963) that these were formed by the blebbing off of peripheral regions of large amyloplasts has not been confirmed. These granules are spherical and remain small (10 μm in diameter) so that, at maturity, both large and small granules are present. In wheat, the large granules, and those which are going to develop into large granules, have a characteristic lenticular shape with a peripheral groove (Evers, 1971). A tubular complex is present inside the amyloplast membrane at each end of

Fig. 10.1 Transverse section through a developing wheat grain 14 days after anthesis at mid-point between base and apex. cr, crease; e, starchy endosperm; v, vascular bundles; fc, funiculus-chalazal region; np, nucellar projection; al, aleurone layer; op, outer pericarp; sc, seed coat; ip, inner pericarp. [Reproduced with permission from Sakri & Shannon (1975).]

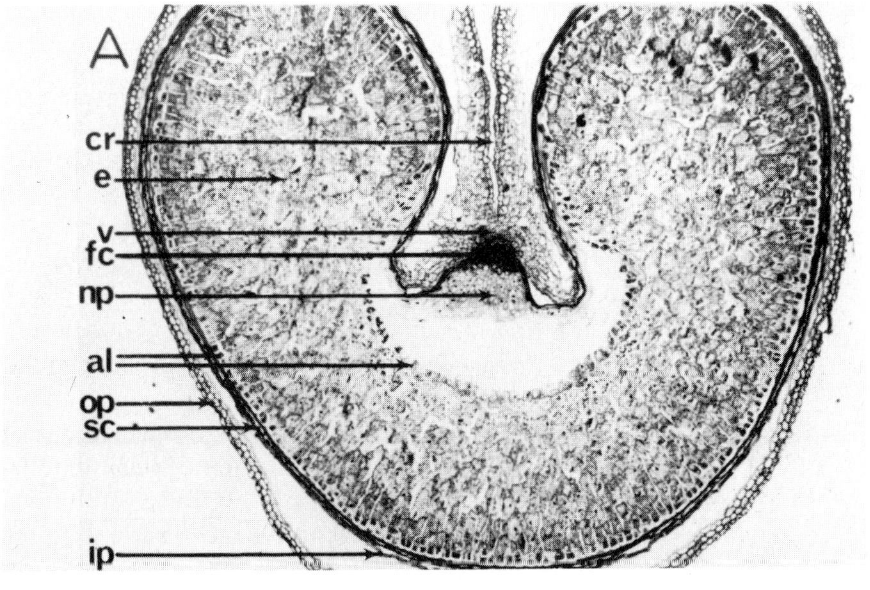

the developing starch granule (M. P. Cochrane, personal communication) (Fig. 10.2). It is possible that this surrounds the granule, possibly in the groove, and may be the site of action of some of the enzymes involved in deposition of starch. It is also possible that the association of plastids with endoplasmic reticulum reported in many types of plant cells is related to synthesis of starch (Thomson & Whatley, 1980).

During much of the developmental period of cereals, mitochondria are closely associated with the amyloplasts (Williams & Duffus, 1977*a*), suggesting that the energy for synthesis of starch may be derived from oxidative respiratory processes. There is much evidence (Duffus, 1979) to suggest that

Fig. 10.2 Electron micrograph of an amyloplast with a tubular complex (t) in the stroma at each end of the starch granule(s). This is part of a transverse section of the starchy endosperm cell immediately inside the sub-aleurone in the flank of a 19 day old field-grown barley grain (micrograph supplied by M. P. Cochrane).

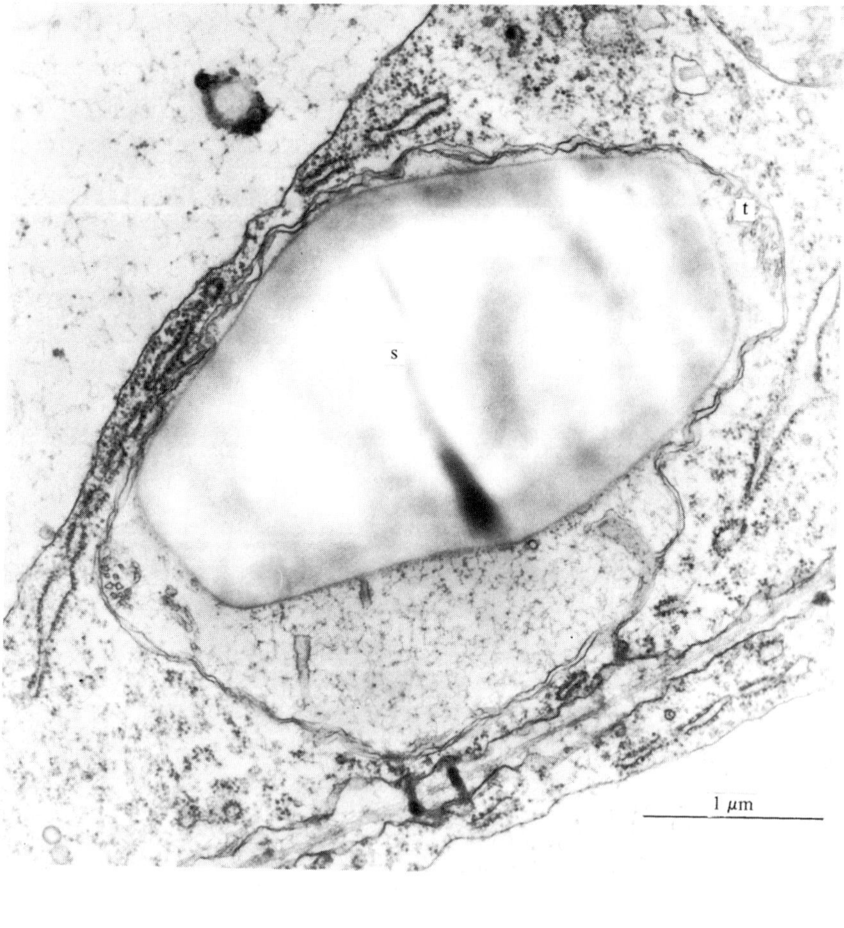

the amylopectin content of the starch synthesized decreases as deposition proceeds. This may relate to changes in the relative numbers of the different size classes of granules present, which show decreasing amylopectin content with increasing size (Duffus & Murdoch, 1979). Matheson (1975) has shown that the very small starch granules, typical of *sugary* endosperms in maize, are made up of phytoglycogen and amylose. As granule size increases, the molecular weight of the phytoglycogen deposited also increases. Hence it may be that the very high molecular weight of amylopectin is a factor in the initiation of starch granule formation. (See Chapter 1 for a brief account of differences between these components of starch.) The morphology, initiation and growth of starch granules have been extensively reviewed by Jenner (1982).

As maturation proceeds, the rate of deposition of starch decreases, dry weight reaches a maximum and dehydration sets in. By maturity, a typical cereal grain contains 60–70% starch, 8–12% protein, 1–3% mineral elements, 2–5% lipid, 12–15% water and 3–11% fibre (Duffus & Slaughter, 1980).

Legumes. Many legume seeds store large amounts of starch. For example, lentils contain over 50% starch. In the garden pea, which may contain as much as 30% starch, the main phase of accumulation commences once the cotyledons have reached their maximum size. The plastids contain only single starch granules which grow to around 10 μm in diameter (Pate & Flinn, 1977). The developing soybean has the curious property of accumulating starch as a transient reserve material, i.e. concentrations of starch increase to mid-development and then decrease (Adams, Rinne & Fjerstad, 1980), presumably as a result of increased amylase activity. In a similar manner, a transient starch-like polysaccharide is synthesized during the initial stages of germination in the commercial lupin variety, *Lupinus luteus* (Saini & Matheson, 1981). As with cereals, it appears that the first insoluble product to accumulate may be branched, being followed by increasing deposition of amylose as starch accumulates.

Tubers. The morphological changes taking place during growth of the potato tuber have been described by Ohad, Friedberg, Ne'eman & Schramm (1971) as well as Reeve, Hautala & Weaver (1969). During all stages of the tuber's development, the starch granules are surrounded by a double membrane and, in most cases, no cytoplasmic or plastid matrix was observed between the granule and the surrounding membrane. The purified membranes are similar in most characteristics to those from other types of plastid (Fishwick & Wright, 1980).

Origin of starch precursors

Sucrose is generally considered to be the primary substrate for the synthesis of reserve starch. Firm evidence for this is lacking and the conclusion is based on a number of more or less indirect observations: firstly,

Fig. 10.3 Hypothetical scheme for conversion of sucrose to α-(1→4)-D-glucan assuming selective permeability of the amyloplast inner membrane. (1), sucrose synthase; (2), invertase; (3), ADP-glucose and UDP-glucose pyrophosphorylases; (4), hexokinase; (5), phosphoglucomutase; (6), phosphohexoisomerase; (7), phosphofructokinase; (8), aldolase; (9), triosephosphate isomerase; (10) fructose-1,6-bisphosphatase; (11), starch synthase. UDP-Glc, UDP-glucose; ADP-Glc, ADP-glucose; Glc-6-P, glucose 6-phosphate; Fru-6-P, fructose 6-phosphate; Fru-1,6-BP, fructose, 1,6-bisphosphate; Glc-1-P, glucose 1-phosphate; G-3-P, glyceraldehyde 3-phosphate; DHAP, dihydroxyacetone phosphate; PPi, pyrophosphate.

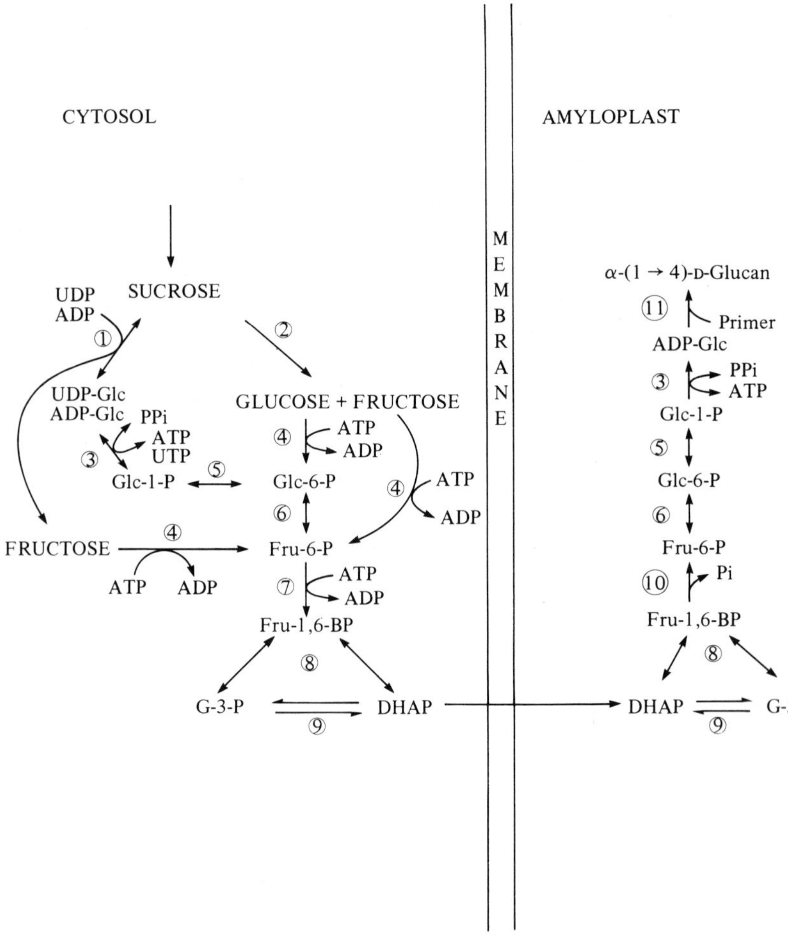

sucrose is the main translocatory material of vascular plants, secondly, enzymes of sucrose metabolism are present in starch-storing tissues and, thirdly, sucrose is generally present in such tissues. Its likely origins and subsequent metabolism are described in recent reviews (Duffus, 1979; Duffus & Cochrane, 1982; Duffus & Duffus, 1984). The main products, which are derived from sucrose in developing cereal endosperms (Duffus, 1979) and potato tubers (Sowokinos, 1973), and which relate to starch synthesis, appear to be UDP-glucose and fructose, together with some ADP-glucose. The developing amyloplast, like the chloroplast, is bounded by a double membrane. If the amyloplastic inner membrane is as highly selective in its permeability as that of the chloroplast, then there may be restrictions on the uptake of the precursors of starch. It may be that many of the early reactions of starch synthesis take place within the developing amyloplast, for example hexoses derived from sucrose could be converted to dihydroxyacetone phosphate *via* glycolysis. After uptake into the amyloplast, this could be converted to ADP-glucose *via* gluconeogenesis and ADP-glucose pyrophosphorylase (Fig. 10.3).

Enzymology of starch synthesis

Formerly starch phosphorylase (EC 2.4.1.1) was believed to catalyse the synthesis of the α-(1,4)-glucosidic linkage:

$$\text{glucose-1-phosphate} + (\text{glucosyl})_n \rightleftharpoons \text{Pi} + (\text{glucosyl})_{n+1}$$

However in 1961, Leloir, de Fekete & Cardini showed that the synthesis of starch could be catalysed by starch synthase (EC 2.4.1.21; EC 2.4.1.11):

$$\text{ADP (UDP)-glucose} + (\text{glucosyl})_n \rightarrow \text{ADP(UDP)} + (\text{glucosyl})_{n+1}$$

and, also in 1961, Recondo & Leloir showed that ADP-glucose rather than UDP-glucose was the preferred substrate, at least for the enzyme in dwarf string beans. It is now considered that the bulk of α-(1,4)-glucan synthesis is catalysed by starch synthase and that the major role of phosphorylase is most likely to be degradatory (Preiss & Levi, 1980). It is nevertheless possible that phosphorylase may be involved in synthesis of α-(1,4) primer i.e. $(\text{glucosyl})_n$ of the equation above (Baxter & Duffus, 1973*b*). The synthesis of the α-(1,6) D-glucosidic linkages in amylopectin is catalysed by the action of a branching enzyme (Q-enzyme) (EC 2.4.1.18) on linear chains of α-(1,4)-linked D-glucose residues. Starch synthases are present either bound to the starch granule or soluble in the cytoplasm. Both can utilize UDP-glucose and ADP-glucose although the latter is generally preferred (Duffus, 1979). In cereals such as rice (Perez *et al.* 1975) and in potatoes (Hawker, Marschner & Krauss, 1979) or barley (Baxter & Duffus, 1973*c*), as deposition of starch proceeds, activity is initially soluble but becomes progressively granule-bound as the cells fill up with starch. Little is known of the regulatory properties of the starch synthases. They are, in general, stimulated by potassium but the effect depends on the source of enzyme. They do not appear to be affected by glycolytic intermediates (Preiss & Levi, 1980). The soluble starch synthases exist in a number of forms and one, starch synthase I, from developing maize kernels has been obtained free from activity of branching enzyme by using a mutant maize endosperm lacking branching enzyme. This enzyme can synthesize starch in the absence of added primer and in the presence of 0.5 M citrate (Boyer & Preiss, 1979). The enzyme itself contained 1.7 mg of anhydroglucose units per enzyme unit and thus reaction mixtures contained endogenous primer. The molecular weight as determined by sucrose density ultracentrifugation was around 70000 (Hawker *et al.* 1974). Little is known of the properties of branching enzyme or its relationship to the starch synthases. In particular, it is still a mystery how the cells produce such constant proportions of the two polymers, amylose and amylopectin. Part of the difficulty stems from problems of assay in the presence of interfering amylases. Some progress, however, has recently been made in this direction by Borovsky *et al.* (1979), using a branching enzyme from potato which is free from α-amylase, D-enzyme and phosphorylase. The substrate used was an amylodextrin containing a single branched linkage. It was suggested that the enzyme acts at random on two chains, long enough to form a stable complex, with branches being formed in both directions. The complex may consist of two α-(1,4) glucan chains lined in parallel and possibly as a double helix.

Use of mutants

One of the problems in understanding how starch is synthesized is to determine how branching enzyme and starch synthase, together no doubt with debranching enzymes, co-operate in the formation of a starch granule within the amyloplast's double membrane. One way of solving this and other problems is to make use of tissues in which mutations affect the synthesis of reserve glucans. The best known mutants of maize include *amylose-extender* (*ae*), hybrids of which have starches with 80% amylose, *waxy* (*wx*) in which, when homozygous, accumulation of amylose is completely blocked and *sugary*-1 which produces relatively large quantities of the water-soluble glucan, phytoglycogen. The range of mutant genes and their associated biochemical lesions present in a number of cereal endosperms has recently been described by Nelson (1980). Like starch synthase, branching enzyme can exist in multiple forms (Boyer & Preiss, 1978*b*; Preiss & Levi, 1980). The physiological role of two of these in developing mutant endosperms of maize, synthesizing both amylopectin and phytoglycogen, has been neatly explained, since one could synthesize a phytoglycogen-like product from either amylose or amylopectin and the other catalyse the branching of amylose to form an amylopectin-type product. The role of those forms present in normal endosperms (Boyer & Preiss, 1978*a*) remains unknown. Further work, using maize mutants, by Boyer & Preiss (1979) has explored further the relationship between the branching enzymes and starch synthases. For example, in *ae* mutants, there is an 80% decrease in one of the branching enzymes. This correlates with the fewer branch points and lower amylopectin content in the *ae* starch. In this case, it proved possible to isolate starch synthase free from branching enzyme. When branching enzyme from normal maize was added back to this fraction, unprimed citrate-stimulated starch synthesis was increased about 3.5 to 4 times. This then indicates that the formation of amylopectin may occur by the combined and simultaneous action of both starch synthase and branching enzyme. That is, branching enzyme may catalyse the formation of an increasing number of non-reducing chain ends in the growing glucan which are then available to accept glycosyl residues from ADP-glucose (Hawker *et al.* 1974). Since activities of starch synthase remained relatively unchanged in comparison with normal maize, the increased amylose content of the mutant starch may be due mainly to the decreased action of branching enzyme. Care must be taken, however, in interpreting the results from mutants, since the precise nature of the lesion is rarely fully described. Alterations in other enzyme systems previously not considered may contribute to a number of the observed effects. It is likely, for example, that synthesis of starch in both normal and mutant endosperms is a function of a range of synthetic and degradative processes whose overall activities may vary independently during the period of deposition of starch and hence control the proportions of amylose and amylopectin synthesized.

The way forward

Since the rate of starch synthesis is a major limiting factor in agricultural production, it is surprising how little is known about the mechanisms involved. As in most plant biochemical research, studies are made on a range of plants and tissues under a variety of environmental conditions where there is little hope of comparing results. At a fundamental level, in order to describe the control mechanisms involved, it is necessary to record *in vivo* concentrations and intracellular locations of both substrates and products, simultaneously with measurements of relevant enzyme activities. Furthermore, the response of these enzymes to endogenous effectors and inhibitors throughout the period of starch biosynthesis should be evaluated before any real understanding can be achieved. So far, an integrated study of this type has not been reported. In most cases, therefore, the enzymes studied and the substrates used are based on analogy, guesswork and convenience.

More specifically, little is known of the location and inter-relationship of the enzymes and substrates concerned in starch granule formation. It is not known, for example, whether or not the permeability characteristics of the inner amyloplastic membrane are as highly selective as those reported for the inner chloroplastic membrane (Walker & Heber, 1979). If they are, then there is little possibility that even sugar nucleotides are transported into the amyloplast and the enzymes of starch synthesis may be located within the organelle. In sum, the precise nature of the genetic and biochemical factors which control the number, size, shape and composition of starch granules remain a mystery.

Degradation of starch

Most of the available information on the mechanisms involved in degradation of reserve starch is derived from studies of germination in starch-storing seeds. The process is complex and is catalysed by a number of different enzymes.

Before degradation of starch can proceed, however, the amyloplast membrane must presumably be broken down or at least rendered permeable to the enzymes of starch hydrolysis. Immature starch granules of barley are resistant to amylases derived from germinated grain and also appear resistant to the amylases known to be present in developing grain (Williams & Duffus, 1977*b*). This indicates that, in the period between maturity and subsequent germination, some barrier to degradation is removed. It is well known, for example, that storage of potato tubers at 4 °C results in degradation of starch and accumulation of sugars whereas storage at 25 °C minimizes these changes. This can be explained by the observation that, during tuber storage at 4 °C, the amyloplast membrane disintegrates and disappears, whereas, at 25 °C, the majority of membranes remain intact (Ohad *et al.* 1971).

Enzymes which degrade starch

α-Amylase(EC 3.2.1.1). It has been suggested that α-amylase is responsible for the initiation of dissolution of granules (Dunn, 1974). Certainly, after incubation with α-amylase, the surface of the large starch granules becomes selectively pitted (Evers & McDermott, 1970). Once the enzyme gains access to the innermost region, digestion proceeds rather faster than at the periphery. Often the two halves of large granules of barley split, showing a hollow layered structure. Similar results were observed during germination (Fig. 10.4*a*) (Sergeant, 1980). The small granules were degraded only slowly and did not show internal cavitation (Fig. 10.4*b*).

α-Amylase degrades both amylose and amylopectin by random hydrolysis of non-terminal α-(1, 4) glucosidic linkages. The enzyme exhibits selectivity in that those linkages adjacent to non-reducing end groups or to α-(1, 6) branch-points are not hydrolysed. The end products of hydrolysis of amylose are maltose and glucose with some maltotriose. Degradation of amylopectin

Fig. 10.4 Scanning electron micrographs of starch granules of wheat during germination. (*a*) Large granules after 5 days' germination. (*b*) Small granules after 4 days' germination. [Reproduced with permission from Sergeant (1980).]

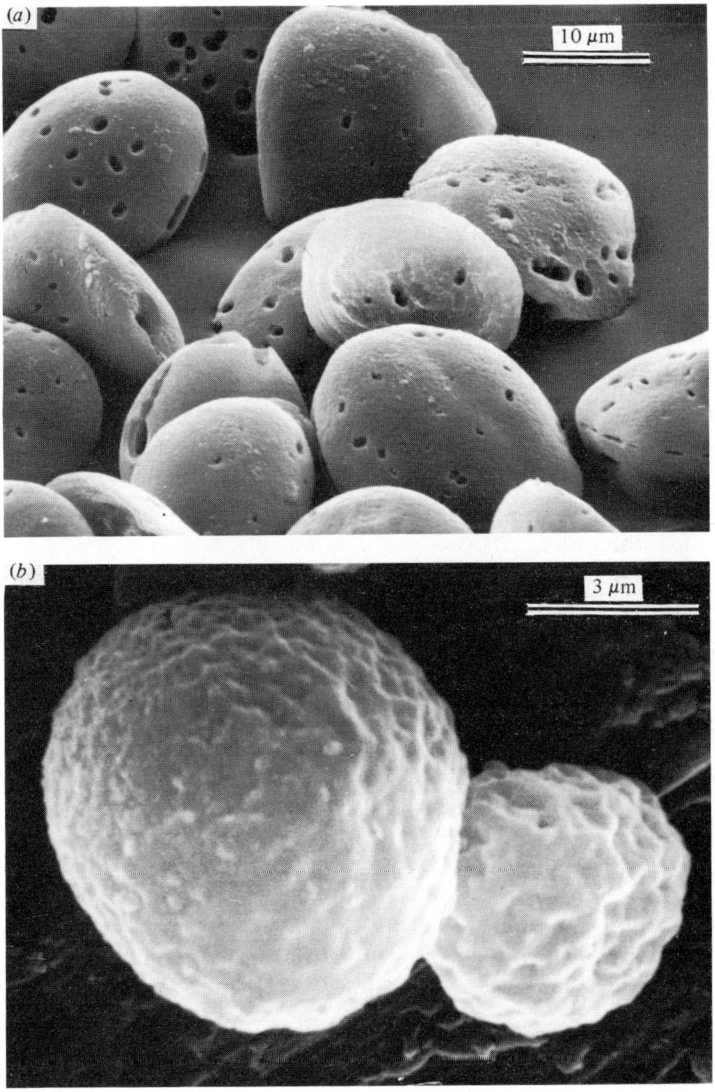

gives the same sugars, together with branched oligosaccharides (α-limit dextrins) derived from the region of the branch-points. The enzyme is a calcium-requiring metallo-enzyme with a pH optimum between 4.5 and 6.0. In common with many other starch-metabolizing enzymes, α-amylase may exist in a number of forms (Manners, 1974). However, the significance of such observations can only be gauged when it can be shown how they relate to the metabolism of starch. For example, they may have different catalytic properties or perhaps be separately localized within the cell.

β-Amylase (*EC 3.2.1.2*). β-Amylase hydrolyses α-(1,4) glucan linkages in both amylose and amylopectin with the successive release of maltose units from the non-reducing ends of the chains. With amylopectin, the hydrolysis is incomplete, stopping at the (1,6) branch-points. The products are then maltose and a β-limit dextrin in approximately equal amounts. The enzyme is a sulphydryl-containing enzyme and can exist in a free, or soluble, form as well as in a latent, or insoluble, form. The active enzyme can be released from the bound form after incubation with the protease, papain (Ford & Guthrie, 1908).

Phosphorylase (*EC 2.4.1.1*). Starch phosphorylase catalyses a reversible reaction which, in the degradative direction, catalyses the formation of glucose 1-phosphate from starch and inorganic phosphate (see above).

In spite of the enormous interest shown by biochemists in the properties of the plant enzyme, notably from potatoes and cereal grains, little is known of its physiological role in starch-storing tissues. The enzyme from sweet corn is a dimer containing pyridoxal phosphate and readily degrades amylopectin *in vitro* (Lee & Braun, 1973). Pea phosphorylase has been described by Matheson & Richardson (1978) and an inverse relationship between the concentration of Pi and the K_m for amylopectin has been demonstrated.

α-Glucosidase (*EC 3.2.1.20*). Maltose and related oligosaccharides are hydrolysed to glucose by α-glucosidase. Since the major activity appears to be with maltose (Hutson & Manners, 1965), the enzyme is often referred to as maltase. However, the enzyme can also hydrolyse maltotriose, maltotetraose and other dextrins.

D-enzyme (*EC 2.4.1.25*). This glycosyl transferase can transfer part of an α-(1,4) glucan chain to a new 4 position. The smallest donor substrate is maltotriose, in which case a maltose unit can be transferred to an acceptor, e.g. another maltotriose, with the formation of free glucose and one molecule of maltopentaose. It is present in extracts of carrots and tomatoes (Manners

& Rowe, 1969) but not, so far as is known, in germinating barley. A glycosyl transferase with some similarity to D-enzyme has been identified in an α-glycosidase preparation from potato tubers free of D-enzyme itself (Killilea & Clancy, 1978). When it was incubated with high concentrations of maltose, glucosyl transferase activity was observed. Paper chromatography of the mixture resolved unhydrolysed maltose, glucose, panose (6-α-glucosyl-maltose) and an unidentified, slowly migrating spot.

Debranching enzyme (*limit dextrinase*) (*EC 3.2.1.10*). The α-(1, 6) inter-chain glucosidic linkage is hydrolysed by specific α-(1, 6)-glucosidases also referred to as debranching enzymes or limit dextrinases. The action of debranching enzymes has been reviewed by Manners (1974) and their specificity and nomenclature are the subject of some controversy. It appears that some debranching enzymes may degrade amylopectin and its β-limit dextrin and also α-dextrins. That from germinating barley (Manners & Hardie, 1977) (limit dextrinase) readily degrades α-dextrins, in all probability the natural substrate, but attacks amylopectin to only a limited extent. Maeda & Nikuni (1978) have shown that debranching enzymes isolated from a number of sources cannot, on their own, degrade intact starch granules. While a partially purified fraction accelerated the rate of digestion by purified α- and β-amylases, a highly purified fraction had little effect. Hence, it seems that a factor, or factors, other than these three enzymes may also be required for complete dissolution of granules. These factors presumably include an α-glucosidase as well as other enzymes and cofactors.

Mobilization of reserve starch

Cereal grains. A key event in the degradation of endospermic starch in germinating cereal grains is the synthesis of gibberellic acid in the imbibed embryo and its subsequent diffusion to the aleurone layer where it triggers the synthesis and release of a number of hydrolytic enzymes into the starchy endosperm. The subsequent structural and physiological changes taking place in the endosperm have been reviewed in relation to malting by Palmer (1980). Dissolution of granules is a relatively late event in germination, beginning around 3–4 days after imbibition (depending on temperature and other variables). This is hardly surprising since the granules are embedded in a matrix of reserve proteins – glutelin and hordein in the case of barley – and contained within cells whose walls consist primarily of β-glucan but with some arabinoxylan (20–25%), protein and mannan. It has been suggested that only after dissolution of the cell walls by β-glucanases and pentosanases (Palmer, 1980), and removal of the protein by endoproteases (hordeinases) (Baxter, Booer & Wainwright, 1978; Slack, Baxter & Wainwright, 1979), are the starch granules accessible to α-amylolytic degradation. Of course, it is important at this stage to point out that the changes observed during malting may not be directly comparable with those occurring during germination under field conditions. In the malting process, for example, grains steeped to a moisture content of 40–45%, are allowed to germinate under controlled conditions of temperature (18 °C) and humidity. True germination, on the other hand, takes place under a wide range of often unpredictable environmental conditions. That factors other than amylolytic and proteolytic activity may be involved in degradation was further demonstrated by Palmer (1975) who showed that, in three cultivars of barley with similar activities of β-glucanase and protease, one (cv. Julia) gave significantly less extract during the subsequent brewing process. ('Extract' is a trade description for removal of storage material from grains during the mashing stage of the brewing process.) Data derived from microscopic and biochemical studies suggested that the cell walls of Julia were more rigid than those of the other two cultivars and reduced migration of the hydrolytic enzymes in the endosperm. That α-amylase may be responsible for the initiation of dissolution has been noted above.

β-Amylase has no effect on maize starch partially digested with bacterial α-amylase or even on the rate of starch digestion by α-amylase. It is not clear, however, in these and other experiments what relationship the starch substrate has to the *in vivo* starch granule. It may be, for example, that β-amylase, which is present in high activity during germination, attacks the starch granules during their isolation. If this is the case, it might explain its apparent inability to degrade the isolated substrate. However, the changes taking place during germination, which include a decrease in starch content, a probable increase in the amylose content of the starch, a small reduction in the β-amylolysis limit and average chain length of the amylopectin, as well as in molecular size of amylose, are similar to those produced by exposing starch granules to limited α-amylolysis (Manners, 1974). Small granules, which account for about 10% of the total starch, are less easily degraded by α-amylase than large granules (Slack *et al.* 1979). That this may not be entirely a consequence of their higher protein content has been suggested by Slack & Wainwright (1980) and may rather be a function of their ease of gelatinization. In experiments with isolated large granules from a range of barleys, those most resistant to α-amylolysis had higher temperatures of gelatinization, as measured by granule swelling. Subsequent degradation of the products is presumably catalysed by β-amylase, α-glucosidase and debranching enzymes.

Although there is no direct evidence that β-amylase is involved in starch degradation, recent results with rice (Okamoto & Akazawa, 1979) using immunofluorescence microscopy indicate that β-amylase is tightly bound to the periphery of starch granules. This was assumed to be the latent, or insoluble, form of the enzyme. Further work from this laboratory (Okamoto & Akazawa, 1980), using incorporation of ^{35}S into β-amylase and identification by quantitative immuno-precipitation, has shown that, at the onset of germination, some β-amylase may be synthesized *de novo* in the scutellum but that, at later stages in germination, the latent form becomes activated.

Tissues other than cereal grains. In a germinating pea seed, energy and metabolites required for initial elongation of the axis are derived from its own reserves. The mobilization of cotyledonary starch which follows is characterized by an increase in α-amylase activity, a decrease in intrinsic viscosity of the amylopectin and an accumulation of β-limit dextrins (Juliano & Varner, 1969; Tárrago & Nicolás, 1976). The decrease in viscosity of the amylopectin is consistent with α-amylolysis. Although activity of phosphorylase in the cotyledons increases before α-amylase, there is no evidence that degradation of starch is a consequence of phosphorolysis. It seems likely that, in peas as in cereals, the initiation of starch degradation is due to α-amylase. A limit dextrinase capable of hydrolysing amylopectin and its β-limit dextrin, present in ungerminated peas, may be involved in the subsequent degradation of the products of α-amylolysis (Yellowlees, 1980). An unusual, but experimentally attractive, tissue to use in studies of degradation of starch is the club of the spadix of *Arum maculatum* (Simon, 1958). At one stage of development, the clubs contain as much as 220–300 mg starch g^{-1} fresh weight and, at a subsequent stage, called thermogenesis, which is a period of around 8 h, almost all of this is consumed (Bulpin & ap Rees, 1978). It appears that starch degradation is catalysed, not by a conventional α-amylase, but rather by an endoamylase with rather different properties. It was concluded from estimations of catalytic activity that phosphorylase does not play a major degradative role (see Chapter 9). That of β-amylase in soybean has been questioned since its activity continues to increase throughout development and into germination before declining, although amounts of starch decrease after mid-development, increasing later during seed germination (Adams *et al.* 1980).

Fate of the products of hydrolysis of starch

In germinating cereals, the major product of hydrolysis of starch is glucose. This is absorbed by the scutellum and converted to sucrose (Edelman, Shibko & Keys, 1959) which is subsequently transported to the axis where it is utilized by the growing roots and shoots (see Chapter 2 for synthesis of sucrose). Eventually the young seedling becomes self-sufficient and fully independent of the seed carbohydrate reserves. The metabolic fate of the products of amylolysis in germinating peas and beans is not clear. It is likely that the conversion of degraded starch into dextrins, maltosaccharides, maltotriose, maltose and ultimately free glucose as well as translocation from the cotyledons must be highly efficient, since no appreciable accumulation of these products is observed (Juliano & Varner, 1969). In germinating lupin seeds, it appears that monosaccharides produced in excess of growth requirements are converted to a branched α-glucan (Saini & Matheson, 1981). Glucose 1-phosphate, derived from cotyledonary starch by phosphorylase action, is probably converted to sucrose via UDP-glucose and sucrose

Fig. 10.5 Some enzyme-catalysed reactions that could be involved in the degradation of starch during cereal grain germination. (1), α-amylase; (2), hexokinase; (3), α-glucosidase; (4), limit dextrinase; (5), β-amylase; (6), phosphorylase; (7), phosphoglucomutase; (8), glucosephosphate isomerase; (9), UDP-glucose pyrophosphorylase; (10), sucrose-phosphate synthase; (11), phosphatase; (12), sucrose phosphatase; (13), sucrose synthase.

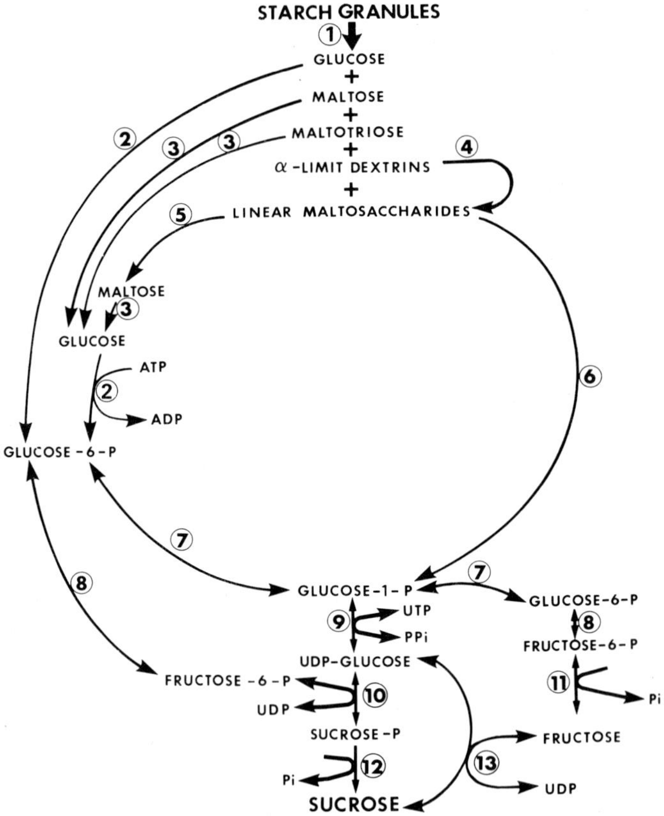

phosphate (Fig. 10.5). The sucrose is presumably then transported to the axis where it acts as a carbon and energy source for the growing seedling.

In the club of *Arum maculatum*, the products of hydrolysis enter the glycolytic pathway via hexokinase (Bulpin & ap Rees, 1978) and subsequently the reactions of the tricarboxylic acid cycle. In this case, the role of starch degradation is not simply as a source of intermediates for use in subsequent biosynthetic reactions. The considerable heat produced volatilizes a number of chemicals such as indoles and amines, which attract particular insects which, in turn, effect pollination.

Degradation of starch – conclusions

A number of enzymes are known to be involved in the degradation of starch granules. Generally, however, the enzymes involved can be investigated only in the absence of the others and using solubilized substrates. Hence, although much is known of the actions of the individual enzymes, little is known of the sequence of events involved in the progressive dissolution of reserve starch under *in vivo* conditions. Figure 10.5 shows some enzyme-catalysed reactions that could be involved in degradation of starch. Moreover, the factors regulating such degradation are still incompletely understood. It is not known, for example, why the susceptibility of starches from different plants to attack by mammalian digestive enzymes varies so greatly such that potato starch (22% amylose) is poorly hydrolysed by pancreatic amylases, whereas corn starch (25% amylose) is easily hydrolysed (Sandstedt, 1965). On the other hand, damaged starch is easily digested and this might explain why starch from *sugary*-2 mutants of maize (40% amylose) is also easily digested, since the large amyloplasts contain starch granules with characteristic extensive external splits. In general, however, the differences in ease of degradation cannot be related to amylose content, temperature of gelatinization or the ability of the granule to absorb cold water. Additionally, it appears that the ease of degradation during germination and malting is further related to the nature of the associated proteins and lipids as well as of the surrounding cell walls. Clearly α-amylolysis is an important site of regulation. The enzyme, α-amylase, which initiates starch degradation is synthesized *de novo* in germinating cereals following release of gibberellic acid (Varner & Chandra, 1964). In turn, it has been suggested that cAMP may be implicated as an intermediate in the action of this hormone (Duffus & Duffus, 1969). Whether or not this control mechanism operates *in vivo* is still open to question.

In addition to regulation by control of protein synthesis, enzyme activity may be regulated by the availability of free calcium, since the α-amylases require calcium for activity. Since most calcium in ungerminated seeds is present in the bound form as calcium phytate, the synthesis of α-amylase requires the presence of an active phytase in order to ensure a supply of available calcium. Changes in environmental conditions may influence enzyme activity directly. For example, the low pH of germinating cereal endosperms correlates well with the pH optimum of α-amylase (4–5). In contrast, the developing endosperm, where presumably starch degradation is inhibited, has a significantly higher pH (around 6.5). Less speculative, perhaps, is the observation that γ-amino butyric acid may be involved in the regulation of activity of α-amylase (Duffus, Duffus & Slaughter, 1972). In isolated aleurone tissue from germinating barley, activity of glutamate decarboxylase declined as α-amylase activity increased. Furthermore, the product of decarboxylase activity, γ-amino butyric acid, inhibited α-amylase activity *in vitro*.

Obviously, many of the other starch-degrading enzymes may be subject to metabolic regulation. For example, β-amylase is unusual in that it can be released from a bound form during germination but the regulatory properties of the protease responsible are unknown. Some of the starch-degrading enzymes shown in Fig. 10.5 are present in developing grains (Duffus & Rosie, 1973) and it is clear that some system of regulation must exist in order to prevent degradation of starch at that time but to switch it on during germination. The observation that starch granules within a given tissue may vary in starch composition further poses an obstacle to any studies of starch degradation. Thus, only very indefinite conclusions can be made about the sequence of events involved in the *in vivo* degradation of reserve starches. In particular much remains to be learnt about the molecular and physiological basis of the control mechanisms concerned.

I would like to thank Dr M. P. Cochrane and Professor D. J. Manners for help in the preparation of this manuscript.

References

Adams, C. A., Rinne, R. W. & Fjerstad, H. C. (1980). Starch deposition and carbohydrase activities in developing and germinating soya bean seeds. *Annals of Botany*, **45**, 577–82.

Baxter, E. D., Booer, C. D. & Wainwright, T. (1978). Degradation of hordein by the proteolytic enzymes of barley and malt. *Journal of the Institute of Brewing*, **84**, 30–3.

Baxter, E. D. & Duffus, C. M. (1973*a*). Enzymes of carbohydrate metabolism in developing *Hordeum distichum* grain. *Phytochemistry*, **12**, 1923–8.

Baxter, E. D. & Duffus, C. M. (1973*b*). Phosphorylase activity in relation to starch synthesis in developing *Hordeum distichum* grain. *Phytochemistry*, **12**, 2321–30.

Baxter, E. D. & Duffus, C. M. (1973*c*). Starch synthetase: comparison of UDPG and ADPG as glucosyl donors in immature barley endosperm. *Planta*, **114**, 195–8.

Borovsky, D., Smith, E. E., Whelan, W. J., French, D. & Kikumoto, S. (1979). The mechanism of Q-enzyme action and its influence on the structure of amylopectin. *Archives of Biochemistry and Biophysics*, **198**, 627–31.

Boyer, C. D. & Preiss, J. (1978*a*). Multiple forms of (1-> 4)-α-D-glucan, (1-> 4)-α-D-glucan-6-glycosyl transferase from developing *Zea mays* L. kernels. *Carbohydrate Research*, **61**, 321–34.

Boyer, C. D. & Preiss, J. (1978*b*). Multiple forms of starch branching enzyme of maize: evidence for independent genetic control. *Biochemical and Biophysical Research Communications*, **80**, 169–75.

Boyer, C. D. & Preiss, J. (1979). Properties of citrate-stimulated starch synthesis catalysed by starch synthase I of developing maize kernels. *Plant Physiology*, **64**, 1039–42.

Brink, R. A. & Cooper, D. C. (1947). The endosperm in seed development. *The Botanical Review*, **13**, 423–541.

Bulpin, P. V. & ap Rees, T. (1978). Starch breakdown in the spadix of *Arum maculatum*. *Phytochemistry*, **17**, 391–6.

Buttrose, M. S. (1963). Ultrastructure of the developing wheat endosperm. *Australian Journal of Biological Sciences*, **16**, 305–17.

Cochrane, M. P. & Duffus, C. M. (1980). The nucellar projection and modified aleurone in the crease region of developing caryopses of barley (*Hordeum vulgare* L. var. *distichum*). *Protoplasma*, **103**, 361–75.

Duffus, C. & Duffus, J. H. (1984). *Carbohydrate Metabolism in Plants*. London: Longman.

Duffus, C. M. (1979). Carbohydrate metabolism and cereal grain development. In *Recent Advances in the Biochemistry of Cereals*, ed. D. L. Laidman & R. G. Wyn Jones, pp. 209–38. London: Academic Press.

Duffus, C. M. & Cochrane, M. P. (1982). Carbohydrate metabolism during cereal grain development. In *The Physiology and Biochemistry of Seed Development, Dormancy and Germination*, ed. A. A. Khan, pp. 43–66. Amsterdam: Elsevier Biomedical Press.

Duffus, C. M. & Duffus, J. H. (1969). A possible rôle for cyclic AMP in gibberellic acid triggered release of α-amylase in barley endosperm slices. *Experientia*, **25**, 581.

Duffus, C. M., Duffus, J. H. & Slaughter, J. C. (1972). Glutamate decarboxylase in barley aleurone and its relationship to α-amylase development during germination. *Experientia*, **28**, 635.

Duffus, C. M. & Murdoch, S. M. (1979). Variation in starch granule size distribution and amylose content during wheat endosperm development. *Cereal Chemistry*, **56**, 427–9.

Duffus, C. M. & Rosie, R. (1973). Starch hydrolysing enzymes in the developing barley grain. *Planta*, **109**, 153–60.

Duffus, C. M. & Slaughter, J. C. (1980). *Seeds and Their Uses*. Chichester: John Wiley and Sons.

Dunn, G. (1974). A model for starch breakdown in higher plants. *Phytochemistry*, **13**, 1341–6.

Edelman, J., Shibko, S. I. & Keys, A. J. (1959). The rôle of the scutellum of cereal seedlings in the synthesis and transport of sucrose. *Journal of Experimental Botany*, **10**, 178–89.

Evers, A. D. (1971). Scanning electron microscopy of wheat starch. III. Granule development in the endosperm. *Stärke*, **23**, 157–62.

Evers, A. D. & McDermott, E. E. (1970). Scanning electron microscopy of wheat starch. *Stärke*, **22**, 23–6.

Fishwick, M. J. & Wright, A. J. (1980). Isolation and characterization of amyloplast envelope membranes from *Solanum tuberosum*. *Phytochemistry*, **19**, 55–9.

Ford, J. S. & Guthrie, J. M. (1908). Contribution to the biochemistry of barley. *Journal of the Institute of Brewing*, **14**, 61–87.

Hawker, J. S., Ozbun, J. L., Ozaki, H., Greenberg, E. & Preiss, J. (1974). Interaction of spinach leaf adenosine diphosphate glucose α-1,4 glucan α-4-glucosyl transferase and α-1,4-glucan, α-1,4-glucan-6-glycosyl transferase in synthesis of branched α-glucan. *Archives of Biochemistry and Biophysics*, **160**, 530–51.

Hawker, J. S., Marschner, H. & Krauss, A. (1979). Starch synthesis in potato tubers. *Physiologia Plantarum*, **4**, 25–30.

Hutson, D. & Manners, D. J. (1965). Studies on carbohydrate-metabolising enzymes. The hydrolysis of α-glucosides, including nigerose, by extracts of alfalfa and other higher plants. *Biochemical Journal*, **94**, 783–9.

Jenner, C. F. (1982). Storage of starch. In *Encyclopaedia of Plant Physiology New Series* (*Plant Carbohydrates*), vol. 13A, ed. F. A. Loewus & W. Tanner, pp. 700–47. Berlin: Springer-Verlag.

Juliano, B. O. & Varner, J. E. (1969). Enzymic degradation of starch granules in the cotyledons of germinating peas. *Plant Physiology*, **44**, 886–92.

Killilea, S. D. & Clancy, M. J. (1978). Properties of potato α-glucosidase. *Phytochemistry*, **17**, 1429–31.

Lee, E. Y. C. & Braun, J. J. (1973). Sweet corn phosphorylase: purification and properties. *Archives of Biochemistry and Biophysics*, **156**, 276–86.

Leloir, L. F., de Fekete, M. A. R. & Cardini, C. E. (1961). Starch and oligosaccharide synthesis from uridine diphosphate glucose. *Journal of Biological Chemistry*, **236**, 636–41.

Luxová, M. (1967). Formation of barley (*Hordeum distichum* L.). *Biologia Plantarum*, **9**, 301–7.

Maeda, I. & Nikuni, Z. (1978). Purification of a debranching enzyme (R-enzyme) from malted barley, and the rôle of the enzyme in the digestion of starch granules during the germination of barley seeds. *Carbohydrate Research*, **61**, 309–20.

Manners, D. J. (1974). Some aspects of the enzymic degradation of starch. In *Plant Carbohydrate Biochemistry*, ed. J. B. Pridham, pp. 109–25. London: Academic Press.

Manners, D. J. & Bathgate, G. N. (1969). α-1,4-Glucans. Part XX. The molecular structure of the starches from oats and malted oats. *Journal of the Institute of Brewing*, **75**, 169–75.

Manners, D. J. & Hardie, D. G. (1977). Studies on debranching enzymes.

Part VI. The starch-debranching system of germinated barley. *Master Brewers Association of America Technical Quarterly*, **14**, 120–5.

Manners, D. J. & Rowe, K. L. (1969). Studies on carbohydrate-metabolising enzymes. Part XXI. The α-glucosidase and D-enzyme activity of extracts of carrots and tomatoes. *Carbohydrate Research*, **9**, 441–50.

Matheson, N. K. (1975). The α(1–4) (1–6) glucans from sweet and normal corns. *Phytochemistry*, **14**, 2017–21.

Matheson, N. K. & Richardson, R. H. (1978). Kinetic properties of two starch phosphorylases from pea seeds. *Phytochemistry*, **17**, 195–200.

May, L. H. & Buttrose, M. S. (1959). Physiology of cereal grain. II. Starch granule formation in the developing barley kernel. *Australian Journal of Biological Sciences*, **12**, 146–59.

Morrison, I. N., Kuo, J. & O'Brien, T. P. (1975). Histochemistry and fine structure of developing wheat aleurone cells. *Planta*, **123**, 105–16.

Nelson, O. E. (1980). Genetic control of polysaccharide and storage protein synthesis in the endosperms of barley, maize and sorghum. In *Advances in Cereal Science and Technology*, vol. III, ed. Y. Pomeranz, pp. 41–71. St Paul, Minnesota: American Association of Cereal Chemists.

Ohad, I., Friedberg, I., Ne'eman, Z. & Schramm, M. (1971). Biogenesis and degradation of starch. I. The fate of the amyloplast membranes during maturation and storage of potato tubers. *Plant Physiology*, **47**, 465–77.

Okamoto, K. & Akazawa, T. (1979). Enzymic mechanism of starch breakdown in germinating rice seeds. 8. Immunohistochemical localization of β-amylase. *Plant Physiology*, **64**, 337–40.

Okamoto, K. & Akazawa, T. (1980). Enzymic mechanism of starch breakdown in germinating rice seeds. 9. *De novo* synthesis of β-amylase. *Plant Physiology*, **65**, 81–4.

Palmer, G. H. (1975). Influence of endosperm structure on extract development. *Proceedings of the American Society of Brewing Chemists, Inc.*, **33**, 174–80.

Palmer, G. H. (1980). The morphology and physiology of malting barleys. In *Cereals for Food and Beverages*, ed. G. Inglett & L. Munck, pp. 301–38. London, New York: Academic Press.

Pate, J. S. & Flinn, A. M. (1977). Fruit and seed development. In *The Physiology of the Garden Pea*, ed. J. F. Sutcliffe & J. S. Pate, pp. 431–68. London: Academic Press.

Perez, C. M., Perdon, A. A., Resurreccion, A. P., Villareal, R. M. & Juliano, B. O. (1975). Enzymes of carbohydrate metabolism in the developing rice grain. *Plant Physiology*, **56**, 579–83.

Preiss, J. & Levi, C. (1980). Starch biosynthesis and degradation. In *The Biochemistry of Plants*, vol. 3, ed. P. K. Stumpf, pp. 371–423. London, New York: Academic Press.

Recondo, E. & Leloir, L. F. (1961). Adenosine diphosphate glucose and starch synthesis. *Biochemical and Biophysical Research Communications*, **6**, 85–8.

Reeve, R. M., Hautala, E. & Weaver, M. L. (1969). Anatomy and compositional variation within potatoes. I. Developmental histology of the tuber. *American Potato Journal*, **46**, 361–73.

Saini, H. S. & Matheson, N. K. (1981). α-Glucan synthesis in the cotyledons of germinating lupins. *Phytochemistry*, **20**, 641–5.
Sakri, F. A. K. & Shannon, J. C. (1975). Movement of ^{14}C-labeled sugars into kernels of wheat (*Triticum aestivum* L.). *Plant Physiology*, **55**, 881–9.
Sandstedt, R. M. (1965). Fifty years of progress in starch chemistry. *Cereal Science Today*, **10**, 305–16.
Sergeant, J. G. (1980). α-Amylase isoenzymes and starch degradation. *Cereal Research Communications*, **8**, 77–86.
Simon, E. W. (1958). Respiration rate and mitochondrial oxidase activity in *Arum* spadix. *Journal of Experimental Botany*, **10**, 125–33.
Slack, P. T., Baxter, E. D. & Wainwright, T. (1979). Inhibition by hordein of starch degradation. *Journal of the Institute of Brewing*, **85**, 112–14.
Slack, P. T. & Wainwright, T. (1980). Amylolysis of large starch granules from barleys in relation to their gelatinisation temperatures. *Journal of the Institute of Brewing*, **86**, 74–7.
Sowokinos, J. R. (1973). Maturation of *Solanum tuberosum*. I. Comparative sucrose and sucrose synthetase levels between several good and poor processing varieties. *American Potato Journal*, **50**, 234–47.
Tárrago, J. F. & Nicolás, G. (1976). Starch degradation in the cotyledons of germinating lentils. *Plant Physiology*, **58**, 618–21.
Thompson, W. P. & Johnston, D. (1945). The cause of incompatibility between barley and rye. *Canadian Journal of Research*, **23**(C), 1–15.
Thomson, W. W. & Whatley, J. M. (1980). Development of non-green plastids. *Annual Review of Plant Physiology*, **31**, 375–94.
Varner, J. E. & Chandra, G. R. (1964). Hormonal control of enzyme synthesis in barley endosperm. *Proceedings of the National Academy of Sciences*, **52**, 100–6.
Walker, D. A. & Heber, U. (1979). The chloroplast envelope – barrier or bridge? *Trends in Biochemical Sciences*, **4**, 252–6.
Williams, J. M. & Duffus, C. M. (1977*a*). The development of endosperm amyloplasts during grain maturation in barley. *Journal of the Institute of Brewing*, **84**, 47–50.
Williams, J. M. & Duffus, C. M. (1977*b*). Separation and some properties of large and small amyloplasts throughout development in barley endosperm. *Plant Physiology*, **59**, 189–92.
Yellowlees, D. (1980). Purification and characterization of limit dextrinase from *Pisum sativum* L. *Carbohydrate Research*, **83**, 109–18.

PETER HALMER

11 Synthesis and degradation of extracellular storage polysaccharides

Synthesis

Introduction

This section deals with the synthesis of the extracellular polysaccharides of plants as a general subject, rather than just of storage polysaccharides. After all, the distinction drawn here is one of physiological function, not of biochemistry. This large subject can only be discussed in a relatively brief way in a chapter such as this. For the reader who wishes to pursue an aspect in greater depth, other reviews, cited here, should be consulted.

Only certain aspects of polysaccharide synthesis are understood well at present – notably the metabolism of sugar nucleotides and the cellular organization of synthesis. About many facets of the mechanism of polymerization, we are still ignorant. How is polysaccharide synthesis initiated and terminated? Are sugar nucleotides the immediate precursors of the glycosidic bond or are other sugar intermediates involved? Is the main chain synthesized in one continuous piece or from preformed oligosaccharide units? Are side chains synthesized en bloc before being added to the main chain, or in part, or not at all? What determines the degree and pattern of side chain substitution, and overall molecular size?

The general model, which seems applicable to all organisms, is that polysaccharide synthesis involves assemblies of membrane-associated *glycosyl transferases*, each specific for a donor sugar nucleotide and acceptor saccharide. In this way, sugar sequence and bond linkage is determined by the specificity of transferases, and sugar composition by the supply of precursors and the relative activities of the enzymes.

This mechanism has been demonstrated most clearly to hold in the glycosylation of proteins and lipids, both by direct biochemical studies (Elbein, 1979), and by analysis of genetic variation, e.g. of human blood group substances. In the simplest version of this model, the glycosidic bond is formed between the 'reducing' (hemiacetal) OH group of the added sugar residue and one of the available 'non-reducing' OH groups in the acceptor saccharide.

Polysaccharides are thus said to be synthesized by growth at their *non-reducing* ends (of which there are potentially very many in a polymer; see Chapter 1):

Sugar Sugar Sugar . . . . . Sugar-OH
Sugar*-Nucleotide
———————————————————————
Sugar* Sugar Sugar Sugar . . . . . Sugar-OH (+ Nucleotide)
Non-reducing end *Reducing end*

However, synthesis at the *reducing* end also occurs in the situation where polysaccharides are synthesized from prefabricated oligosaccharide units. In this case, each oligosaccharide unit is synthesized at its non-reducing end in the above way. However, there is also a 'donor' group attached at the reducing end of each unit which provides the energy for the glycosidic bonds that link the units together in the polymer. The growing polysaccharide likewise carries a donor group, and this is lost as the next unit is added at its reducing end:

Unit Unit Unit . . . . . Unit-Donor
Unit*-Donor*
———————————————————————
Unit Unit Unit . . . . . Unit Unit*-Donor* (+ Donor)
Non-reducing end *Reducing end*

Long-chain polysaccharides (e.g. bacterial cell-wall lipopolysaccharides and peptidoglycans) are synthesized in this way. One point of particular interest about this mechanism is that both precursor and product are covalently bonded to another molecule, such as a protein or a lipid, which permits 'tethering' of the process of polysaccharide synthesis to cell membranes.

Another form of polysaccharide growth would be where a prefabricated side chain is added to another polysaccharide. (This mechanism appears to operate during synthesis of pectinic acid – see later.) Depending on one's point of view, this would be growth at a non-reducing end of the polysaccharide or at the reducing end of the side chain:

Sugar Sugar . . . . . Sugar Sugar Sugar Sugar . . . . . Sugar-OH
Oligosaccharide-Donor
———————————————————————
Sugar Sugar . . . . . Sugar Sugar Sugar Sugar . . . . . Sugar-OH
|
Oligosaccharide (+ Donor)

The main experimental approach to the study of the mechanism of polysaccharide synthesis has been to obtain cell-free preparations of synthetases, incubate them with various radioactively-labelled sugar nucleotides and characterize the products. Such synthetases are found in particulate fractions of cell extracts, from which in some cases they can be solubilized by treatment with detergents. This suggests that polysaccharide synthetases are closely associated with membranes in intact cells. A big problem, however, has been that these preparations rapidly lose activity, so that the radioactive products represent the addition of only a few residues to a pre-existing chain or other acceptor. Complete polysaccharides cannot yet be synthesized *in vitro* by preparations from vascular plants (although this is now possible in fungal extracts, and possibly in an algal extract too – see below). It is difficult to characterize unambiguously the new saccharide sequences (e.g. is there one type or a mixture of several?) and even more so the acceptor polysaccharide (if one is involved). It is therefore not clear how closely the *in vitro* products resemble 'natural' polysaccharides. The usual tactic is to analyse sugar composition and bond linkage in the radioactive fragments, using techniques such as partial degradation by acetolysis, acid hydrolysis or digestion by specific hydrolases. Simpler criteria, such as the solubility of products in warm water or dilute alkali were used formerly but these methods by themselves are now regarded as inadequate. Another problem with cell-free preparations is that the extracts are crude, and may contain mixtures of different polysaccharide synthetases, along with enzymes of sugar nucleotide interconversion and protein glycosylation etc. Membrane association makes purification difficult and this has not been achieved yet with any such enzymes from plants.

Virtually all the synthetases studied in vascular plants have been prepared from tissues that are synthesizing structural cell-wall polysaccharides. As mentioned above, we can reasonably assume that extracellular storage polysaccharides are synthesized by fundamentally the same mechanism. Most attention has been given to the synthesis of β-glucan. Synthetases have been found in cell-free extracts of several seedling shoots or pollen tubes, and also in semi-intact tissue preparations, such as root slices, broken cotton fibres and suspension-callus tissue (Clark & Villemez, 1972; Larsen & Drummond, 1974; Storm & Hassid, 1974; Southworth & Dickinson, 1975; Helsper, Veerkamp & Sassen, 1977; Heineger & Delmer, 1977; Anderson & Ray, 1978; Brett, 1978; Raymond, Fincher & MacLachlan, 1978). All are specific for UDP-Glc. Where products have been well characterized, they have been found in every case to consist largely of $(1\rightarrow3)$-β-linkages, with a low proportion of $(1\rightarrow4)$-β-linked residues. The product(s) therefore resemble callose, the polysaccharide produced as a rapid wound response in plants, rather than cellulose (whose synthesis most of these authors were aiming at first to study). A common observation is that the proportion of $(1\rightarrow3)$-β-linkages increases with concentration of UDP-Glc. It is still unclear whether this represents the presence of separate $(1\rightarrow3)$-β and $(1\rightarrow4)$-β-glucan synthetases, which have different optimal substrate concentrations, or whether the preparations are synthesizing a mixed (1 → 3), (1 → 4)-β-glucan. Whatever the case, (1 → 3)-β-glucans are not normally synthesized in such quantities by the tissues from which the preparations have been made: their production *in vitro* is apparently due to artefacts, such as tissue-shock, arising during extraction. There have been reports too of the synthesis of various 'hemicelluloses' by cell-free preparations, e.g. arabinoxylans, galactans, galacturonans and xyloglucans, from UDP-sugar precursors (Villemez, Lin & Hassid, 1965; Odzuck & Kauss, 1972; Heller & Villemez, 1972; Panayotatos & Villemez, 1973), and glucanmannan from GDP-sugar precursors (e.g. Hinman & Villemez, 1975). For a detailed review of this work, see Ericson & Elbein (1980), or Franz & Heineger (1982).

At present, the most promising experimental systems for the study of polysaccharide synthesis are fungal cell-free preparations. Conditions have now been devised that allow the synthesis of milligram quantities (enough to be visible!) of chitin (2-acetamido-2-deoxy-(1 → 4)-β-glucan) and (1 → 3)-β-glucan in extracts of *Mucor rouxii* and *Phytophthora cinnamomi* (Ruiz-Herrera, Sing, Van der Woude & Bartnicki-Garcia, 1975; Wang & Bartnicki-Garcia, 1976; Duran & Cabib, 1978). These preparations, once activated by protease treatment, are reported to produce complete fibrils and do not require exogenous primers. The stage is now therefore set for an *in vitro* biochemical assault on the problem of extracellular polysaccharide synthesis.

Synthesis and interconversion of nucleoside-diphosphate sugars – 'sugar nucleotides'

In contrast to the relative lack of progress in research into the polymerization of polysaccharides, the metabolism of sugar nucleotides, the ubiquitous biological precursors for the synthesis of the glycosidic bond, has been thoroughly investigated over the past 30 years and few puzzles remain. This subject is well covered in textbooks and many reviews (see e.g. Karr, 1976; Feingold, 1982), and only a summary is given here.

Sugar nucleotides may be synthesized directly from free monosaccharides, ATP and a nucleoside triphosphate, by the sequential action of a kinase, a phosphoglycomutase and a nucleoside-diphosphate-sugar pyrophosphorylase (Fig. 11.1). All have the sugar in the α-configuration. Thus, for example, glucose is converted to uridine-diphosphate glucose (UDP-Glc) *via* Glc-6-P and Glc-1-P. (Glucose phosphates are also, of course, produced by gluconeogenesis, photosynthesis and starch phosphorolysis.) In vascular plants, there is good evidence that free glucuronic acid, an intermediate in the *myo*-inositol pathway produced from Glc-6-P and *myo*-inositol, is an

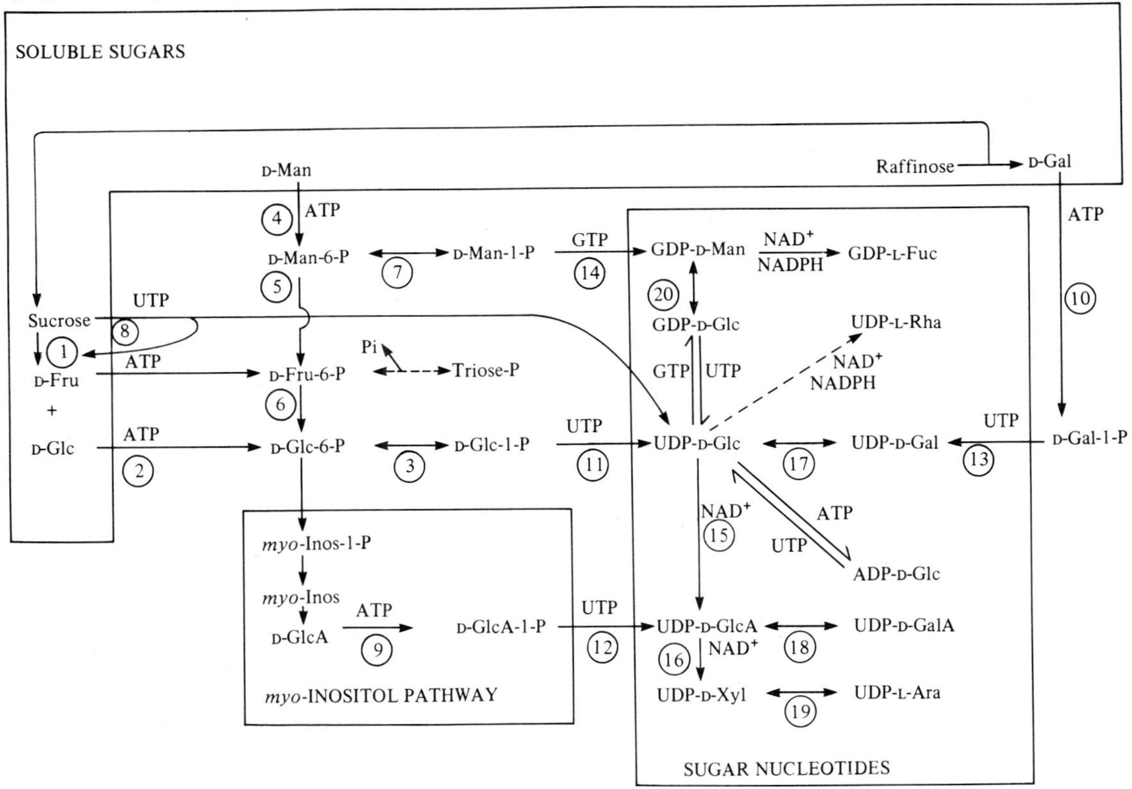

Fig. 11.1 Pathways in vascular plants that may be operative in the synthesis and interconversion of nucleoside-diphosphate-sugars (sugar nucleotides) from free sugars and sugar phosphates. Enzymes: (1) invertase; (2) D-glucokinase; (3) phospho-D-glucomutase; (4) D-mannokinase; (5) phospho-D-mannoseisomerase; (6) phospho-D-glucoseisomerase; (7) phospho-D-mannomutase; (8) sucrose synthetase; (9) D-glucuronokinase; (10) D-galactokinase; (11) UDP-D-glucose pyrophosphorylase; (12) UDP-D-glucuronic acid pyrophosphorylase; (13) UDP-D-galactose pyrophosphorylase; (14) GDP-D-mannose pyrophosphorylase; (15) UDP-D-glucose dehydrogenase; (16) UDP-D-glucuronic acid decarboxylase; (17) UDP-D-galactose-4-epimerase; (18) UDP-D-galacturonic acid-4-epimerase; (19) UDP-L-arabinose-4-epimerase; (20) GDP-D-mannose-4-epimerase.

important precursor for the synthesis of uronic acid and pentose nucleotides (Fig. 11.1 and Chapter 6). Similar pyrophosphorylases could participate in the synthesis of sugar nucleotides other than those of glucose and glucuronic acid. However, in vascular plants, these are likely to operate when storage polysaccharides are being *mobilized*, rather than during their synthesis.

Another important source of UDP-glucose is sucrose, via the action of sucrose synthetase. This seems to provide an important supply of precursors for cell-wall polysaccharide synthesis in young seedlings (Rollit & MacLachlan, 1974; Dick & ap Rees, 1976) and may well do the same during the formation of seeds and other storage organs, which are supplied with translocated sucrose (see Chapter 2).

A key feature of sugar nucleotide metabolism is that it is mainly as their nucleotides that sugar residues are interconverted. A central set of pathways linking Glc, Gal, GlcA, GalA, Xyl and Ara involves UDP-Glc:NAD^+ dehydrogenase, UDP-GlcA decarboxylase and three C-4 epimerases (Fig. 11.1). By pathways connected to these, GDP-D-Glc is converted to GDP-D-Man and then to GDP-L-Fuc, and UDP-L-Rha is formed from UDP-D-Glc (Barber, 1963; Liao & Barber, 1971).

These enzymes are recovered in the soluble fraction of cell extracts and are thus considered to be located in the cytoplasm. This raises the question of how sugar nucleotides are conveyed across the membranes to the compartments where polysaccharide synthesis occurs (see below).

Although the basic biochemistry of these pathways of interconversion is now established, there is still much to be learnt about how the pathways are regulated and about the relative importance of different routes to the same sugar nucleotide in individual tissues.

Intermediates in polysaccharide synthesis

Polyisoprenol-linked saccharides are formed as intermediates during the synthesis of cell-wall and capsular polysaccharides in bacteria and during synthesis of glycoprotein in a wide range of eukaryotes, e.g. fungi, birds, mammals and vascular plants. This participation of lipids reflects the association of glycosyl transferases with membranes. Presumably, the lipid moieties allow the positioning of intermediates in membranes or their transport through membranes between cellular compartments. In recent years, there has been a tremendous increase in knowledge about lipid–saccharide intermediates, particularly in the synthesis of glycoproteins. Whether lipids are involved in a similar way in the synthesis of plant extracellular polysaccharides is not known at present. Nevertheless, the subject is worth discussing here in view of the light it might throw on possible synthetic mechanisms.

Polyisoprenols have the general formula:

$$H-(CH_2-\overset{\displaystyle CH_3}{\overset{|}{C}}=CH-CH_2)_n-OH$$

In bacteria, *n* is 11 – an undecaprenol. In eukaryotes, *n* is variable; 14 in yeast,

Fig. 11.2 Comparison of the role of lipid-linked saccharides in the synthesis of (A) the O-antigen of *Salmonella* lipopolysaccharide, (B) Asn-linked side chains in eukaryotic glycoproteins (simplified: see e.g. Tonn & Gardner, 1979; Parodi & Leloir, 1979).

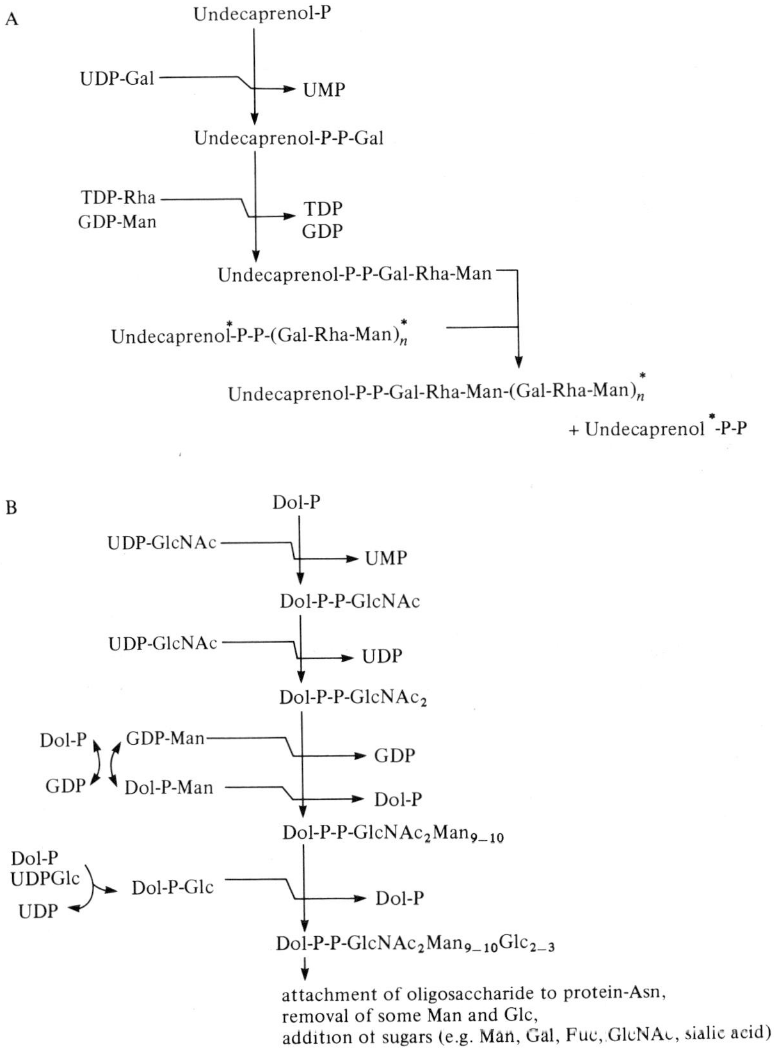

18–22 in mammalian tissue and 20 in vascular plants. Also, in eukaryotes the α-residue (nearest the OH group) is saturated – constituting the dolichol class of polyisoprenols. Polyisoprenols function as coenzymes by forming lipid-(pyro)phosphate-sugar esters, which serve in two ways: (1) to transfer individual sugars from sugar nucleotides to acceptor saccharides; and (2) as chain initiators and carriers during the synthesis of oligosaccharide units, which are then assembled into larger polymers.

In bacteria, for instance, lipopolysaccharides in the outer cell-wall membrane contain O-antigen polysaccharides, which are made up of repeating sequences of sugars, such as A-B-C-A-B-C-A-B-C- etc. Each unit sequence is assembled on an undecaprenyl carrier by successive transfer of sugars from sugar nucleotides and the units are then joined together (Fig. 11.2A). Chain initiation is by the formation of an undecaprenol-pyrophosphate-sugar linkage. This linkage remains intact as subsequent sugars are added and also as the completed unit is transferred to the growing polysaccharide chain. Likewise, the peptidoglycan in bacterial cell walls, which contains the repeating sequence (MurNAc-GlcNAc), is polymerized from peptidyl-disaccharide units synthesized on undecaprenyl carriers (Tonn & Gardner, 1979). Also, there is evidence that lipid intermediates, including polyprenyl-pyrophosphate cellobiose, are involved in the synthesis of bacterial cellulose in *Acetobacter xylinum* (Garcia, Recondo & Dankert, 1974). A similar situation may exist in some eukaryotes. Mammalian proteoglycans contain polysaccharides such as heparan and hyaluronic acid which have repeating disaccharide sequences, and there are clues that dolichol-pyrophosphate-linked saccharides are involved in their synthesis (see Parodi & Leloir, 1979).

Dolichol-linked intermediates certainly participate in the synthesis of the main type of glycosylation of eukaryotic proteins, in which an oligosaccharide is linked via a glycosidic bond to the amide-N of an asparaginyl residue. The structure and size of these side chains is very variable but all seem to share a $Man_3GlcNAc_2$ pentasaccharide sequence at the reducing end. This is thought to be because all side chains are synthesized by the attachment of a common prefabricated core oligosaccharide ($Glc_3Man_9GlcNAc_2$). The core is then processed in many different ways by the removal of some sugars and the addition of others (see review by Parodi & Leloir, 1979; Elbein, 1979). Dolichols function as chain initiators and carriers during the synthesis of these core oligosaccharides (Fig. 11.2B), to which they are linked by a pyrophosphate group in an analogous way to undecaprenols in bacteria. Also dolichol-phosphate-sugars, formed from dolichol phosphate and sugar nucleotides, are precursors for the addition of some Man and Glc residues. (Other sugars are donated directly by their nucleotides.)

Mechanisms like this might operate during synthesis of the main chains of extracellular polysaccharides in plants, which typically have a simple structure consisting of one or two sugar residues and linkage bonds. Also, side chains (typically one to four sugar residues) might be formed from presynthesized 'core' units, followed by processing after they are joined to the main chain.

At present though, there is very little evidence that dolichol-saccharides, specifically, do in fact intervene during the synthesis of extracellular polysaccharides in plants. Several workers have observed dolichol-like lipids amongst the radioactive products when cell-free polysaccharide synthetase preparations are incubated with nucleotides of labelled sugars (e.g. Jung & Tanner, 1973; Green & Northcote, 1979). However, such dolichol saccharides could well have been formed by glycoprotein-synthesizing enzymes also present in the crude extracts used. Helsper (1979) has examined the issue more carefully by measuring the kinetics of pulse-labelling of the various products when a cell-free (1→3)-β-glucan synthetase preparation from *Petunia* pollen tubes was incubated first with UDP-[^{3}H]-Glc and then with additional UDP-[^{14}C]-Glc (twice the radioactivity). If dolichol-linked intermediates were involved in the glucan synthetase, then the ^{14}C:^{3}H ratio should increase to two, first in the lipid fraction, and then a matching increase should be seen in the fraction containing (1→3)-β-glucan. This, however, was not the case: the ratio in the glucan fraction increased and reached the steady-state value sooner than in the lipid fraction.

Dolichol-precursors do not seem to be required for yeast (1→3)-β-glucan synthesis either. In a cell-free preparation, the rate of synthesis of glucan is unaffected by addition or removal of dolichol-phosphate-Glc (Shematek, Braatz & Cabib, 1980).

In contrast, there is some evidence that both lipid- and protein-linked intermediates are involved in the synthesis of β-glucan in the alga, *Prototheca zopfii*. In the presence of UDP-Glc, a cell-free preparation synthesizes a water soluble protein-linked glucan intermediate, apparently via a dolichol-linked oligosaccharide intermediate. In the additional presence of GDP-Glc, the water-soluble glucan is incorporated into a (1→4)-β-glucan, which, like cellulose, is insoluble in hot alkali. These intriguing observations suggest that polysaccharides are formed in several stages: on lipid and then protein carriers from UDP-Glc, and then into larger polymers using GDP-Glc (Hopp, Romero, Daleo & Pont-Lezica, 1978). Protein-saccharide (i.e. glycoprotein) intermediates have also been detected during the synthesis of intracellular (1→4)-α-glucans in vascular plants (Tandecarz, Lavintman & Cardini, 1975; Pisigan & Rosario, 1976).

Some 'pectins' and mucilages have different structures from other cell-wall polysaccharides, in the sense that very large side chains are present, e.g. arabinogalactan blocks attached to the polygalacturonate main chain in 'pectinic acid'. In pectinic acid, it appears that the main chain of polygalacturonate and the side chains of arabinogalactan are synthesized separately and then joined together. Evidence comes from experiments in which ^{14}C-Glc was fed in a short pulse to sycamore callus tissue culture. Radioactivity was recovered first in methylated polyGalA chains without side chains. Later this radioactive component disappeared and radioactivity became associated with a pectinic acid fraction, in which both the GalA residues and the Ara and Gal residues in the side chains were labelled (Stoddart & Northcote, 1967).

This observation has not been followed up and the mechanism by which the two polysaccharides are joined remains to be explored.

Cellular sites of synthesis

Synthesis of extracellular polysaccharides is a highly co-ordinated process, involving the simultaneous deposition in different cellular locations of mixtures of polysaccharides. The pattern of deposition and the types of polysaccharide change as the cell ages.

In brief, the consensus is that cellulose and callose are synthesized at the plasmalemma in vascular plants. Cell-wall matrix polysaccharides ('pectins' and 'hemicelluloses') and mucilages, however, are initiated within the endoplasmic reticulum, completed within the cisternae of the dictyosomes (Golgi bodies) and transported from there within vesicles to the plasmalemma or tonoplast for secretion into the cell wall. This flow of membrane, and of enzymes, polysaccharides and other material associated with membranes, through the cell is an aspect of the 'endomembrane' concept of eukaryotic cellular organization (see Hanke, 1977, section 8.4.2, for a review) in which plastid outer membranes, endoplasmic reticulum, plasmalemma and tonoplast are functionally linked. The evidence for this has come from ultrastructural and cell-fraction studies in a variety of plant (and animal) tissues, and particularly from following the distribution of radioactivity after feeding labelled sugars to intact tissues (reviewed critically in detail by Robinson, 1977).

Only one example of the organization of synthesis of plant storage extracellular polysaccharides has been studied – in the formation of the endosperm in leguminous seeds. There, the deposition of galacto-mannan does not involve dictyosomes at all. Instead, the polysaccharide is formed within stacks of linked, and grossly distended, rough endoplasmic reticulum (Meier & Reid, 1977). This abnormal state of affairs is associated with the massive amount of deposition taking place. In fenugreek endosperms, for example, deposition during six weeks results in cells that are almost completely occluded by the cell wall and, apart from the aleurone layer, dead. In other legumes, such as guar and carob, the process does not go quite as far. A proportion of the endosperm cells remain alive and indeed participate in the mobilization of their own much thickened cell walls (McClendon, Nolan & Wenzler, 1976; Seiler, 1977; Meier & Reid, 1977). In seeds of *Tropaeolum*, the thickening of cotyledonary cell walls is very rapid indeed, involving the deposition of large amounts of xylo-(1 → 4)-β-glucans during the last week or days of seed maturation (Koimann, 1957).

Maturing seeds such as these, in which metabolism is given over to the production of a single polysaccharide, would seem to offer useful experimental systems for the study of extracellular polysaccharide synthesis in plants. Another possibility is to use protoplasts in the process of regenerating cell walls (see Hanke, 1977). Here the very absence of the cell wall allows the use of relatively gentle methods of cell extraction, which might circumvent problems of membrane disruption or induction of wound responses.

Degradation

Introduction

Virtually all the work on the mobilization of extracellular polysaccharides of vascular plants has used endosperms. Species studied include the galacto-(1 → 4)-β-mannan storing legumes, fenugreek, carob and guar; the (1 → 4)-β-mannan storing celery, lettuce and date palm; and the starchy cereals, wheat and barley, where cell-wall (1 → 3),(1 → 4)-β-glucans and arabino-(1 → 4)-β-xylans provide a secondary carbohydrate reserve. In these seeds, degradation to oligosaccharides and free sugars occurs in the endosperm and these saccharides are then absorbed by the cotyledons, where further hydrolysis and utilization occurs.

Enzymes

Breakdown of polysaccharides involves the co-operation of two classes of hydrolases (EC 3.2.1.), which break the glycosidic bond between the C1 and O atoms. *Endo-glycanases* cleave the main chain to produce both large fragments and short oligosaccharides (2 to 4 residues long): *glycosidases* attack poly- and oligosaccharides by the removal of single residues from the non-reducing ends of main chains, side chains or fragments. Complete mobilization typically involves sets of these enzymes, each specific for a certain sugar and bond-linkage and, in some cases, sequence of sugar and molecular size of substrate.

Leguminous seeds offer an example of the co-operation of different enzymes in mobilization of polysaccharides. In these seeds, the binding of endo-(1 → 4)-β-mannanase to its galacto-mannan substrate requires two contiguous unsubstituted mannose residues in the main chain (Dey, 1980). Accordingly, in highly substituted galacto-mannans, with Gal: Man ratios in the range 1:1.5–2.0, a preliminary attack by α-galactosidase is necessary to clear regions of the main chain. This preferential mobilization of galactose residues is detectable in seeds of guar and lucerne, where the Gal:Man ratio decreases markedly as mobilization proceeds (McCleary & Matheson, 1975; McCleary, Matheson & Small, 1976). Mobilizing enzymes seem to be tailored to their natural substrates. Endo-β-mannanases, and α-galactosidases specific for galacto-mannans, tend to be less active on substrates from 'foreign' species (McCleary & Matheson, 1975; Williams, Villaroya & Petek, 1977; Dey, 1980). It is this specificity of α-galactosidases that has made them so useful as tools for the structural analysis of galacto-mannans. The endo-glycanases and glycosidases, found in many plant tissues, are not only involved in the mobilization of intra- and extracellular polysaccharides, but also in cell-wall modifications, abscission, the softening of fleshy fruits, and certain defence mechanisms. Apart from the enzymes of leguminous seeds, endoglycanases of plants have not been well characterized. Most of our knowledge of these enzymes is based on work using plant pathogenic fungi and bacteria (Dekker & Richards, 1976).

The production of mobilizing enzymes in germinated seeds

Substantial differences between species occur in the sites of origin and action of the endo-glycanases and glycosidases involved in mobilization of endospermic reserves.

In leguminous seeds, the enzymes are produced mostly in the endosperm itself. In fact, isolated endosperms, prepared by the 'de-embryonation' of dry seeds, can mobilize galactomannan at comparable rates to intact germinated seeds under similar conditions (Reid & Meier, 1972; Seiler, 1977). Likewise, β-mannosidase and α-galactosidase are produced in 'normal' amounts by de-embryonated seeds (Fig. 11.3A, B; Seiler, 1977). Moreover, glycosidases are also produced in germinating embryos (Fig. 11.3A; McCleary & Matheson, 1974, 1975; Seiler, 1977), indicating that products of breakdown from the endosperm may be absorbed into the embryo as galactosylated oligo-mannans and then hydrolysed completely. In fenugreek endosperm, hydrolases are produced by the aleurone layer, the only living tissue present. Mobilization begins in the outer part of the endosperm, next to it (Reid, 1971). In guar and carob, however, some 'internal' endospermic cells are alive and produce substantial amounts of hydrolases too (McClendon *et al.* 1976; Seiler, 1977). In carob, endospermic mobilization starts in the cells nearest the centrally-placed embryo, as well as next to the outer aleurone layer.

In cereal grains, the aleurone layer is also the major source of cell-wall hydrolases. In barley, for instance, it secretes endo-$(1\rightarrow3)$, $(1\rightarrow4)$-βglucanases, endo-(1→4)-β-xylanases and α-arabinosidase (Bennett & Chrispeels, 1972; Taiz & Honigman, 1976; Dashek & Chrispeels, 1977).

In some seeds the endosperm and cotyledons supply different enzymes. In lettuce for example, the endosperm, which consists of an aleurone layer two cells thick, synthesizes endo-β-mannanase, whereas β-mannosidase and α-galactosidase are produced entirely by the cotyledons (Halmer & Bewley, 1979; Leung, Reid & Bewley, 1979). α-Galactosidase is secreted into the endosperm in substantial amounts, where it perhaps participates in the breakdown of the endospermic wall, which contains galactose residues. However, β-mannosidase remains in the cotyledons (Leung *et al.* 1979).

In seeds where the endosperm is entirely dead, mobilizing enzymes come from the embryo. In palm seeds, mobilization occurs by the ingrowth of a haustorial cotyledon, which secreted endo-β-mannanase and β-mannosidase (Keusch, 1968; Tampion, 1972).

Little is known of the mobilization of polysaccharides that are stored in cotyledons. An enzyme capable of degrading xylo-(1→4)-β-glucan is said to

Fig. 11.3 Changes in α-galactosidase (○ ●), β-mannosidase (△ ▲) and endo-(1→4)-β-mannanase (□ ■) activities in (A) endosperms (● ▲ ■) and embryos (○ △ □) of intact fenugreek seed and (B) endosperms isolated from dry seeds. Incubations were by floating seeds/seed parts on 2.0 ml water (after Reid & Meier, 1973; Reid, Davies & Meier, 1977). Isolated endosperms produce similar amounts of the glycosidases, but reduced amounts of mannanase, compared to endosperms in germinated seeds. Large activities of glycosidases are also found in germinating embryos.

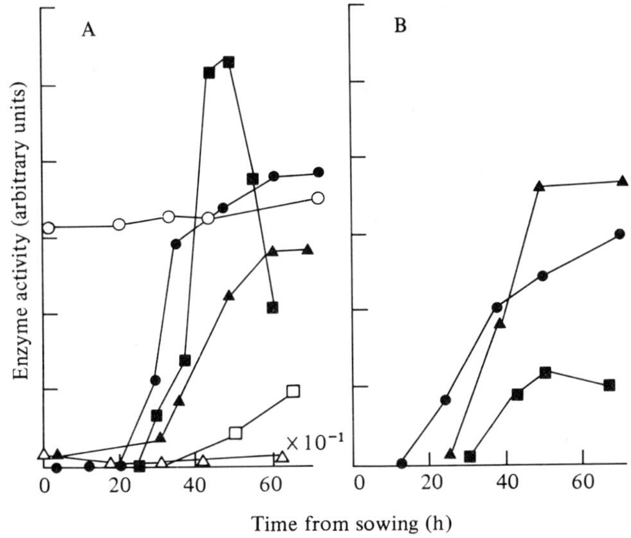

be present in the cotyledons of germinated *Tropaeolum* seeds, as would be expected (Koimann, 1957), but this seems to be the extent of information on the subject.

The uptake and fate of endospermic cell-wall breakdown products

The cotyledon, rather than the axis of the embryo, is the organ of uptake of the products of endospermic degradation. Indeed mobilization typically begins only after germination, when the axis has often emerged entirely from the seed.

Sugar uptake has been studied only in cereals. Glucose is absorbed into the scutellum (a modified cotyledon) by a combination of active and passive transport, whereas maltose is hydrolysed in the process by a cell-surface α-glucosidase (Humphrey, 1975).

Generally, sugars once absorbed are rapidly metabolized. In the germinated endospermic legumes, for example, concentrations of free mannose and galactose remain low in the cotyledons at all times. In cereal grains and lettuce seeds, mobilized sugars are converted to sucrose and translocated to the growing axis (Bewley & Black, 1978; Halmer, Bewley & Thorpe, 1978) whereas, in seeds of carob and fenugreek, much of the incoming carbohydrate is converted to starch as a temporary reserve (Reid, 1971; Seiler, 1977). A certain amount of sugar is also consumed in internal cotyledonary metabolism. For example, in guar embryos fed with ^{14}C-labelled hexoses, radioactivity is recovered as $^{14}CO_2$ and in cell-wall polysaccharides, as well as in glucose, fructose and sucrose (McCleary & Matheson, 1976). The first metabolic step in the assimilation of sugars is probably phosphorylation *via* kinases. In legumes, Gal and Man are thought to be thus converted to Gal-1-P and Man-6-P: the latter is converted into Fru-6-P by phosphomannoisomerase (Fig. 11.1). This enzyme of widespread distribution in these seeds declines in activity as galactomannan is consumed (McCleary *et al.* 1976). Sugar phosphates might conceivably be converted *via* their sugar nucleotides, too (Fig. 11.1).

Control of production of hydrolases during endospermic mobilization

Major mobilization of reserves only occurs after germination is completed or, to put it another way, after growth has begun. In imbibed dormant or germinating seeds very little mobilization occurs. Evidently, the production and distribution of key mobilizing enzymes are controlled so that reserves are not supplied until they are needed.

The main such control mechanism operating in cereal grains has been well documented over the past 20 years. In these seeds, the production and secretion of many of the hydrolases involved in endospermic degradation (e.g.

starch and protein) depends upon the translocation and diffusion of gibberellins from the embryo – in essence a direct, positive control over mobilization. Endospermic cell walls provide only a minor carbohydrate reserve – about an eighth of that supplied by starch in wheat and barley – but in some grasses, like *Bromus*, the cells walls are much thickened and provide a more substantial reserve. Wall breakdown is also necessary for the release of hydrolases from the aleurone layer and for their diffusion within the endosperm. Amongst the cell-wall hydrolases, the following are controlled by gibberellins: the production and release of endo-xylanase (synthesized *de novo*), β-xylosidase and α-arabinosidase, and also the release of endo-$(1\rightarrow3),(1\rightarrow4)$-$\beta$-glucanase (Bennett & Chrispeels, 1972; Taiz & Honigman, 1976). (Interestingly, gibberellins also inhibit the *synthesis* of cell-wall arabino-xylan in the aleurone layer, which otherwise continues at a low rate (Johnson & Chrispeels, 1973). Gibberellins thus control a switch between synthesis and degradation of walls.)

A somewhat similar control by gibberellins may operate in the mobilization of endospermic cell walls in celery seeds. In this species, the embryo grows considerably at the expense of the endosperm before the radicle emerges through the seed coat. Isolated celery endosperms are not degraded but can be induced to do so by gibberellins (Jacobsen & Pressman, 1979). The endospermic walls probably contain $(1\rightarrow4)$-β-mannan (see Chapter 1) but the enzymes that degrade them have not been investigated.

The mechanism of control in endospermic legumes is quite different, however. As already mentioned, breakdown of endospermic cell walls and production of hydrolases are more or less unaffected by the removal of the embryo (Fig. 3A, B; Seiler, 1977), which indicates that there is no positive control by the embryo over endospermic mobilization. Nor is there any evidence of feedback control mechanisms for the rate of hydrolysis of galacto-mannan is unaffected by the accumulation of free mannose and galactose, which reach high concentrations in isolated endosperms (Reid & Meier, 1972; Seiler, 1977).

However, the situation in leguminous seeds may not be as simple as it seems. The picture may be influenced by the choice of experimental conditions, as some recent work on lettuce endosperms has revealed. In this tissue, the *de novo* synthesis of endo-$(1\rightarrow4)$-β-mannanase is prevented by an (unidentified) inhibitor, which is removed from the endosperm by the cotyledons under the control of the axis (Table 11.1). As germination is completed, the concentration of inhibitor falls and endo-mannanase synthesis starts (Halmer & Bewley, 1979). This amounts to a form of negative control – an inhibitor removed by the embryo – rather than the positive control seen in the cereal grains. Production of endo-mannanase can occur in isolated endosperms in certain experimental conditions, namely when the incubation volume used is sufficiently large (2 ml) to leach out the inhibitor and allow enzyme synthesis to begin (Table 11.1). In fact, the inhibitor can be detected by bioassay of the endospermic leachate. Thus, in these circumstances, mannanase synthesis *appears* to be independent of embryonic control, though this is not actually the case. It follows that a similarly misleading state of affairs might arise in experimental work using isolated leguminous endosperms. Certainly, although isolated carob endosperms must be washed thoroughly with water to remove inhibitory substances before mobilization can take place (Seiler, 1977), washing is not a requisite for mobilization in intact seeds. Some mechanism must presumably exist for the removal or inactivation of the inhibitor during

Table 11.1 *Summary of evidence that endo-(1→4)-β-mannanase production in the lettuce endosperm is controlled by the removal/inactivation of an inhibitor*

Incubation (all in 0.1 ml, except where stated)	Units mannanase × 10^{-3} produced per 10 seed parts
Experiment 1	
1. Endosperms in 2.0 ml water	2.4
2. Endosperms	0.1
3. Prewashed endosperms	2.4
4. Prewashed endosperms in 2.0 ml leachate from other endosperms	0.4
Experiment 2	
5. Seeds 'de-axised' before germination	0.1
6. Seeds 'de-axised' during germination but before enzyme production	3.6
7. Seeds 'de-axised' before germination, incubated with dissected axes	2.9
8. Endosperms, incubated with dissected axes	0.1
9. Intact germinated seeds	4.0

Endosperms isolated from non-germinating seeds produce mannanase when incubated in a 'large' volume (2.0 ml) but not in a limited volume (0.1 ml) of water (lines 1 & 2). However, prewashed endosperms can produce the enzyme under the latter conditions (line 3). An inhibitory substance can be detected in the leachate of endosperms (line 4). In intact seeds, the inhibition is relieved by a mechanism, not yet fully explained, involving the cotyledons under the control of the axis. Some of the evidence is as follows. Mannanase production requires the presence of the germinating axis for the first part of the germination phase, after which the axis may be removed without subsequent effect (lines 5 & 6). De-axised seeds can also produce the enzyme if dissected axes are incubated alongside them (line 7), indicating a hormonal control by the axis. Endosperms alone do not respond to the axis (line 8), indicating that the cotyledons are responsible for relieving the inhibition in the endosperm, in response to the signal from the axis. (Summarized from Halmer & Bewley, 1980.)

germination. Thus, further investigation may reveal that a negative control over endospermic mobilization operates in leguminous seeds too.

Financial support during the writing of this chapter by a Science Research Council Post Doctoral Research Fellowship is gratefully acknowledged.

References

Anderson, K. L. & Ray, P. M. (1978). Labelling of the plasma membrane of pea cells by a surface-localized glucan synthetase. *Plant Physiology*, **61**, 723–30.

Barber, G. A. (1963). The formation of uridine diphosphate L-rhamnose by enzymes of the tobacco leaf. *Archives of Biochemistry and Biophysics*, **103**, 276–82.

Bennett, P. A. & Chrispeels, M. J. (1972). De novo synthesis of ribonuclease and β-1,3-glucanase by aleurone cells of barley. *Plant Physiology*, **49**, 445–7.

Bewley, J. D. & Black, M. (1978). *Physiology and Biochemistry of Seeds in Relation to Germination*, vol. I, *Development, Germination and Growth*. Berlin, Heidelberg, New York: Springer-Verlag.

Brett, C. T. (1978). Synthesis of β-(1 → 3)-glucan from extracellular uridine diphosphate glucose as a wound response in suspension cultured soybean cells. *Plant Physiology*, **62**, 377–82.

Clark, A. F. & Villemez, C. L. (1972). The formation of β-(1 → 4)-glucan from UDP-α-D-glucose catalyzed by a *Phaseolus aureus* enzyme. *Plant Physiology*, **50**, 371–4.

Dashek, W. V. & Chrispeels, M. J. (1977). Gibberellic acid-induced synthesis and release of cell-wall-degrading exoxylanase by isolated aleurone layers of barley. *Planta*, **134**, 251–6.

Dekker, R. F. H. & Richards, G. N. (1976). Hemicellulases: their occurrence, purification, properties and mode of action. *Advances in Carbohydrate Chemistry & Biochemistry*, **32**, 277–352.

Dey, P. M. (1978). Biochemistry of plant galactomannans. *Advances in Carbohydrate Chemistry & Biochemistry*, **35**, 341–76.

Dey, P. M. (1980). Biochemistry of α-D-galactosidic linkages in the plant kingdom. *Advances in Carbohydrate Chemistry and Biochemistry*, **37**, 283–372.

Dick, P. S. & ap Rees, T. (1976). Sucrose metabolism by roots of *Pisum sativum*. *Phytochemistry*, **15**, 255–9.

Duran, A. & Cabib, E. (1978). Solubilization and partial purification of yeast chitin synthetase. Confirmation of the zymogenic nature of the enzyme. *Journal of Biochemistry*, **253**, 4419–25.

Elbein, A. D. (1979). The role of lipid-linked saccharides in the biosynthesis of complex carbohydrates. *Annual Review of Plant Physiology*, **30**, 239–72.

Ericson, M. C. & Elbein, A. D. (1980). Biosynthesis of cell wall polysaccharides and glycoproteins. In *The Biochemistry of Plants*, vol. 3, ed. J. Preiss, pp. 589–616. New York, San Francisco, London: Academic Press.

Feingold, D. S. (1982). Aldo (and keto) hexoses and uronic acids. In *Encyclopedia of Plant Physiology*, New Series, vol. 13A, Plant Carbohydrates I, Intracellular Carbohydrates, ed. F. A. Loewus & W. Tanner, pp. 3–76. Berlin, Heidelberg, New York: Springer-Verlag.

Franz, G. & Heineger, U. (1982). Biosynthesis and metabolism of cellulose and non-cellulosic cell wall glucans. In *Encyclopedia of Plant Physiology*, New Series, vol. 13B, Plant Carbohydrates II, Extracellular Polysaccharides, ed. W. Tanner & F. A. Loewus, pp. 47–67. Berlin, Heidelberg, New York: Springer-Verlag.

Garcia, R. C., Recondo, E. & Dankert, M. (1974). Polysaccharide biosynthesis in *Acetobacter xylinum*. Enzymatic synthesis of lipid diphosphate and monophosphate sugars. *European Journal of Biochemistry*, **43**, 93–105.

Green, J. R. & Northcote, D. H. (1979). Polyprenol phosphate sugars synthesised during slime-polysaccharide production by membranes of the root-cap cells of maize (*Zea mays*). *Biochemical Journal*, **178**, 661–71.

Halmer, P. & Bewley, J. D. (1979). Mannanase production by the lettuce endosperm: control by the embryo. *Planta*, **144**, 333–40.

Halmer, P., Bewley, J. D. & Thorpe, T. A. (1978). Degradation of the endosperm cell walls of *Lactuca sativa* L., cv. Grand Rapids. Timing of mobilization of soluble sugars, lipid and phytate. *Planta*, **139**, 1–8.

Hanke, D. E. (1977). The endomembrane system and the integration of cellular activities. In *The Molecular Biology of Plant Cells*, ed. H. Smith, pp. 182–207. Oxford: Blackwell Scientific Publications.

Heiniger, U. & Delmer, D. P. (1977). UDP-glucose: glucan synthetase in developing cotton fibres. II. Structure of the reaction product. *Plant Physiology*, **59**, 719–23.

Heller, J. S. & Villemez, C. L. (1972). Interaction of soluble glucosyl- and mannosyl-transferase enzyme activities in the synthesis of a glucomannan. *Biochemical Journal*, **129**, 645–55.

Helsper, J. P. F. G. (1979). The possible role of lipid intermediates in the synthesis of β-glucans by a membrane fraction from pollen tubes of *Petunia hybrida*. *Planta*, **144**, 443–50.

Helsper, J. P. F. G., Veerkamp, J. H. & Sassen, M. M. A. (1977). β-Glucan synthetase activity in Golgi vesicles of *Petunia hybrida*. *Planta*, **133**, 303–8.

Hemming, F. W. (1974). Lipids in glycan biosynthesis. *MTP International Reviews of Science, Biochemistry, Series 1*, **4**, 39–97.

Hinman, M. B. & Villemez, C. L. (1975). Glucomannan biosynthesis catalysed by *Pisum sativum* enzymes. *Plant Physiology*, **56**, 608–12.

Hopp, H. E., Romero, P. A., Daleo, G. R. & Pont-Lezica, R. (1978). Synthesis of cellulose precursors. The involvement of lipid-linked sugars. *European Journal of Biochemistry*, **84**, 561–71.

Humphrey, T. E. (1975). Maltose uptake in *Zea mays* scutellum. *Phytochemistry*, **14**, 333–40.

Jacobsen, J. V. & Pressman, E. (1979). A structural study of germination in celery (*Apium graveolens* L.) seed with emphasis on endosperm breakdown. *Planta*, **144**, 241–8.

Johnson, K. D. & Chrispeels, M. J. (1973). Regulation of pentosan biosynthesis in barley aleurone tissue by gibberellic acid. *Planta*, **111**, 353–64.

Jung, P. & Tanner, W. (1973). Identification of the lipid intermediate in yeast mannan biosynthesis. *European Journal of Biochemistry*, **37**, 1–6.

Karr, A. L. (1976). Cell wall Biogenesis. In *Plant Biochemistry*, 3rd edn, ed. J. Bonner & J. E. Varner, pp. 405–26. New York, San Francisco, London: Academic Press.

Keusch, L. (1968). Mobilisation of reserve mannan in germinating date seeds. *Planta*, **78**, 321–50.

Koimann, P. (1957). Amyloids of plant seeds. *Nature*, **179**, 107–9.

Larsen, G. L. & Drummond, D. O. (1974). (1–4)-D-Glucan synthesis from UDP-[^{14}C]-D-glucose by a solubilised enzyme from *Lupinus albus*. *Phytochemistry*, **13**, 361–6.

Leung, D. W. M., Reid, J. S. G. & Bewley, J. D. (1979). Degradation of the endosperm cell walls of *Lactuca sativa* L., cv. Grand Rapids in relation to the mobilisation of proteins and the production of hydrolytic enzymes in the axis, cotyledons and endosperm. *Planta*, **146**, 335–41.

Liao, T. H. & Barber, G. A. (1971). The synthesis of guanosine-5′-diphosphate L-fucose by enzymes of a higher plant. *Biochimica et Biophysica Acta*, **230**, 64–71.

McCleary, B. V. & Matheson, N. K. (1974). α-D-Galactosidase activity and galactomannan and galactosylsucrose oligosaccharide depletion in germinating legume seeds. *Phytochemistry*, **13**, 1747–57.

McCleary, B. V. & Matheson, N. K. (1975). Galactomannan structure and β-mannanase and mannosidase activity in germinating legume seeds. *Phytochemistry*, **14**, 1187–94.

McCleary, B. V. & Matheson, N. K. (1976). Galactomannan utilization in germinating legume seeds. *Phytochemistry*, **15**, 43–7.

McCleary, B. V., Matheson, N. K. & Small, D. M. (1976). Galactomannans and a galactoglucomannan in legume seed endosperms – structural requirements for β-mannanase hydrolysis. *Phytochemistry*, **15**, 1111–17.

McClendon, J. H., Nolan, W. G. & Wenzler, H. F. (1976). Role of endosperm in germination of legumes – galactomannan, nitrogen and phosphorus changes in germination of guar (*Cyamopsis tetragonoloba – Leguminosae*). *American Journal of Botany*, **63**, 790–7.

Meier, H. & Reid, J. S. G. (1977). Morphological aspects of the galactomannan formation in the endosperm of *Trigonella foenum-graecum* L. (Leguminosae). *Planta*, **133**, 243–8.

Odzuck, W. & Kauss, H. (1972). Biosynthesis of pure araban and xylan. *Phytochemistry*, **11**, 2489–94.

Panayotatos, N. & Villemez, C. L. (1973). Formation of a β-(1–4) D-galactan chain catalysed by a *Phaseolus aureus* enzyme. *Biochemical Journal*, **33**, 263–71.

Parodi, A. J. & Leloir, L. F. (1979). The role of lipid intermediates in the glycosylation of proteins in the eucaryotic cell. *Biochimica et Biophysica Acta*, **559**, 1–37.

Pisigan, R. A. & Rosario, E. J. (1976). Polysaccharide biosynthesis. Isoenzymes of soluble starch synthetase from *Oriza sativa* grains. *Phytochemistry*, **15**, 71–3.

Raymond, Y., Fincher, G. B. & MacLachlan, G. A. (1978). Tissue slice and particulate β-glucan synthetase activities from *Pisum* epicotyls. *Plant Physiology*, **61**, 938–42.

Reid, J. S. G. (1971). Reserve carbohydrate metabolism in germinating seeds of *Trigonella foenum-graecum* L. (Leguminosae). *Planta*, **100**, 131–42.

Reid, J. S. G., Davies, C. & Meier, H. (1977). Endo-β-mannanase the leguminous aleurone layer and the storage galactomannan in germinating seeds of *Trigonella foenum-graecum* L. *Planta*, **133**, 219–22.

Reid, J. S. G. & Meier, H. (1972). The function of the aleurone layer during galactomannan mobilisation in germinating seeds of fenugreek (*Trigonella foenum-graecum* L.), crimson clover (*Trifolium incarnatum* L.) and lucerne (*Medicago sativa* L.): a correlative biochemical and ultrastructural study. *Planta*, **106**, 44–60.

Reid, J. S. G. & Meier, H. (1973). Enzymic activities and galactomannan mobilisation in germinating seeds of fenugreek (*Trigonella foenum-graecum* L. Leguminosae). Secretion of α-galactosidase and β-mannosidase by the aleurone layer. *Planta*, **112**, 301–8.

Robinson, D. G. (1977). Plant cell wall synthesis. *Advances in Botanical Research*, **5**, 89–151.

Rollit, J. & MacLachlan, G. A. (1974). Synthesis of wall glucan from sucrose by enzyme preparations from *Pisum sativum*. *Phytochemistry*, **13**, 167–374.

Ruiz-Herrera, J., Sing, V. O., Van der Woude, W. J. & Bartnicki-Garcia, S. (1975). Microfibril assembly by granules of chitin synthetase. *Proceedings of the National Academy of Sciences of the USA*, **72**, 2706–10.

Seiler, A. (1977). Galaktomannanabbau im Keimenden Johannisbrotsamen (*Ceratonia siliqua* L.). *Planta*, **134**, 209–21.

Shematek, E. M., Braatz, J. A. & Cabib, E. (1980). Biosynthesis of the yeast cell wall. L. Preparation and properties of β-(1–3)-glucan synthetase. *Journal of Biological Chemistry*, **255**, 888–94.

Southworth, D. & Dickinson, D. B. (1975). β-1, 3-glucan synthase from *Lilium longiflorum* pollen. *Plant Physiology*, **56**, 83–7.

Stoddart, R. W. & Northcote, D. H. (1967). Metabolic relationships of the isolated fractions of the pectic substances of actively growing sycamore cells. *Biochemical Journal*, **105**, 45–59.

Storm, D. L. & Hassid, W. Z. (1974). Partial purification and properties of a β-D-glucosyl transferase occurring in germinating *Phaseolus aureus* seeds. *Plant Physiology*, **54**, 840–5.

Sugiyama, N., Shimagara, H., Andoh, T. & Takemoto, M. (1973). Studies on mannan and related compounds. 3. Konjac mannanase from tubers of *Amorphophallus konjac* (C. Koch.). *Agricultural and Biological Chemistry (Tokyo)*, **37**, 9–17.

Taiz, L. & Honigman, W. A. (1976). Production of cell wall hydrolysing enzymes by barley aleurone layers in response to gibberellic acid. *Plant Physiology*, **58**, 380–6.

Tampion, J. (1972). A viscometric assay for β-mannanase. *Enzymologia*, **42**, 253–5.

Tandecarz, J., Lavintman, N. & Cardini, C. E. (1975). Formation of a glucoproteic acceptor by a potato nonsedimentable preparation. *Biochimica et Biophysica Acta*, **399**, 345–55.

Tonn, S. J. & Gardner, J. E. (1979). Biosynthesis of polysaccharides by prokaryotes. *Annual Review of Microbiology*, **33**, 169–99.

Villemez, C. L., Lin, T. Y. & Hassid, W. Z. (1965). Biosynthesis of polygalacturonic acid chain of pectin by a particulate enzyme preparation from *Phaseolus aureus* seedlings. *Proceedings of the National Academy of Sciences of the USA*, **54**, 1626–33.

Wang, M. C. & Bartnicki-Garcia, S. (1976). Synthesis of β-1,3-glucan microfibrils by a cell-free extract from *Phytophthora cinnamoni*. *Archives of Biochemistry and Biophysics*, **175**, 351–4.

Williams, J., Villaroya, H. & Petek, F. (1977). Purification and properties of an α-D-galactoside galactohydrolase from seeds of *Trifolium repens* (white clover). *Biochemical Journal*, **161**, 509–15.

INDEX

(*Where vernacular names of plants are followed by botanical names in parentheses, the botanical names do not appear in the text – see Preface.*)